A DICTIONARY OF
OIL & GAS
INDUSTRY TERMS

A DICTIONARY

—— OF ——

OIL & GAS
INDUSTRY
TERMS

SECOND EDITION

PETER ROBERTS

OXFORD
UNIVERSITY PRESS

OXFORD
UNIVERSITY PRESS

Great Clarendon Street, Oxford, OX2 6DP,
United Kingdom

Oxford University Press is a department of the University of Oxford.
It furthers the University's objective of excellence in research, scholarship,
and education by publishing worldwide. Oxford is a registered trade mark of
Oxford University Press in the UK and in certain other countries

Published in the United States of America by Oxford University Press
198 Madison Avenue, New York, NY 10016, United States of America

British Library Cataloguing in Publication Data
Data available

Library of Congress Control Number is on file at the Library of Congress

ISBN 978-0-19-287346-0

DOI: 10.1093/oxfordhb/9780192873460.001.0001

Printed in the UK by
Ashford Colour Press Ltd, Gosport, Hampshire

Links to third party websites are provided by Oxford in good faith and
for information only. Oxford disclaims any responsibility for the materials
contained in any third party website referenced in this work.

INTRODUCTION TO THE
FIRST EDITION

This dictionary is a guide to the abbreviations, acronyms, terms, and phrases which are most commonly encountered in the legal, regulatory, technical, commercial, and financial aspects of the international oil and gas industry today. It consists of the following parts:

Part A: abbreviations and acronyms

Part B: terms and phrases

Part C: appendices

Appendix 1: petroleum units and conversions

Appendix 2: technical elements:
    2A—petroleum formation;
    2B—exploration drilling;
    2C—drilling a wellbore;
    2D—petroleum production; and
    2E—refining

Appendix 3: promote levels

Appendix 4: geological time periods

Appendix 5: well symbols

Appendix 6: API gravity

Appendix 7: SPE PRMS

An inescapable exposure applies to the author of a dictionary. It is the accusation of a reader that the dictionary misses or misrepresents something and so is incomplete or inaccurate.[1] In anticipated defence of such a charge the following points can be made:

Technical terms: the oil and gas industry is highly technical and the key technical terms, which are most commonly encountered in practice, are referenced in this dictionary. But this dictionary is not intended to recite absolutely every technical term, however obscure, used in every element of the industry today. To do so would

---

[1] A sentiment best expressed by the great lexicographer Dr Samuel Johnson in the preface to his *Dictionary of the English Language* in 1775: 'It is the fate of those who toil at the lower employments of life ... to be exposed to censure, without hope of praise; to be disgraced by miscarriage, or punished for neglect. ... Among these unhappy mortals is the writer of dictionaries ... Every other author may aspire to praise; the lexicographer can only hope to escape reproach.'

simply swamp the dictionary. There are several excellent online technical directories which can be used to find an outlying term not covered in this dictionary.

Multiple usages: many of the terms which are defined in this dictionary can be represented in different ways. Synonyms are referenced, but would that the lexicon of the oil and gas industry is completely settled; many of these synonyms are not universally agreed and many are used ambiguously. Also, some defined terms have several (often unrelated) meanings. A consistent and ubiquitous usage of commonly agreed terms is not a feature of the industry.

Depth of reference: this dictionary is intended to offer a concise explanation of key oil and gas industry terms. It is a starting point for the acquisition of knowledge, from which further research can be conducted if required. It is not intended to be an encyclopaedia which contains lengthy narratives on each topic but a partial exception to this principle comes in Appendix 2, where a lengthier explanation of certain matters is intended to save a reader from patching together a number of related definitions in the search for an answer.

The intended purpose of this dictionary is to be concise, comprehensive, and useful. Some readers will surely have opinions about unreferenced terms which they think should be included; others about terms which they think should be defined differently (or even not at all). The publishers welcome the communication of these opinions, not least because the intention is that this dictionary will be the subject of revision and further issue as the industry continues to evolve.

Thanks are due to the Energy Institute for its review of the technical terms in this dictionary. Thanks are also due to Katie Hooper and Zoe Tustin of the Oxford University Press for their help in making this project a reality and to my super secretary Jackie Harper for the heavy lifting on the manuscript.

The information contained in this dictionary is provided for general information purposes only. Whilst the author, the OUP, and the Energy Institute have each applied reasonable care in developing this dictionary, no representations or warranties, express or implied, are made by any of the author, the OUP, or the Energy Institute concerning the applicability, suitability, accuracy, or completeness of the information contained herein. Each of the author, the OUP, and the Energy Institute accept no responsibility whatsoever for the use of this information and shall have no liability in any way for any liability, loss, cost, or damage incurred as a result of the receipt or use of the information contained herein.

Finally, I would respectfully disagree with Dr Johnson's assertion that 'to make dictionaries is dull work'. It has been anything but that.

Peter Roberts
Guildford, Surrey
October 2018

INTRODUCTION TO THE SECOND EDITION

The structure of the first edition of this dictionary is repeated in this second edition, and the ambitions expressed and the caveats made in the introduction to the first edition apply equally here too.

The second edition applies a significant number of new definitions (largely drawn from the worlds of clean energy, energy transition, and energy financing), takes account of various review comments and suggestions received, and makes a number of cross-referencing improvements.

As with the first edition, I welcome more feedback (which can be sent to me at petro@crosskeysenergy.com), and I also wish to express my thanks to Fay Gibbons of the Oxford University Press.

Peter Roberts
Guildford, Surrey
February 2023

PART A

ABBREVIATIONS AND ACRONYMS

1C	*see* one C
1P	*see* proved reserves
2C	*see* two C
2D	*see* two D
2P	*see* probable reserves
3C	*see* three C
3D	*see* three D
3P	*see* possible reserves
4D	*see* four D
A&D	*see* acquisition and divestment
A&E	*see* amend and extend
A&R	*see* amendment and restatement
A&W	*see* abandoned and whipstocked
AAPG	*see* American Association of Petroleum Geologists
AAPL	*see* American Association of Professional Landmen
AAV	*see* ambient air vaporiser
ABC	*see* anti-bribery and corruption
ABS	*see* American Bureau of Shipping
ACCC	Australian Competition & Consumer Commission
ACQ	*see* annual contract quantity
AD	*see* anaerobic digestion; authorised distributor
ADI	all dates inclusive
ADP	*see* annual delivery programme
ADR	*see* alternative dispute resolution
AECO	*see* Alberta Energy Company
AF	*see* Arctic floater
AFC	*see* approved for construction
AFD	*see* approved for design
AFE	*see* authorisation for expenditure
AFR	*see* air–fuel ratio
AFRA	*see* average freight rate assessment
AG	*see* associated gas
AGI	*see* above-ground installation
AGO	*see* automotive gas oil
AGR	*see* above-ground review; acid gas removal
AGW	*see* agricultural grade water; all going well
AHC	*see* active heave compensation; ad hoc charter

AHD	*see* along-hole depth; along-hole distance
AHV	*see* anchor handling vessel
AI	*see* artificial island
AIEN	Association of International Energy Negotiators; *see* AIPN
AIPN	*see* Association of International Petroleum Negotiators
AIS	*see* automatic identification system
ALARP	*see* as low as reasonably practicable
ALP	*see* articulated loading platform
ALS	*see* accidental limit state
AMA	*see* asset management agreement
AMI	*see* area of mutual interest
AMIA	*see* area of mutual interest agreement
AMM	*see* abandoned mine methane
AMPLA	*see* Australian Mining Petroleum Law Association
AMR	*see* automated meter reading
ANGA	*see* American Natural Gas Association
ANSCO	*see* Alaskan North Slope crude oil
ANSI	*see* American National Standards Institute
AO	*see* asset optimisation; associated oil
AOA	*see* add on afterwards
AOL	*see* arrival on location
AOP	annual operating plan
AP	*see* accounting procedure
API	*see* American Petroleum Institute
APO	*see* after payout
APPEA	*see* Australian Petroleum Production & Exploration Association
APRV	*see* annular pressure relief valve
ARA	*see* Amsterdam-Rotterdam-Antwerp
ARCA	*see* advanced reservation of capacity agreement
ARO	*see* abandonment and reclamation obligation
ASOG	*see* activity-specific operating guidelines
ASP	*see* average system pressure
ASTM	*see* American Society for Testing and Materials
ASV	*see* accommodation support vessel
ASWP	*see* any safe world port
ATDNSHINC	any time day or night Sundays and holidays included
ATG	*see* aviation turbine gasoline
ATK	*see* aviation turbine kerosene

Atm	*see* atmosphere
ATSBE	*see* all time saved both ends
AUV	*see* autonomous underwater vehicle
AVO	*see* amplitude variation with offset
AVR	*see* ad valorem royalty
AVTAG	*see* aviation turbine gasoline
AVTUR	*see* aviation turbine kerosene
AWP&B	annual work programme and budget; *see* work programme and budget
B	*see* barrel; billion
B2B	*see* business-to-business
B2C	*see* business-to-consumer
BA	breathing apparatus
BAF	*see* bunker adjustment factor
BAFO	best and final offer
BAT	*see* best available technology/technique
BATNA	*see* best alternative to a negotiated agreement
BATNEEC	*see* best available technology not entailing excessive cost
BAU	*see* business as usual
BBC	*see* bareboat charter
Bbl	*see* barrel
BBQ	*see* Bonny, Brass River, and Qua Iboe
BBSM	*see* behaviour-based safety management
BCGA	*see* basin-centred gas accumulation
BCO	*see* beneficial cargo owner
BCP	*see* business continuity plan
BD	barrels per day
BDI	both dates inclusive
BE	best endeavours
BECCS	*see* bioenergy carbon capture and storage
BEP	*see* best environmental practice
BER	beyond economic repair
BEWL	best estimate without liability
BFG	*see* boiler fuel gas
BFOE	*see* Brent Forties Oseberg Ekofisk
BHA	*see* bottomhole assembly
BHL	*see* bottomhole location
BI	business interruption

BIAPO	*see* back-in after pay-out
BII	*see* business interruption insurance
BIMCO	*see* Baltic International Maritime Council
BIS	*see* bought-in services
BIT	*see* bilateral investment treaty
BL	*see* battery limit
BLEVE	*see* boiling liquid expanding vapour explosion
BLPD	*see* barrels of liquid per day
BNOC	*see* British National Oil Corporation
BOD	*see* basis of design; biochemical oxygen demand
BOE	*see* barrel of oil equivalent
BOEPD	BOE per day
BOG	boil-off gas
BOGCS	*see* boil-off gas combustion system
BOGRS	*see* boil-off gas recovery system
BOO	*see* build own operate
BOOT	*see* build own operate transfer
BOP	*see* blowout preventer
BOPE	*see* blowout preventer equipment
BOR	*see* boil-off rate
BPO	*see* before payout
BPS	*see* basis point
BPSD	*see* barrels per stream day
BRC	*see* bid round circular
BRDP	*see* bid round data package
BRINDEX	*see* Association of British Independent Oil Exploration Companies
BROA	*see* British Rig Owners' Association
BROB	bunkers remaining on board
BRT	*see* below rotary table
BRV	*see* bulk rock volume
BS	*see* bottom sediment; British Standards
BS&W	*see* basic sediment and water; bottom settlings and water
BSI	*see* British Standards Institution
BSODU	*see* bottom-supported offshore drilling unit
BTEX	benzene, toluene, ethylbenzene, and xylene
BTL	*see* biomass to liquid
Btu	*see* British thermal unit
BTX	*see* benzene, toluene, xylene

BW	*see* barrels of water
BWEAS	*see* ballast water exchange at sea
BWP	*see* benchmark weighted price
BWPD	BW per day
BWPH	BW per hour
C1	methane (referencing one carbon atom)
C2	ethane (referencing two carbon atoms)
C3	propane (referencing three carbon atoms)
C4	butane (referencing four carbon atoms)
C5	pentane (referencing five carbon atoms)
C6	hexane (referencing six carbon atoms)
C&C	*see* circulating and conditioning; contract and collaboration
C&F	*see* cost and freight
C&M	*see* care and maintenance
C&P	*see* cased and perforated
C&S	*see* cased and suspended
CA	*see* confidentiality agreement
CAF	*see* currency adjustment factor
CAGR	*see* compound annual growth rate
CALM	*see* catenary anchor leg mooring
CAM	*see* capacity allocation mechanism
CAP	*see* competence assurance programme; condition assessment programme
CAPEX	*see* capital expenditure
CAPL	*see* Canadian Association of Petroleum Landmen
CAR	*see* construction all risks
CBD	*see* cross-block development
CBM	*see* coal bed methane; condition-based monitoring; conventional buoy mooring
CBOB	*see* conventional blendstock for oxygenate blending
CBR	*see* crude by rail
CBW	*see* clay-bound water
CCC	*see* care, custody, control; command, control, communication
CCGT	*see* combined cycle gas turbine
CCIP	*see* contractor-controlled insurance programme
CCR	central control room; *see* Conradson carbon residue
CCS	*see* carbon capture and storage; cargo containment system
CCSU	*see* carbon capture and storage unit
CCUS	carbon capture, utilisation, storage

CDA	*see* community development agreement
CDD	*see* community driven development
CDU	*see* conical drilling unit; crude dehydration unit; crude distillation unit
CEGH	*see* Central European Gas Hub
CEIMS	*see* crisis, emergency, incident management system
CEMP	*see* construction environmental management plan
CEMS	*see* continuous emissions monitoring system
CEP	*see* certified emissions profile
CFAR	*see* cash flow at risk
CFD	*see* contract for differences
CFG	cubic feet of gas
CFPP	*see* cold filter plugging point
CGL	*see* commercial general liability
CGR	*see* condensate–gas ratio
CGT	capital gains tax
CHOPS	*see* cold heavy oil production with sand
CHOPT	*see* charterer's option
CHOTO	*see* commissioning, handover, and takeover
CHP	*see* combined heat and power
CI	*see* carbon intensity
CIA	*see* conflict impact assessment
CIC	commercial in confidence
CIDS	*see* concrete island drilling system
CIF	*see* cost insurance freight
CIM	*see* confidential information memorandum
CISG	*see* Convention on the International Sale of Goods
CJV	*see* contractual joint venture
CLL	*see* concentrated live load
CLNG	*see* compact LNG
CM	*see* conditioning monitoring
CM/OA	*see* carbon mitigation/offset agreement
CMA	*see* construction management agreement
CMM	*see* coal mine methane
CMS	*see* condition monitoring system
CMT	*see* crisis management team
CN	*see* cetane number; confirmation notice
CNG	*see* compressed natural gas
CNS	Central North Sea; *see* United Kingdom continental shelf
CO2e	*see* carbon dioxide equivalent

CO	*see* change order
COA	*see* contract of affreightment
COB	*see* clean on board; close of business
COC	*see* carbon offset credit
COCO	*see* company-owned company-operated
COD	*see* commencement of deliveries; commercial operations date
COFO	*see* company-owned franchisee-operated
COFR	*see* Certificate of Financial Responsibility
COLA	*see* crude oil lifting agreement
Colreg	*see* collision regulation
Con-A	*see* Condition A
Con-B	*see* Condition B
Con-C	*see* Condition C
Con-O	*see* other condition materials
COO	*see* certificate of origin
COP	*see* cessation of production
COPAS	*see* Council of Petroleum Accountants Societies
CoQ	certificate of quality; certificate of quantity
CORAN	*see* core analysis report
COS	*see* chance of success; cost of service
COSPA	*see* crude oil sale and purchase agreement
COSS	*see* company-owned service station
COTA	*see* coal trading agreement; crude oil transportation agreement
COU	*see* conditions of use
COW	*see* crude oil washing
CP	*see* cathodic protection
CPOC	*see* Conoco Phillips optimised cascade
CPP	*see* central processing platform; clean petroleum products
CPR	*see* competent person's report
CPT	*see* cone penetration test
CQQO	certificate of quantity, quality and origin
C-risk	*see* completion risk
CRA	*see* corrosion resistant alloy
CRINE	*see* cost reduction in the new era
CRISTAL	*see* contract regarding an interim supplement to tanker liability
CS	*see* continental shelf
CSA	*see* credit support agreement; common stream agreement
CSEM	*see* controlled source electro-magnetic
CSG	*see* coal seam gas

CSI	*see* commercially sensitive information
CSO	*see* complete shut-off
CSP	*see* collector and separation platform
CSR	*see* corporate and social responsibility
CSS	*see* cyclic steam stimulation
CSU	*see* commissioning and start-up
CSV	*see* construction support vessel
CT	*see* coiled tubing
CTA	*see* commodity trading advisor
CTIA	*see* construction and tie-in agreement
CTMS	*see* custody transfer measurement system
CWD	*see* commercial well declaration
D&A	*see* dry and abandoned; drug and alcohol
D&C	*see* drilling and completion
D:E	*see* debt to equity
DAMFORDET	*see* damages for detention
DAO	*see* deasphalted oil
DAP	*see* delivered at place
DAPS	*see* days all purposes
DAS	*see* day ahead scheduling; data acquisition system
DAT	*see* delivered at terminal
DBBV	*see* double block and bleed valve
DBC	*see* depletion-based contract
DBO	*see* double banking operation
DBT	*see* double bottom tank
DCC	*see* deep catalytic cracking
DCP	damage control plan
DCQ	*see* daily contract quantity
DCS	Danish continental shelf
DCT	*see* double containment tank
DDA	*see* depreciation, depletion, and amortisation
DDC	*see* daily drilling costs
DDDC	*see* dedicated design day capacity
DDPPI	*see* dollar-denominated petroleum production interest
DDR	*see* daily drilling report
DEA	*see* drag embedment anchor
DERV	*see* diesel-engined road vehicle
DES	*see* delivered ex-ship

DFDE	*see* dual-fuel diesel electric
DFI	*see* development finance institution
DFIT	*see* diagnostic fracture injection test
DFQ	*see* downward flexibility quantity
DFS	*see* depleted field storage
DG	*see* decision gate
DH	*see* drilling history
DHC	*see* dry hole costs
DHI	*see* direct hydrocarbon indicator
DHO	*see* domestic heating oil
DHSV	*see* downhole safety valve
DIF	*see* diversified investment fund
DIK	*see* dollar inch kilometre
DIP	*see* developed in production
DLB	*see* ductility level blast
DLC	*see* documentary letter of credit
DLS	*see* dog-leg severity
DMA	*see* dynamic mooring analysis
DME	*see* dimethyl ether
DML	*see* deliverables master list
DMO	*see* domestic market obligation
DMR	*see* dual mixed refrigerant
DNP	*see* developed not producing
DNV	*see* development not viable
DOC	*see* daily operating costs
DOD	*see* drill-on date
DODO	*see* dealer-owned dealer-operated
DOF	*see* digital oilfield
DOH	*see* development on hold
DOP	*see* deliver or pay
DOSS	*see* dealer-owned service station
DP	*see* development pending; dynamic positioning
DPP	*see* dirty petroleum products
DPSA	*see* development and production sharing agreement
DPT	*see* downhole pressure and temperature
DPR	*see* daily production rate
DPW	*see* deep pool well
DR	*see* developed reserves
DRA	*see* drag-reducing agent

DRO	*see* discovered resources opportunities
DRP	*see* drill ready prospect
DRR	*see* dismantling, removal, and restoration
DSA	*see* decommissioning security agreement
DSCR	*see* debt service coverage ratio
DSRA	*see* debt service reserve account
DSS	*see* days since spud
DST	*see* drill stem testing
DSU	*see* debt service undertaking
DSV	*see* downhole safety valve
DSVC	*see* deep sea vessel capacity
DSWL	*see* design still water level
Dth	*see* dekatherm
DTW	*see* dealer tank wagon
DW	*see* deep water
DWP	deep water port
DWT	*see* deadweight tonnage
DYNPOS	*see* dynamic positioning
E&A	*see* exploration and appraisal
E&C	engineering and construction; ethics and compliance
E&E	ecology and environment; electrical and electronics
E&P	*see* exploration and production
E&S	environmental and social
EBB	*see* electronic bulletin board
EBITDA	*see* earnings before interest, tax, depreciation, and amortisation
EBITDAX	*see* EBITDA exploration expenses
ECA	*see* emission control area; export credit agency
ECD	*see* equivalent circulating density
ECS	*see* elemental capture spectroscopy; Energy Charter Secretariat
ECT	*see* Energy Charter Treaty
ED	*see* external diameter
EDC	*see* equivalent distillation capacity
EDI	*see* electronic data interchange
EDS	*see* emergency disconnect sequence
EEA	*see* exclusive exploitation authorization
EED	energy exploration and development
EEMUA	*see* Engineering Equipment and Material Users Association
EEZ	*see* exclusive economic zone

EFET	*see* European Federation of Energy Traders
EFM	*see* electronic flow measurement
EHS	extremely hazardous substance
EI	*see* Energy Institute
EIA	*see* earn-in agreement/earn-out agreement; Energy Information Administration; environmental impact assessment
EIS	East Irish Sea; *see* United Kingdom continental shelf
EITI	*see* Extractive Industries Transparency Initiative
EJV	*see* equity joint venture
EMF	*see* extensible modelling framework
EMT	*see* environmental management team
ENTSOG	*see* European Network of Transmission System Operators for Gas
EOA	*see* earn-in agreement/earn-out agreement
EOD	*see* end of day
EOFL	*see* end of field life
EOP	*see* enhanced oil production
EOR	*see* enhanced oil recovery; exporter of record
EOS	*see* east of Suez
EOSP	*see* end of sea passage
EOWR	*see* end of well report
EPCI	*see* engineering procurement construction installation
EPF	*see* early production facility
EPIRB	*see* emergency position-indicating radio beacon
EPM	*see* ethane propane mix
EPR	*see* explosion protection review
EPS	electric power sector
EPSA	*see* exploration and production sharing agreement
EPT	*see* electromagnetic propagation tool; extended production test
ERC	*see* emergency release coupling
ERD	*see* extended reach drilling
ERP	*see* economically recoverable petroleum
ERR	*see* economically recoverable reserves; effective royalty rate
ERRV	*see* emergency response rescue vessel
ERT	*see* emergency response training
ESD	*see* emergency shut-down
ESG	*see* environmental, social, and governance
ESIA	*see* environmental and social impact assessment

ESP	*see* electric submersible pump
ESS	*see* expandable sand screen
ESSA	*see* emergency systems sustainability analysis
ETA	*see* estimated time of arrival; event tree analysis
ETBE	*see* ethyl tertiary butyl ether
ETD	*see* estimated time of departure
ETRM	*see* energy trading risk management
ETS	*see* emissions trading scheme
EUR	*see* estimated ultimate recovery
EV	*see* expected value
EVA	*see* extreme value analysis
EVL	*see* erosional velocity limit
EWT	*see* extended well test
ExplrPSA	*see* exploration PSA
ExpltPSA	*see* exploitation PSA
F&G	*see* fire and gas
FAC	*see* first aid case
FAF	*see* fuel adjustment factor
FAME	*see* fatty acid methyl ester
FAOP	*see* full away on passage
FARO	flowed at rate of
FAT	*see* factory acceptance test
FCC	*see* fluid catalytic cracking
FCT	*see* full containment tank
FDP	*see* field development plan
FDPSO	*see* floating drilling production storage and offtake
FE	*see* formation evaluation
FEA	*see* fire and explosion analysis
FEED	*see* front-end engineering design
FEP	*see* flexible energy purchasing
FEPC	*see* FEED EPC
FEWD	*see* formation evaluation while drilling
FFM	*see* full-field model
FFO	*see* funded from operations
FFP	fit for purpose
FFS	fit for service

FGD	*see* flue gas desulphurisation
FGS	*see* field gathering station
FGSS	*see* fuel gas supply system
FIA	*see* farm-in agreement/farm-out agreement
FID	*see* final investment decision
FIDIC	*see* Fédération Internationale Des Ingénieurs-Conseils
FIFO	*see* first-in first-out
FILO	*see* first-in last-out
FIP	*see* free in pipe
FIT	*see* formation integrity test
FL&U	*see* fuel lost and unaccounted for
FLCR	*see* field life coverage ratio
FLE	*see* full lifecycle emissions
FLNG	*see* floating LNG
FLNGV	*see* floating LNG vessel
FLS	*see* fatigue limit state
FM	flare measurement; flow meter; *see* force majeure
FME	*see* force majeure excepted
FMEA	*see* failure mode and effects analysis
FMP	*see* first marketable product
FMPER	*see* force majeure pre-emption right
FMRQ	*see* force majeure restoration quantity
FMV	*see* fair market value
FOA	*see* farm-in agreement/farm-out agreement
FOB	*see* free on board
FOC	fibre optic cable
FOD	*see* fuel oil domestique
FONSI	*see* finding of no significant impact
FPF	*see* floating production facility
FPSO	*see* floating production, storage, and offtake
FPW	flowback and produced water
FQG	*see* feed quality gas
FRC	*see* fast rescue craft
FS	*see* fail-safe
FSO	*see* floating storage offshore
FSRU	*see* floating storage and regasification unit
FSS	*see* firm storage service
FSU	*see* floating storage unit
ft3	*see* cubic foot

FTE	*see* full-time employee
FTGG	*see* full tensor gravity gradiometry
FTP	*see* first tranche petroleum
FVF	*see* formation volume factor
FW	fresh water
FWKO	*see* free water knock-out
G&A	*see* general & administrative overhead
G&G	*see* geological and geophysical
G&OCM	*see* gas and oil-cut mud
G&P	*see* gathering and processing
G&POFP	*see* good and prudent oilfield practice
G&T	*see* gathering and transportation
GAAP	*see* generally accepted accounting principles
GAR	*see* gas analysis recorder
GBA	*see* gas balancing agreement
GBS	*see* gravity-based structure
GCL	*see* glycol carriage line
GCM	*see* gas-cut mud
GCMS	*see* gas chromatography mass spectrometry
GCU	*see* gas containment unit
GDP	*see* gas disposal project
GDPP	*see* general development and production plan
GDT	*see* gas down-to
GDU	*see* gas dehydration unit
GECF	*see* Gas Exporting Countries Forum
GEF	*see* gas expansion factor
GERM	*see* gas export regulation manifold
GFBE	good faith best estimate
GFE	good faith estimate
GGG	*see* grid-grade gas
GHG	*see* greenhouse gas
GIIGNL	*see* Groupe International des Importateurs de Gaz Natural Liquefié
GIIP	*see* gas initially in place
GIS	*see* geographic information system
GISB	*see* Gas Industry Standards Board
GJ	*see* gigajoule
GLC	*see* government-linked company
GMDSS	*see* global maritime distress safety system

GNL	*see* gaz natural liquefié
GN/WM	*see* gross negligence/wilful misconduct
GOC	*see* gas-oil contact; government-owned company
GOM	Gulf of Mexico
GOR	*see* gas-oil ratio; gross overriding royalty
GOSP	*see* gas/oil separation plant
GOST	*see* Gosudarstvennyy standart
GOV	*see* gross observed volume
GP	*see* general purpose
GPA	*see* Gas Processors Association
GPM	*see* gallons per minute
GPP	*see* gas processing plant; guilty party pays
GPW	*see* gross product worth
GRT	*see* gross registered tonnage
GRV	*see* gross rock volume
GSA	*see* gas sales agreement
GSE	*see* government-sponsored enterprise
GSPA	*see* gas sale and purchase agreement
GSV	*see* gross standard volume
GT	gas turbine
GTA	*see* gas transportation agreement
GTCs	*see* general terms and conditions
GTD	*see* general technical description
GTL	*see* gas-to-liquids
GTP	gas treatment platform; *see* gas-to-power
GTS	*see* geologic time scale
GWC	*see* gas-water contact
GWP	*see* global warming potential
HAT	*see* highest astronomical tide
HAZID	hazard identification
Hazmat	hazardous material
HAZOP	*see* hazard and operability
HBL	*see* hydrostatically balanced load
HBP	*see* held-by production
HC	*see* human capital
HCHP	*see* high consequence high potential
HCLP	*see* high consequence low potential

HCWC	*see* hydrocarbon water contact
HD5	consumer-grade propane
HD10	propane which is graded below HD5
HFO	*see* heavy fuel oil
HFS	*see* hydraulic fracture stimulation
HGA	*see* host government agreement
HHV	*see* higher heating value
HIP	*see* hydrate inhibitor pipeline
HiPo	high potential
HIPPS	*see* high integrity pressure protection system
HLO	*see* helicopter landing officer
HLS	*see* Heavy Louisiana Sweet
HLV	heavy lift vessel
HOCM	*see* heavy oil-cut mud
HP	high-pressure
HP/HT	*see* high-pressure/high-temperature
HPU	*see* hydraulic power unit
HSD	*see* high-speed diesel
HSE	*see* health, safety, and environment
HSES	health, safety, environment, and security
HSFO	*see* high-sulphur fuel oil
HSI	*see* high-stability instrumentation
HSP	*see* hydraulic submersible pump
HSV	*see* hyperbaric support vessel
HT	high temperature
HUTIC	*see* hook-up, tie-in, and commissioning
HVAC	heating, ventilation, air conditioning
HVR	*see* Hague-Visby Rules; *see* hydrocarbon value realization
HX	*see* heat exchanger
I&A	inspection and acceptance
I&CN	*see* industrial and cultural noise
I&R	inspection and rejection
IA	*see* implementation agreement
IACS	*see* International Association of Classification Societies
IADC	*see* International Association of Drilling Contractors
IBA	*see* impact and benefits agreement
IBP	*see* initial boiling point; Irish Balancing Point
ICC	*see* International Chamber of Commerce

ICE	*see* Intercontinental Exchange
ICOP	*see* infrastructure code of practice
ICS	Irish continental shelf; *see* International Commission on Stratigraphy
ICSID	*see* International Centre for Settlement of Investment Disputes
ICSS	*see* integrated control and safety system
ICV	*see* inflow control valve
ID	*see* internal diameter
IDD	*see* internal due diligence
IDHL	immediately dangerous to health and life
IEA	*see* International Energy Agency
IEM	*see* integrated energy market
IFO	*see* intermediate fuel oil
IGA	*see* inter-government agreement
IGC	*see* IGC Code; internal gas consumption
IGCC	*see* integrated gasification combined cycle
IGF	*see* IGF Code
IGS	*see* inert gas system
IGU	*see* International Gas Union
IHUC	*see* installation, hook-up, and commissioning
ILS	*see* industry-led solution
ILX/ILEx	*see* infrastructure-led exploration
IM	*see* information memorandum
IMHH	*see* industry mutual hold harmless
IMO	*see* International Maritime Organisation
ImpG	Imperial gallon
IMR	*see* inspection, maintenance, and repair
INL	*see* International Navigating Limits
INOC	*see* international national oil company
IO	*see* integrated operations
IOC	*see* international oil company
IOGP	*see* International Association of Oil and Gas Producers
IOR	*see* importer of record; improved oil recovery
IPIECA	*see* International Petroleum Industry Environmental Conservation Association
IPMT	*see* integrated project management team
IPP	*see* independent power project
IPR	*see* initial production rate; intellectual property rights

IRR	*see* internal rate of return
IS	Irish Sea
ISBL	*see* inside battery limit
ISD	*see* inherently safer design
ISDA	*see* International Swaps and Derivatives Association
ISM	*see* international safety management certificate
ISO	*see* independent system operator; International Organization for Standardization
ISP 98	*see* International Standby Practices 98
ISPS	*see* International Ship and Port Facility Security Code
ISS	*see* interruptible storage service
ITM	*see* in-the-money
ITO	*see* independent transmission operator
ITOPF	*see* International Tanker Owners Pollution Federation
ITT	*see* invitation to tender
IWL	*see* Institute Warranty Limits
IWS	*see* in-water survey
J&A	*see* junked and abandoned
J&S	*see* joint and several
J&WO	*see* jettison and washed overboard
JCC	*see* Japan Customs-cleared Crude
JDZ	*see* joint development zone
JET A	*see* aviation turbine kerosene
JET B	*see* aviation turbine gasoline
JFD	*see* justified for development
JIB	*see* joint interest billings
JIP	*see* joint industry project
JKM	*see* Japan Korea Marker
JOA	*see* joint operating agreement
JODI	*see* joint oil data initiative
JSBA	*see* joint study and bidding agreement
JTLTS	*see* Joule Thompson low temperature separator
J-UB	*see* jack-up barge
JVCo	*see* joint venture company
K4K	*see* knock-for-knock
KB	*see* kelly bushing
KBD	thousands of barrels per day

KBOE	one thousand BOE
KD	*see* knocked down
Kh	*see* permeability height
KJ	*see* kilojoule
KPI	*see* key performance indicator
L&O	*see* lease and operate
LACT	*see* lease automatic custody transfer
LAER	*see* lowest achievable emissions rate
LASROB	*see* liquids and solids remaining on board
LAT	*see* lowest astronomical tide
LBG	*see* liquefied biogas
LC	*see* letter of credit; local content
LCA	*see* lifecycle assessment
LCHP	*see* low consequence high potential
LCIPS	*see* lowest carbon intensity project solution
LCLP	*see* low consequence low potential
LCS	lowest cost solution
LCZ	*see* lost circulation zone
LDs	*see* liquidated damages
LDC	*see* local distribution company
LDD	*see* loss damage delay
LDZ	*see* local distribution zone
LEG	*see* liquefied ethylene gas
LEL	*see* lower explosive limit
LHV	*see* lower heating value
LIFO	*see* last-in first-out
LILO	*see* last-in last-out
LIP	*see* lifting in place
LKG	lowest known gas
LKO	lowest known oil
LLC	*see* late life compression
LLCR	*see* loan life coverage ratio
LLI	*see* long-lead item
LLS	*see* Light Louisiana Sweet
LMRP	*see* lower marine riser package
LMS	*see* legal minimum stock
LNA	*see* learning needs assessment
LNG	*see* liquefied natural gas

LNGC	*see* LNG carrier
LOA	*see* length overall
LOC	*see* local oil company
LOGIC	*see* Leading Oil and Gas Industry Competitiveness
LOPA	*see* layer of protection analysis
LOPI	*see* loss of production income insurance
LOS	*see* line of sight
LOT	*see* loaded on top
LOTO	*see* lock-out tag-out
LOTOTO	*see* lock-out tag-out try-out
LP	liquid propane; low-pressure; *see* linear programme
LPFO	*see* low pour fuel oil
LPG	*see* liquefied petroleum gas
LQ	*see* living quarters
LQE	*see* letter of quiet enjoyment
LR1	*see* large range 1
LR2	*see* large range 2
LRG	*see* liquefied refinery gas
LRS	*see* long-range storage
LRV	*see* liquefied natural gas regasification vessel
LSA	*see* low specific activity
LSC	*see* longshore current
LSFO	*see* low sulphur fuel oil
LSMF	*see* low sulphur marine fuel
LSTK	*see* lump sum turnkey
LSWR	*see* low sulphur waxy residue fuel oil
LT	*see* long ton; low temperature
LTBP	*see* London Tanker Brokers' Panel
LTC	*see* long term contract
LTFQ	*see* long term fixed quantity
LTI	*see* long term interruptible; lost time injury
LTM	*see* last twelve months
LTMEBITDA	*see* last twelve months EBITDA
LTMEBITDAX	*see* last twelve months EBITDA exploration expenses
LTO	*see* light tight oil
LTSA	*see* long term sales agreement; long term services agreement
LTT	*see* lock-out tag-out try-out
LTX	*see* low temperature separator

LUSC	*see* liquefied natural gas unit shipping cost
LWD	*see* logging while drilling
M	one thousand
M&S	*see* marketing and sales
M&T	*see* measurement and testing
M-100	*see* mazut
m3	*see* cubic metre
Ma	millions of years
MAASP	*see* maximum allowable annulus surface pressure
MAC	*see* material adverse change; mobile Arctic caisson
MAE	*see* material adverse event
MAF	*see* materials unaccounted for
MAH	major accident hazard
MAOP	*see* maximum allowable operating pressure
MARPOL	*see* International Convention for the Prevention of Pollution from Ships
MASP	*see* maximum allowable surface pressure
M-bal	*see* material balance equation
MBBL	one thousand barrels
MBL	*see* mid-body length; minimum breaking load
MCal	*see* megacalorie
MCHE	*see* main cryogenic heat exchanger
MCR	*see* maximum continuous rating
MCV	*see* modular containment vessel
MD	*see* measured depth; measured distance
MDEA	*see* methyl diethanolamine
MDL	*see* methane drainage licence
MDO	*see* marine diesel oil
MDS	*see* middle distillate synthesis
MDT	*see* modular dynamic testing
MEFC	*see* maximum economic finding cost
MEG	*see* mono-ethylene glycol
MER	*see* maximising economic recovery; maximum efficient rate/ most efficient recovery; merchant–equity ratio
MFDT	*see* modular formation dynamics tester
MFN	*see* most-favoured nation
MFO	*see* marine fuel oil
MFP	*see* minimum facility platform

MGO	*see* marine gas oil
MGPS	*see* marine growth prevention system
MHH	*see* mutual hold harmless
MHHW	*see* mean higher high water
MHWN	*see* mean high water neaps
MHWS	*see* mean high water springs
MICP	*see* mercury injection capillary pressure
MILOS	*see* middle distillates lower olefins selective
Mini-man	*see* minimised manning
MIR	moving in rig
MIT	*see* mechanical integrity test; multilateral investment treaty
MIYP	*see* minimum internal yield pressure
MLA	*see* marine loading arm
MLLW	*see* mean lower low water
MLWN	*see* mean low water neaps
MLWS	*see* mean low water springs
MM	one million
MMBBL	one million barrels
MMLS	*see* moveable modular liquefaction system
MMybp	*see* millions of years before present
MOB	*see* man overboard boat
MOC	*see* management of change
MODU	*see* mobile offshore drilling unit
MOF	*see* material offloading facility
MOIC	*see* multiple on invested capital
MON	*see* motor octane number
MOOIP	*see* moveable oil originally in place
MOPO	*see* manual of permitted operations
MOPU	*see* mobile offshore production unit
MOR	*see* means of rescue
MPa	*see* megapascal
MPD	*see* managed pressure drilling; multiple product dispenser
MPFM	*see* multiphase flow meter
MPOP	*see* maximum pipeline operating pressure
MPPF	*see* multi-pad production facility
MR	*see* medium range
MRL	*see* mid-reach lateral
MRS	*see* medium-range storage
MRV	*see* measurement, reporting, verification

MSA	*see* master sales agreement; master services agreement
MSDS	*see* material safety data sheet
MSF	*see* multi-stage fracking
MSL	*see* mean sea-level; minimum stock to load
MSPA	*see* master sale and purchase agreement
MT	*see* metric tonne
MTBE	*see* methyl tertiary butyl ether
MTC	*see* mass-transport complex
MTD	*see* mooring tension damper
MTM	*see* mark-to-market
MTO	*see* materials take off
MTPA	metric tons per annum
MTTF	*see* mean time to failure
MUI	*see* maintenance of uniform interest
MVOP	multi-vessel operation
MWC	*see* minimum work commitment
MWD	*see* measurement while drilling
MWE	*see* minimum work expenditure
MWO	*see* minimum work obligation
NAESB	*see* North American Energy Standards Board
NAG	*see* non-associated gas
NBP	*see* National Balancing Point
NBP 1997/ 2015	*see* short term flat agreement
NCI	*see* Nelson complexity index
NCG	*see* NetConnect Germany
NCS	Norwegian continental shelf
NDA	*see* non-disclosure agreement
NDT	*see* non-destructive testing
NFI	*see* no further investment
NFW	*see* new field wildcat
NGLs	*see* natural gas liquids
NGO	*see* non-governmental organisation
NGV	*see* natural gas vehicle
NLPM	*see* normal litre per minute
NMI	*see* near miss incident
NMOC	*see* non-methane organic compounds
NMR	*see* nuclear magnetic resonance

NNS	Northern North Sea; *see* United Kingdom continental shelf
NOC	*see* national oil company
NOMI	*see* non-operated minority interest
NOR	*see* notice of readiness
NORA	*see* notice of readiness accepted
NORM	*see* naturally occurring radioactive material
NORR	*see* notice of readiness rejected
NORSOK	*see* Norwegian offshore operational standards
NORT	*see* notice of readiness tendered
NOV	*see* non-operated venture
NOX	*see* nitrogen oxides
NPAI	*see* not permanently attended installation
NPD	*see* Norwegian Petroleum Directorate
NPF	*see* non-producing facility
NPG	*see* non-petroleum gases
NPI	*see* net profit interest
NPRI	*see* non-participating royalty interest
NPT	*see* non-productive time
NPU	*see* non-petroleum use
NPV	*see* net present value
NPW	*see* net product worth; new pool wildcat; non-producing wellbore
NRA	*see* national regulatory authority
NRT	*see* net registered tonnage
NRU	*see* nitrogen rejection unit
NSV	*see* net standard volume
NTGR	*see* net–gross ratio
NTM	*see* next twelve months
NTPA	*see* negotiated third party access
NTRs	*see* non-technical risks
NTS	*see* National Transmission System; non-technical summary
NUI	*see* normally uninhabited (unattended) installation
NYMEX	*see* New York Mercantile Exchange
O&M	*see* operating (operation) and maintenance
O&MA	*see* operating (operation) and maintenance agreement
OBA	*see* operational balancing agreement
OBC	*see* ocean bottom current
OBE	*see* open book estimate; operating basis earthquake
OBL	*see* oil-backed loan

OBM	*see* obsolescing bargain model; oil-based mud
OBN	*see* ocean bottom node
OBO	*see* operated-by-other
OBQ	*see* on-board quantity
OCGT	*see* open cycle gas turbine
OCIMF	*see* Oil Companies International Marine Forum
OCIP	*see* owner-controlled insurance programme
OCM	*see* oil-cut mud; operating committee meeting
OCO	*see* one-cancels-other
OCS	*see* outer continental shelf
ODT	*see* oil down-to
OE	operating efficiency/operational efficiency; *see* oil equivalent
OEE	*see* operator's extra expense
OEG	*see* oil-equivalent gas
OEL	*see* Oil Exploration Licence
OEM	*see* original equipment manufacturer
OFID	*see* OPEC fund for international development
OFO	*see* operational flow order
OFS	*see* oil field services
OGM	*see* on ground maximum
OGSE	*see* oil and gas services and equipment
OGUK	*see* Oil and Gas UK Limited
OH	*see* open hole
OHADA	*see* Organisation pour l'Harmonisation en Afrique du Droit des Affaires
OHGP	*see* open hole gravel pack
OIIP	*see* oil initially in place
OIM	*see* offshore installation manager
OLR	*see* onshore licensing round
OML	*see* Oil Mining Licence
OMS	*see* operations management system
OOC	office/occupancy costs
OOL	*see* oil offloading line
OOS	*see* oil offloading system
OpCom	*see* operating committee
OPE	*see* oil price escalation
OPEC	*see* Organisation of Petroleum Exporting Countries
OPEC+	*see* Organisation of Petroleum Exporting Countries +
OPEX	*see* operating expenditure

OPF	*see* open flow potential
OPL	*see* Oil Prospecting Licence
OPM	*see* other peoples' money
OPOL	*see* offshore pollution liability agreement
ORA	*see* operational risk assessment
ORRI	*see* overriding royalty interest
ORV	*see* open rack vaporiser
OS&D	*see* over short and damage
OSBL	*see* outside battery limit
OSC	*see* oilfield service company
OSCP	oil spill contingency plan
OSP	*see* official sales price
OSPAR	*see* Oslo and Paris Conventions
OSR	*see* offshore service rig
OSV	*see* offshore support vessel; offshore supply vessel
OTM	*see* out-of-the-money
OTR	*see* on the rig
OTT	*see* over the tide
OTTP	*see* offshore title transfer point
OU	*see* ownership unbundling
OVSP	*see* offset vertical seismic profile
OWC	*see* oil–water contact
P10	*see* possible reserves
P50	*see* probable reserves
P90	*see* proved reserves
P&A	*see* plug and abandon
P&I	*see* protection and indemnity club
P&ID	*see* piping and instrumentation diagram
P&MS	*see* plug and make safe
P&P	*see* porosity and permeability
P&S	*see* produced and saved
Pa	*see* Pascal
PACV	*see* pressure activated circulating valve
PALS	*see* park and loan service
PAM	*see* passive acoustic monitoring
PAR	*see* pre-assembled rack
PAU	*see* pre-assembled unit
P/B	*see* propane/butane ratio

PBE	*see* play-based exploration
PBL	*see* parallel body length
PBM	*see* political bargaining model
PBS	*see* pilot boarding station
PBSI	*see* pilot boarding station inbound
PBSO	*see* pilot boarding station outbound
Pc	*see* chance of commerciality
PC	*see* part cargo
PCG	*see* parent company guarantee
PCOP	*see* permanent cessation of production
PCP	*see* progressive cavity pump
Pd	*see* chance of development
PDH	*see* propane dehydrogenation
PDNP	*see* proved developed not producing
PDP	*see* proved developed producing
PE	petroleum engineer; *see* permanent establishment
PEC	*see* production enhancement contract
PEDL	*see* petroleum exploration development licence
PEG	*see* Point d'Échange de Gaz
PEL	*see* permissible exposure limit
PEMS	*see* predictive emissions monitoring system
PER	*see* process energy requirement
PERC	*see* powered emergency release coupling
PFD	*see* planned for development
PFF	*see* pipes, flanges, and fittings
PFP	*see* passive fire protection
Pg	*see* chance of geologic discovery
PGA	*see* planning gain agreement
PHA	*see* production handling agreement
PI	*see* productivity index
PIANC	*see* Permanent International Association of Navigation Congresses
PIB	*see* product information bulletin
PIC	payment in cash
PIF	*see* petroleum investment fund
PIK	payment in kind; *see* payment in-kind interest
PIP	*see* produced into pipeline
PIR	*see* potential impact radius
PIT	*see* pressure integrity test

PKD	*see* partly knocked down
PLA	*see* port liability agreement
PLCR	*see* project life coverage ratio
PLEM	*see* pipeline end manifold
PLET	*see* pipeline end termination
PLSV	*see* pipelay support vessel
PLT	*see* production logging tool
PLV	*see* pipelay vessel
PMC	project management consultant
PMS	*see* premium motor spirit
PMT	*see* project management team
PND	*see* proved not developed
PNG	*see* pipeline natural gas
PO	*see* purchase order
POA	*see* power of attorney
POB	*see* personnel on board
POC	*see* proof of concept
POD	*see* plan of development
POL	*see* petrol, oil, lubricants
POP	*see* percentage of proceeds contract; periodic operating plan; pipeline operating principles
POS	*see* probability of success
POSA	*see* processing and operating services agreement
POSMOOR	*see* position mooring
POT	*see* proven offshore technology
PPA	*see* power purchase agreement
PPD	*see* pour-point depressant
PPE	*see* personal protective equipment
PPI	*see* percentage participating interest
PPM	parts per million; *see* planned preventative maintenance; predictive preventative maintenance
PPMV	parts per million (measured by) volume
PPMW	parts per million (measured by) weight
PQG	*see* pipeline quality gas
PR	*see* proved reservoir
PRA	*see* price reporting agency
PRG	*see* partial risk guarantee
PRI	*see* political risk insurance
PRICO	*see* poly refrigerant integral cycle operation

PSA	*see* pressure swing adsorption; product sales agreement
PSC	*see* production sharing contract
PSD	*see* planned shut-down
PSDM	*see* pre-stack depth migration
Psi	*see* pound per square inch
PSL	*see* pipeline safety limit
PSM	*see* profit-sharing mechanism
PSO	*see* public service obligation
PSONR	*see* permanent sovereignty over natural resources
PSR	*see* pipeline system rules
PSS	*see* pseudo steady state
PSSR	*see* pre-start-up safety review
PSTM	*see* pre-stack time migration
PSUA	*see* pre-start-up audit
PSV	*see* platform supply vessel; Punto di Scambio Virtuale
PTA	*see* pressure transient analysis
PTL	*see* produce the limit
PTO	*see* permit to operate
PTW	*see* permit to work
PTZ	*see* pan tilt zoom
PU	*see* petroleum use
PUA	*see* pre-unit agreement
PUD	*see* proved undeveloped
PV 10	*see* present value 10
PV	*see* pore volume
PVV	*see* pressure vacuum valve
PVT	*see* pressure, volume, temperature
PW	*see* potable water; produced water; producing wellbore
PWD	*see* pressure while drilling
PWO	*see* produced water overboard
Q88	*see* Questionnaire 88
Q&Q	quality and quantity
QAP	*see* quality activity plan
QA/QC	*see* quality assurance/quality control
QC/DC	quick connection/disconnection coupling
Q-Flex	*see* Qatar Flex

Q-Max	*see* Qatar Max
QRA	*see* quantitative risk assessment
QRMH	*see* quick release mooring hooks
QTF	*see* quadratic transfer function
QU	*see* quarters and utilities
R&D	research and development
R&LE	refurbishment and life extension
R&M	*see* refining and marketing; repair and maintenance
R&R	*see* repeatability and reproducibility
RACC	*see* refinery acquisition cost of crude
RAD	*see* radioactive densitometer
RAM	*see* reliability, availability, maintenance
RAO	*see* response amplitude operator
RBC	*see* regulation below cost
RBI	*see* risk-based inspection
RBL	*see* reserves-based lending
RBOB	*see* reformulated blendstock for oxygenate blending
RC	reinforced concrete
RCA	*see* root cause analysis; routine core analysis
RCM	*see* restricted catenary mooring
RCP	*see* reduced circulating pressure
RCS	*see* regulation cost of service
RD&D	research, design, and development
RE	reasonable endeavours; reservoir engineer
REBCO	*see* Russian export blend crude oil
RES	*see* renewable energy source
RF	*see* recovery factor
RFG	*see* reformulated gasoline
RFO	*see* ready for operation; residual fuel oil
RFT	*see* repeat formation testing
RMMLF	*see* Rocky Mountain Mineral Law Foundation
RMP	*see* reservoir management plan
RNG	*see* renewable natural gas
RO	*see* retail outlet
ROB	*see* retention on board
ROFO	*see* right of first offer
ROFR	*see* right of first refusal

RON	*see* research octane number
ROP	*see* rate of penetration
ROR	*see* rate of return
ROT	*see* retention of title
ROTR	*see* resources other than reserves
ROV	*see* remote operated vehicle
ROZ	*see* residual oil zone
RP	*see* return period
RPG	*see* remote power generation
RPM	*see* resale price maintenance
RPO	*see* reasonable and prudent operator
RPT	*see* rapid phase transition
RQ	*see* reservoir quality
RRF	*see* risk reduction factor
RRR	*see* reserves replacement ratio
RSC	*see* rotary sidewall core
RSD	*see* routine shut-down
RSG	*see* responsibly sourced gas
RSWC	rock samples with cuttings
R/T	*see* royalty/tax
RTD	*see* reserve tail date
RTP	*see* ready to produce
RTPA	*see* regulated third-party access
RU	*see* reservoir
RVP	*see* Reid vapour pressure
S&M	*see* statute and mandate
S&OR	safety and operational risk
S&W	sediment and water; *see* basic sediment and water
SAGD	*see* steam-assisted gravity drainage
SALM	*see* single anchor loading/leg mooring
SAPG	*see* stranded area power generation
SAS	*see* surveyed after shipment
SBC	*see* supply based contract
SBGD	*see* sea-bed gas diverter
SBLC	*see* standby letter of credit

SBM	*see* single buoy mooring; synthetic-based mud
SBS	*see* surveyed before shipment
SBU	*see* standard bundled unit
SC	*see* sidewall core
SCA	*see* structural consequences analysis
SCADA	*see* supervisory control and data acquisition
SCAL	*see* special core analysis
SCAP	*see* supply chain action plan
SCE	*see* safety critical element
scf	standard cubic foot; *see* cubic foot
SCGT	*see* single cycle gas turbine
SCM	*see* supply chain management
SCO	*see* stabilised crude oil
SCT	*see* single containment tank
Scuf	standard cubic foot; *see* cubic foot
SCV	*see* submerged combustion vaporiser
SDA	*see* solvent deasphalting
SDR	*see* shared data repository
SDV	*see* shut-down valve
SEC	US Securities and Exchange Commission
SFO	*see* storage facility owner
SFU	*see* storage facility user
SHE	*see* safety, health, and environment
SHEQ	safety, health, environment, quality
SHEX	*see* Sundays and holidays excluded
SHINC	*see* Sundays and holidays included
SHOPT	*see* shipowner's option
SIA	*see* smoke ingress analysis
SIBHP	*see* shut in bottomhole pressure
SICP	*see* shut in casing pressure
SIDSP	*see* shut in drill stem pressure
SIF	*see* safety instrumented function
SIGTTO	*see* Society of International Gas Tanker and Terminal Operators
SIL	*see* safety integrity level
SIMOPS	*see* simultaneous operations
SIRE	*see* ship inspection report programme
SIS	*see* safety instrumented system; surface to in-seam drilling

SITA	*see* shut in temporarily abandoned
SLA	*see* social lifecycle analysis
SLB	*see* strength level blast
SLL	*see* sustainability linked loan
SLPM	*see* standard litre per minute
SLR	*see* seaward licensing round
SLS	*see* specific lifting schedule
SMOG	*see* standardised measure of oil and gas
SMR	*see* single mixed refrigerant
SMUT	*see* spirit of mutual understanding and trust
SNG	*see* sustainable natural gas; synthetic natural gas
S/NR	*see* signal/noise ratio
SNS	Southern North Sea; *see* United Kingdom continental shelf
SOE	*see* state-owned enterprise
SOLAS	*see* Safety Of Life At Sea
SOLR	*see* supplier of last resort
SOP	standard operating procedure
SOR	*see* statement of requirements
SOX	*see* sulphur oxides
SP	*see* spontaneous potential
SPA	*see* sale and purchase agreement
SPE	*see* Society of Petroleum Engineers
SPE PRMS	*see* Society of Petroleum Engineers—Petroleum Resources Management System
SPEE	*see* Society of Petroleum Evaluation Engineers
SPJ	*see* steel piled jacket
SPMT	*see* self-propelled multi-wheel trailer
SPR	*see* single point responsibility; strategic petroleum reserve
Sprof	*see* seismic profile
SPS	sub-sea pipeline system
SPV	*see* special purpose vehicle
SPW	*see* shallow pool well
SRB	*see* sulphate-reducing bacteria
SRL	*see* standard reach lateral
SRME	*see* surface-related multiple elimination
SRS	*see* short-range storage
SRU	*see* sulphur reduction unit
SRV	*see* shuttle regasification vessel
SS	sub-sea; sub-surface; *see* safety stock

SSB	*see* sun, surf, and breeze
SSC	*see* storage service contract
SSCS	*see* ship-shore compatibility study
SSCV	*see* semi-submersible crane vessel
SSD	*see* safety shut-down; slow-speed diesel; special shut-down
SSE	*see* safe shut-down earthquake
SSG	*see* slight show of gas
SSI	*see* sub-surface installation
SSIV	*see* sub-sea isolation valve
SSLNGC	*see* small-scale LNG carrier
SSO	*see* slight show of oil
SSP	*see* senior supervisory personnel
SSSV	*see* sub-surface safety valve
SSTB	*see* sub-sea tie-back
SSU	*see* Saybolt Second Universal
SSV	*see* seismic survey vessel
ST	*see* short ton
STB	*see* stock tank barrel
STCs	*see* special terms and conditions
STIC	*see* state take in cash
STIK	*see* state take in kind
STL	*see* submerged turret loading
STO	shut-down, turnaround, and outage
STOOIP	*see* stock tank oil originally in place
STP	*see* standard temperature and pressure
STS	*see* ship-to-ship
SU	*see* safety ullage
SUAR	*see* start-up assurance review
SUKO	*see* Shell UK Oil
SURF	*see* sub-sea umbilicals, risers, and flowlines
SURM	*see* service user risk management
SV	*see* support vessel
SW	salt water; *see* shallow water; water saturation
SWDW	*see* salt water disposal well
SWF	*see* sovereign wealth fund
SWHE	*see* spiral wound heat exchanger
SWL	*see* safe working load; still water level
SWOPS	single well oil production system

T	*see* ton; trillion
T&D	*see* transportation and distribution
T&M	*see* time and materials contract
T&P	*see* transportation and processing
T&T	*see* treatment and transportation
TA	*see* temporary abandonment
TAC	technical advisory committee
TAN	*see* total acid number
TAP	*see* take and pay
TAR	*see* true amplitude recovery; turnaround
TC	*see* tidal current
TCM	*see* technical committee meeting
TCom	*see* technical committee
TCP	*see* time charter party
TCV	*see* total calculated volume
TD	*see* time-depth
TDS	*see* total dissolved solids
TEA	*see* technical evaluation agreement
TEG	*see* tri-ethylene glycol
TEMPSC	*see* totally enclosed motor propelled survival craft
TEOR	*see* thermal enhanced oil recovery
TFDE	*see* tri-fuel diesel electric
THA	*see* tophole location
THAI	*see* toe-to-heel air injection
THD	*see* true horizontal distance
TLI	*see* time losing injury
TLIFR	*see* time losing injury frequency rate
TLP	*see* tension leg platform
TOC	*see* total organic carbon
TOE	*see* tonne of oil equivalent
TOP	*see* take or pay
TOS	*see* temporary overnight shelter
TOUT	*see* time of use tariff
TOV	*see* total observed volume
TOVALOP	*see* Tanker Owners Voluntary Agreement concerning Liability for Oil Pollution
TPA	*see* third-party access
TPAA	*see* trustee and paying agent agreement
TPH	tons per hour; *see* total petroleum hydrocarbons

TPI	*see* third party inspection
TPOSA	*see* transportation, processing, and operating services agreement
TRA	*see* temporary refuge area
TSDF	*see* treatment, storage, and disposal facility
TSJ	*see* titanium stress joint
TSO	*see* transmission system operator
TSR	*see* tank storage receipt
TSS	*see* total suspended solids
TT	*see* transit time
TTD	*see* time to depth; time to distance
TTF	*see* Title Transfer Facility
TUA	*see* terminal use agreement
TUTU	*see* topside umbilical termination unit
TVD	*see* true vertical depth
TVDBRT	*see* true vertical depth below rotary table
TVDSS	*see* true vertical depth sub-sea
TVP	*see* true vapour pressure
TVT	*see* true vertical thickness
TX	*see* turboexpander
UAP	*see* unallocated provision
UCG	*see* unconventional gas; underground coal gasification
UCP 600	*see* Uniform Customs and Practices for Documentary Credits
UCS	*see* unconfined compressive strength
UDLL	*see* uniformly distributed live load
UDW	*see* ultra-deep water
UEL	*see* upper explosive limit
UFG	*see* unaccounted-for gas
UFQ	*see* upward flexibility quantity
UGPC	*see* upstream government petroleum contract
UIC	*see* underground injection control
UIOLI	*see* use it or lose it
UKC	*see* under keel clearance
UKCS	*see* United Kingdom continental shelf
ULCC	*see* ultra-large crude carrier
ULEC	*see* ultra-large ethane carrier
ULSD	*see* ultra-low sulphur diesel
UNCITRAL	*see* United Nations Commission on International Trade Law

UNCLOS	*see* United Nations Convention on the Law of the Sea
UNDRIP	United Nations Declaration on the Rights of Indigenous Peoples
UOM	*see* urgent operational matter
UPF	*see* unmanned production facility
UPS	*see* uninterruptible power supply
URDG 758	*see* Uniform Rules for Demand Guarantees
UROA	*see* unconventional resources operating agreement
USG	United States gallon
UST	underground storage tank
UTA	*see* umbilical termination assembly
UTE	*see* Unión Transitoria de Empresas
UTM	*see* Universal Transverse Mercator
UUOA	*see* unitisation and unit operating agreement
UWHP	*see* unmanned wellhead platform
UWI	*see* unique well identifier
UWILD	*see* underwater inspection in lieu of dry-docking
UXO	*see* unexploded ordnance
VAR	*see* value at risk
VBN	*see* votes by notice
VCP	*see* voyage charter party
VDD	*see* vendor due diligence
VDL	*see* variable deck load
VDR	*see* virtual data room
VDU	*see* vacuum distillation unit
VEF	*see* vessel experience factor
VFM	*see* virtual flow meter
VGIP	*see* volumetric gas in place
VGM	*see* viscosity gel meter
VI	*see* viscosity index
VIP	*see* virtual interconnection point
VIV	*see* vortex induced vibration
VLCC	*see* very large crude carrier
VLEC	*see* very large ethane carrier
VLGC	*see* very large gas carrier
VNG	*see* vehicular natural gas
VOC	*see* volatile organic compound
VOIP	*see* volumetric oil in place
VOWD	*see* value of work done

VPPI	*see* volumetric production payment interest
VPR	*see* vessel presentation range
VRL	*see* vapour return line
VRR	*see* voidage replacement ratio
VRS	*see* vapour recovery system
VSD	variable speed drive
VSP	*see* vertical seismic profile
VTP	Virtual Trading Point; *see* Central European Gas Hub
VTV	*see* vessel-to-vessel
VWAP	*see* volume-weighted average price
W2W	*see* walk to work
W&II	*see* warranty and indemnity insurance
W&S	*see* won and saved
WACOG	*see* weighted average cost of gas
WAG	*see* water alternating gas
WAP	*see* weighted average price
WAT	*see* wax appearance temperature
WBM	*see* water-based mud
WBS	*see* work breakdown structure
WCMC	*see* worst credible metocean conditions
WCP	*see* well containment package; well control plan
WD	water depth
WDT	*see* wax disappearance/dissolution temperature
WGR	*see* water–gas ratio
WGS	*see* world geodetic system
WHP	*see* wellhead platform
WHRU	*see* waste heat recovery unit
W/hr	*see* Watt/hour (Appendix 1)
WIB	*see* working interest barrel
WLL	*see* working load limit
WMS	*see* well monitoring system
WO	*see* work order
WOB	*see* weight on bit
WOC	*see* waiting on cement
WOG	without guarantee
WOR	*see* water–oil ratio

WOS	West of Shetland *see* United Kingdom continental shelf; *see* west of Suez
WOW	without warranty; *see* waiting on weather; watered out well
WP&B	*see* work programme and budget
WPB	*see* within-pipe blending
WPC	*see* World Petroleum Council
WQA	*see* Wobbe quality adaptation
WRM	*see* well and reservoir management
WSL	*see* well surface location
WSO	*see* water shut off
WTI	*see* West Texas Intermediate
WTT	*see* well to tank
WVSP	*see* walkaway vertical seismic profile
WW	waste water
WWS	*see* wind, wave, and solar
XL	*see* exploration licence
XLOT	*see* extended leakoff test
X-O	*see* crossing
X-OV	*see* crossover valve
XRD	*see* x-ray diffraction
XRL	*see* extended reach lateral
X-tree	*see* Christmas tree
ybp	years before present
YOY	*see* year-on-year
YP	*see* yield point/yield pressure
ZOVSP	*see* zero offset vertical seismic profile
ZT	*see* zone time
ZTP	*see* Zeebrugge trading point

PART B

TERMS AND PHRASES

A

A-frame   A particular type of lifting gear which is installed on the stern of a pipe-laying or other offshore construction vessel for sub-sea load handling.

Abandoned and whipstocked (A&W)   A cased wellbore which has been plugged and abandoned but which has also been whipstocked, whereby the resultant side-track could be reopened for further development.

Abandoned mine methane (AMM)   Residual methane deposits which are found in and are recoverable from disused coal mine workings. Contrast with **coal mine methane** in active coal mine workings. *See also* **methane drainage**.

Abandoned well   A wellbore which is not in use and has been plugged and abandoned (either because the wellbore was originally a dry hole when it was first drilled or because the production of petroleum through the wellbore has ceased over time).

Abandonment   An alternative (and increasingly unfashionable) term for **decommissioning**.

Abandonment and reclamation obligation (ARO)   An alternative term for a **decommissioning obligation**.

Abandonment pressure   The point where the capillary pressure of natural gas within a formation, and the rate of gas production, has dropped such that the continued production of gas has become no longer economically viable.

Abated petroleum   A quantity of petroleum which has been produced in accordance with defined carbon reduction standards (and which could attract a price premium). *See also* **unabated petroleum.**

Abatement   An activity or a process which reduces the level of pollution from a petroleum facility or the carbon intensity of produced petroleum.

Ablation debris   Small pieces of rock which are broken up by the perforation process.

Abnormal peak day   The highest predictable demand for energy which is forecasted to occur in respect of any day within a defined period of time.

Above-ground installation (AGI)   A petroleum facility which is installed at or above ground level. Contrast with a **sub-surface installation**, which is found sub-surface.

Above-ground review (AGR)   A risk assessment which focuses on social, political, and environmental risks and not on sub-surface issues.

Abrasive blasting   The cleaning of steel with compressed air-propelled abrasives prior to painting.

Absolute forfeiture   Under a JOA, the forfeiture by a defaulting party of the entirety of its petroleum project interests because of that party's payment default. Contrast with a **withering interest provision**, whereby the defaulting party forfeits part of its interests proportionate to the degree of the payment default. *See also* **forfeiture**.

Absolute open flow   A measure of the maximum flow rate from a formation which a wellbore could theoretically deliver through a wellhead at any time.

Absolute ownership   An alternative term for **ownership-in-place**.

Absolute permeability   A measure of the ability of a formation to conduct the flow of a phase when the formation is totally saturated with that phase.

Absolute porosity   A measure of the porosity of a rock sample which is expressed as a percentage of the total volume of the rock sample. *See also* **effective porosity**.

Absolute pressure   A measure of pressure which is determined as the combination of atmospheric pressure plus gauge pressure.

Absorber   An alternative term for an **absorption tower**.

Absorption
(1)   A process by which a gas stream is saturated with water or an absorption oil in order to remove the heavier petroleum fractions from the gas stream. Also called **gas absorption** and **scrubbing**. *See also* **absorption tower**.
(2)   The permeation a gas or a liquid in a mineral deposit (such as petroleum in a sedimentary rock deposit). Contrast with **adsorption**, whereby a gas or a liquid adheres to the surface area of a mineral deposit.

Absorption oil   A light oil which is used in the absorption process. Also called **wash oil**.

Absorption tower   In a refinery, a large vertical steel cylinder which separates gases from liquid fractions within a scrubbing process. Also called an **absorber**.

Absorptive capacity   A measure of the ability of a petroleum-producing state to absorb the investment of domestically generated petroleum project revenues and taxation back into the state's wider domestic economy in order to maintain growth in the economy.

Abstract   A written summary of the historical and/or current ownership interests in a concession or a petroleum project. *See also* **title opinion**.

ACCC netback   A pricing methodology for Australian east coast LNG exports which is published by the Australian Competition & Consumer Commission.

Access   An alternative term for **third-party access**.

Accidental limit state (ALS)   The ability of a petroleum infrastructure item to exceed its limit state because of an accident but still to be operational thereafter.

Accommodation sale   In an integrated company, the purchase by the company's refining function of crude oil from the company's petroleum production function.

Accommodation support vessel (ASV)   An offshore personnel accommodation module which houses workers engaged in petroleum project development and petroleum production operations. Also called a **flotel**.

Accounting date   A date in an asset transfer agreement when the economic benefit and burden associated with the asset transfers to the buyer ahead of the legal transfer of the asset.

Accounting procedure (AP)   A procedure which is appended to a JOA and is intended to regulate the cash flows associated with the conduct of joint operations, the right of the operator to levy costs against the parties, and the management of inventories.

Accounting separation   Unbundling which is realised by the need for different elements of a vertically integrated business to provide separate management accounts, thereby revealing any cross-subsidies which might exist between those elements.

Accrual accounting   An accounting methodology whereby a revenue item is accounted for when earned and an expense item is accounted for when incurred. Contrast with **cash accounting**, whereby a revenue item is accounted for when received and an expense item is accounted for when paid out.

Accumulation   An alternative term for a **reservoir**.

Accumulator   A facility which is used for the temporary storage of gas within a gas pipeline network or a gas processing facility which can release stored gas to ensure the continuous flow of gas through the network or the facility in the event of an imbalance. *See also* **surge tank**.

Acid fracture   An alternative term for **acidisation**.

Acid gas
(1) An alternative term for **sour gas**.
(2) A gas such as hydrogen sulphide or carbon dioxide which forms an acid when mixed with water.

Acid gas removal (AGR)   A process for removing acidic components from a gas stream by the use of aqueous solutions of various alkylamines. Also called **amine scrubbing** and **amine treatment**.

Acid stimulation   An alternative term for **acidisation**.

Acid treatment   The treatment of unfinished products such as gasoline, diesel, naphtha, and lubricants with sulphuric acid to approve appearance and odour. *See also* **sweetening**.

Acid wash   The circulation of acid through a wellbore in order to clean the wellbore of accumulated scale, mud cake, and other impediments which have built up and could constrict the initial or the ongoing flow of petroleum.

Acidisation   A process for increasing the flow of petroleum from a formation by pumping high-pressure hydrochloric (or other) acid into a wellbore in order to increase the porosity of sedimentary rocks in the formation which the wellbore has penetrated. Also called **acid fracture** and **acid stimulation**.

Acidity   A measure of the level of acidic compounds which exist within a petroleum stream, which is measured by the total acid number.

Acoustic logging   A method for determining the petroleum prospectivity of a formation by the transmission of sound waves. Also called **acoustic surveying**. *See also* **seismic survey**.

Acoustic surveying   An alternative term for **acoustic logging**.

Acquisition and divestment (A&D) The activity of buying or selling petroleum businesses, projects, or assets.

Acre A unit of land measurement which is equal to 4,047 square metres or 0.4047 hectares.

Acre foot The volume of a payzone which is defined by a combination of the areal measure of one acre and one foot of depth.

Acreage A geographical area (whether marine or terrestrial) which has been leased by an owner or licensed by a state to a private sector participant for petroleum exploration and/or production activities. Contrast with an **open area**, which is not presently so leased or licensed.

Act of God clause An alternative term for **force majeure**.

Active heave-compensation (AHC) Motion sensors which detect a ship's heave, pitch, and roll movements and compensate crane loads accordingly.

Active mud tank On a drilling unit, a storage tank that is used for circulating mud which is in active use during drilling operations.

Active survey A survey of a petroleum facility which is conducted with active intervention (such as the mechanical inspection of the interior of a pipeline). Contrast with a **passive survey**, as a survey which is conducted without active intervention.

Active well A wellbore which is currently producing petroleum. Contrast with an **inactive well**, as a wellbore which typically has not produced petroleum in a preceding twelve-month period.

Activity-specific operating guidelines (ASOG) Guidelines which are issued by the International Maritime Organisation which relate to the safe operation of dynamically positioned vessels.

Actual recovery The total quantity of petroleum which is actually recovered from a formation over the productive lifetime of the formation. Contrast with **ultimate recovery**, as the total quantity of petroleum which could be recovered from a formation.

Actual total loss The complete destruction and loss of an insured asset. *See also* **constructive total loss** and **total loss**.

Actualisation In a risk analysis process, the application of risk mitigations to gross risks in order to identify accurate net risk positions.

Actuals Petroleum quantities which are traded for physical delivery under a physical contract (in contrast with petroleum which is traded under a notional contract).

Ad coelum doctrine A shorthand term for the Latin term *cuius est solum, eius est usque ad coelum et ad infernos* (literally 'whoever's is the soil is theirs to heaven and to hell') which describes the principle that a landowner owns all of the sub-surface mineral wealth which lies beneath its land. *See also* **ownership-in-place**.

Ad hoc arbitration An arbitration which is conducted without applying a particular body of arbitration rules (such that the parties to the arbitration will need to choose or agree their own procedural rules). Contrast with **institutional**

arbitration, as an arbitration which is conducted according to a particular body of arbitration rules. *See also* **arbitration rules.**

Ad hoc charter (AHC)   A charterparty which is arranged on short notice, often for a single voyage.

Ad valorem   A payment or levy which is calculated as a proportion of a given value or number.

Ad valorem royalty (AVR)   A royalty whereby the amount which is payable to the royalty-holder by the owner of the burdened concession varies according to the monetary value from time to time of the produced petroleum.

Ad valorem tax   A taxation liability whereby the amount which is payable to the taxation authority varies according to the price of the taxable commodity.

Adaptation clause   Under a petroleum sales contract, a provision which allows for the revision (adaptation) of certain commercial terms (such as price) in certain circumstances. *See also* **hardship clause**, **price review clause**, and **spirit of mutual understanding and trust.**

Add on afterwards (AOA)   An infrastructure design philosophy whereby a rudimentary petroleum infrastructure item is installed early on and is later modified to meet emerging operational needs.

Added mass   The volume of water which is displaced by a ship's heave, pitch, and roll movements.

Additional insurance   Items of insurance cover which are beyond a level of base insurance in respect of a petroleum project.

Additional loss payee   An alternative term for a **co-insured.**

Additives   Specialty chemicals which are incorporated into fuel and lubricant products in order to give certain performance enhancements and/or environmental characteristics.

Adhesion contract   A contract for the supply of goods or services which is issued on the supplier's standard terms and offers little or no scope for negotiation by the intended consumer.

Adiabatic change   A change in any of the pressure, temperature, or volume of a quantity of gas which takes place without any accompanying loss or gain of heat.

Adjacency   Under a concession, the principle that the contract area will be extended in the concession-holder's favour to include any unlicensed adjacent (lateral level) areas into which a discovery could extend. *See also* **subjacency.**

Adjustable choke   A choke which can be opened to different sizes. Contrast with a **fixed choke**, which has a one-size opening only.

Adjusted ACQ   Under a petroleum sales contract, in an annually based take or pay calculation the annual contract quantity after the adjustments have been applied (which is relevant to where the take or pay commitment is expressed as a percentage of the adjusted ACQ).

Adjustment   An alternative term for **calibration.**

Adjustments Under a petroleum sales contract, in a take or pay calculation a series of deductions to a defined contract quantity which reflect the circumstances in which petroleum was not delivered to the buyer or was not taken delivery of by the buyer and reduce the amount of the buyer's take or pay commitment. *See also* **adjusted ACQ**.

Admeasurement A process by which the dimensions of a ship (including gross registered tonnage and net registered tonnage) are determined.

Administered arbitration An alternative term for **institutional arbitration**.

Administered price A price for petroleum in a market which is centrally planned, administered, and controlled, rather than being left for determination by the application of market forces, often as a regulatory intervention in order to establish or to stabilise a market.

Admiralty constant A factor which is applied to determine the ability of a new-build to attain a desired speed.

Adrift A ship which is floating but which is not under control.

Adsorption The adherence of a gas or a liquid to the surface area of a mineral deposit (such as in coal bed methane, whereby gas is adsorbed to the surface of the coal). Contrast with **absorption**, whereby a gas or a liquid permeates a mineral deposit.

Advance make good An alternative term for **carry forward**.

Advance payment financing An alternative term for a **forward purchase**.

Advanced recovery An alternative term for **secondary recovery** and **tertiary recovery**.

Advanced reservation of capacity agreement (ARCA) An alternative term for a **capacities-based contract**.

Advantaged LNG An alternative term for **clean LNG**.

Advantaged oil A crude oil development project which benefits from a short cycle time from discovery to production and/or a low project development cost.

Advocacy An alternative term for **energy advocacy**.

Aeolian source The creation of a petroleum deposit which is influenced by the depositional/erosional action of the wind.

Aerated mud Drilling fluid which is injected with air or with another gas in order to reduce its density during drilling operations, often to adjust the degree of over-balance which is being achieved in the wellbore.

Aeration Any injection of air, natural gas, or other gases into a liquid in order to reduce the density of the liquid.

Aero-mag An aerially conducted magnetic survey of the Earth's natural magnetic fields which is intended to reveal anomalies which indicate the potential presence of petroleum in a formation. *See also* **grav-mag**.

Affiliate The relationship between two companies which are any of the parent or the subsidiary or the siblings of each other, which is defined by contract or by applicable law.

Afforestation   A form of carbon offset which is represented by the planting of new trees. *See also* **avoided deforestation**.

Affreightment   *See* **contract of affreightment**.

Aframax   A crude carrier with a deadweight tonnage in the range of 80,000 to 120,000 metric tons.

After payout (APO)   In relation to a back-in after payout provision, the commercial relationship between the parties as it exists after the payout point.

Age   A sub-division of geological time. *See* Appendix 4. Also called a **stage**.

Ageing   An alternative term for **curing**.

Agency agreement   A contract by which an agent is formally appointed to act.

Agency service   An arrangement whereby a petroleum buyer gives authority to an agent to act on the buyer's behalf in arranging or administering the associated pipeline transportation services.

Agent   A person who acts (formally or informally) on behalf of another person (the principal—which could be disclosed or undisclosed to any third parties with whom the agent deals) in the negotiation of a business arrangement which will apply between the third party and the principal. *See also* **commercial agent**, **del credere agent**, and **factor**.

Agglomeration   The formation of larger bubbles or particles from smaller bubbles or particles.

Aggradation   The deposition of sedimentary material by a river or a current. Also called **alluviation**.

Aggregate receipt point   A physical hub at which different supply sources of petroleum intersect on a common pipeline.

Aggregation   The regulatory condition within which an aggregator functions. Also called the **single buyer model**.

Aggregator   A buyer of petroleum from multiple sources, principally for onward resale to end-users rather than for its own consumption, which stands as an intermediary between petroleum producers and end-users and often has statutorily established rights of monopsony and/or monopoly.

Agios   A measure of the difference in value between two or more different currencies.

Agitated boil-off   An increase to a boil-off rate which is caused by the agitation of an LNG cargo in-tank, which is caused for example by sloshing and/or the application of metocean conditions.

Agricultural grade water (AGW)   Water which is liberated from onshore petroleum production operations but which is fit only for irrigation purposes.

Air draught   The vertical distance between the sea surface and the highest point of a ship, an offshore drilling unit, or a petroleum production platform.

Air drilling   A method of rotary drilling which uses compressed air, rather than drilling fluid, as the circulation medium in the wellbore.

Air–fuel ratio (AFR)   The ratio of air to fuel which is present during the combustion of petroleum.

Air gap   The vertical distance between the sea surface and the lowest deck part of an offshore drilling unit or petroleum production platform.

Air injection   The injection of air into a formation as a means of artificial lift.

Air lock   An enclosed space for transiting between a gas-safe zone and a gas-dangerous place, which is intended to maintain a measure of separation between the two places.

Air vapour eliminator   A device which separates gases from liquids in a petroleum stream prior to measurement of the stream to give more accurate measurement readings.

Alaskan North Slope crude oil (ANSCO)   A crude oil blend from the United States.

Alberta Energy Company (AECO)   A benchmark price which is used for determining Canadian natural gas prices.

Albino oil   An alternative term for **white oil**.

Aliphatic hydrocarbon   A petroleum molecule which is formed by a straight chain of carbon atoms. Contrast with **aromatic hydrocarbon**, which is formed by a ring of carbon atoms.

Aliquot   A small part of a whole, such as a petroleum sample, which is taken for analysis.

Alkaline flooding   An enhanced oil recovery technique whereby alkaline chemicals such as sodium hydroxide are injected into a formation as part of a waterflood process.

Alkanes   A generic name for straight chain saturated petroleum with the general formula C_nH_n.

Alkanolamine processing   The use of specific organic compounds in order to remove carbon dioxide and hydrogen sulphide from a gas stream.

Alkylation   A process using sulphuric or hydrofluoric acid in order to produce high octane gasoline blending stocks in a refinery.

All-events   An absolute contractual obligation which is stated as being not capable of being relieved by the application of force majeure or otherwise excused. Also called a **hell-or-high water** obligation.

All-or-nothing   An agreement which is made between a group of concession-holders that if any part of the concession area which they jointly hold is to be made subject to the process of unitisation then all and not only a part of that concession area will be unitised.

All time saved both ends (ATSBE)   The aggregate of running hours which are saved in each function of cargo loading and cargo unloading, thereby reducing overall demurrage liability.

Alliance agreement   An alternative term for a **pairing agreement**.

Alligator grab   A specific tool which is used to recover a fish from a wellbore.

Allision   The situation in which a moving object strikes a stationary object. Contrast with **collision**, in which two moving objects strike each other.

Allocation   A mathematical arrangement by which petroleum which is delivered through a pipeline in a commingled stream from a number of different input sources is identified and is allocated proportionately back to each input source, so that the performance characteristics of each input source can be understood.

Allogeny   A geological term for the parts of a formation which have been formed in one location but which over time have moved to another, such as a sedimentary rock formation. Contrast with **authigeny**, whereby parts of a formation have been formed in-situ and have not moved over time.

Allowable quantity   A defined maximum quantity of petroleum which, for regulatory or commercial purposes, is permitted to be produced through a wellbore. *See also* **overproduction**.

Allowed laytime   A defined notional period of time within which a ship is required to load or unload its cargo (as appropriate), subject to the possibility of certain extensions. *See also* **SHEX** and **SHINC**. Also called **free time**. Contrast with **used laytime**, as the actual amount of time which a ship has taken to load or unload its cargo.

Alluvial plane   A large, level surface (onshore or sub-sea) which has been created over time by the deposition of sediments.

Alluviation   An alternative term for **aggradation**.

Along-hole depth (AHD)   An alternative term for **measured depth**.

Along-hole distance (AHD)   An alternative term for **measured distance**.

Alternative dispute resolution (ADR)   A method for the resolution of a dispute between the parties to a contract which involves the use of a mediation-based approach, rather than the use of a more adversarial approach such as arbitration or litigation.

Ambient air vaporiser (AAV)   A heat exchanger which is used to convert LNG to regas, using the ambient temperature of adjacent air to raise the temperature of the LNG to the point at which vaporisation occurs. *See also* **open rack vaporiser** and **submerged combustion vaporiser**.

Ambient conditions   The prevailing atmospheric conditions (such as humidity, pressure, or temperature) which surround a petroleum measurement unit.

Ambient temperature   The ordinary, prevailing temperature of local air or water.

Ambit claim   In a negotiation, an extravagant opening offer which is made in the expectation of an equally extravagant counter-offer and an eventual compromise in the middle.

Amend and extend (A&E)   Under a loan agreement, an extension of the terms of the facility which is made by the lender in favour of the borrower in order to enable the borrower to meet certain short-term financing needs.

Amendment and restatement (A&R)   An agreement between the parties to a contract for the amendment of the contract and for the restatement of the amended contract as a single new contract between the parties, typically with effect from the date of the amended and restated contract. *See also* **consolidation**.

Amerada bomb A wireline tool which measures bottomhole pressure or temperature.

American Association of Petroleum Geologists (AAPG) A US-based organisation which represents the interests of international petroleum geologists (http://www.aapg.org).

American Association of Professional Landmen (AAPL) A US-based organisation which represents the interests of US landmen (http://www.landman.org).

American Bureau of Shipping (ABS) A US-based ship classification agency (http://www.2.eagle.org).

American National Standards Institute (ANSI) A US-based organisation which defines common measurement standards (http://www.ansi.org).

American Natural Gas Association (ANGA) A US-based organisation for the promotion of natural gas, which became part of the API in 2016.

American option An option agreement which can be exercised by the option-holder at any time during its currency, up to its expiry date. Contrast with a **European option**, which can be exercised by the option-holder only on its expiry date.

American Petroleum Institute (API) A US-based organisation which represents the interests of the US petroleum industry (http://www.api.org).

American Society for Testing and Materials (ASTM) A US-based organisation which defines common performance standards for materials (http://www.astm.org).

Amiable compositeur An alternative term for **ex aequo et bono**.

Amine A compound which is used to treat a natural gas stream as part of a scrubbing process.

Amine scrubbing An alternative term for **acid gas removal**.

Amine treatment An alternative term for **acid gas removal**.

Amortisation Under a loan agreement, a schedule for the repayment of debt (measured as principal and interest) by the borrower to the lender in equal periodic instalments.

Amplitude A seismically determined measure of the difference in properties between two sub-surface layers which indicates the presence of a formation.

Amplitude variation with offset (AVO) An analysis tool which is used to determine the thickness, porosity, density, velocity, lithology, and fluid content of a formation.

Amsterdam—Rotterdam–Antwerp (ARA) A shorthand term for the regional concentration of petroleum storage and refining capacity which is found in this region.

Anaerobic digestion (AD) A process by which microorganisms break down biodegradable material in the absence of oxygen, typically in landfill sites in order to release landfill gas.

Analogue   A producing reservoir which can, because of the similarity of certain properties and characteristics, be used as a prediction tool to assess the likely performance of an as-yet undeveloped reservoir. Also called a **production analogue**.

Analysis   A process for the determination of the composition (quality) of a given quantity of petroleum. Contrast with **metering**, as a determination of the quantity of petroleum.

Anchor buoy   A floating marker which is used in a spread to mark the position of the anchors on a remotely anchored vessel such as an offshore drilling unit.

Anchor buyer   A prospective buyer of petroleum from a petroleum project which makes a commitment (in price and volume) which is sufficient to allow a final investment decision to be taken in favour of the project by the project developers. Also called a **foundation customer**.

Anchor handling vessel (AHV)   A ship which is used specifically to handle anchors for offshore petroleum exploration and production platforms.

Anchor lay   A method of laying a pipeline offshore whereby a pipelay vessel is anchored and is moved along the planned pipeline route by pulling against its anchors as the pipeline is laid.

Anchored storage   An alternative term for **floating storage**.

Ancillary services   Supporting services which are needed for the effective operation of a petroleum facility.

Angle of deflection   The angle, expressed in degrees (°), by which a wellbore deviates from the vertical. *See also* **high-angle well** and **kickoff point**.

Angle of dip   In geological analysis, the measured angle by which a formation slopes downward from the horizontal plane.

Angle of heel   The angle by which a ship or a drilling unit at sea inclines to one side because of prevailing weather and/or tidal conditions. *See also* **loll**.

Angular unconformity   A rock formation whereby rock layers have been tilted away from the horizontal plane by geologic deformation and subsequent rock layers have formed on top of these layers and the resulting formations are not parallel.

Aniline   An organic compound which is produced from coal tar and is used for the manufacture of petrochemicals.

Aniline point   The lowest point of temperature at which equal volumes of aniline and crude oil will mix to form a single phase.

Anisotropy   The ability of a substance to assume different physical or mechanical properties when measured across different axes.

Announcements clause   Under a petroleum project contract, a provision which regulates the ability of a contracting party to freely make public announcements regarding the project or the contract.

Annual contract quantity (ACQ)   Under a petroleum sales contract, a defined quantity of petroleum which is agreed to be delivered by the seller and bought by the buyer in respect of a defined calendar or contract year.

Annual delivery programme (ADP)   Under an LNG sale and purchase agreement, provision for the annual scheduling of LNG cargoes which are to be delivered in respect of a defined contract year.

Annubar   A gas measurement device which calculates a gas volume based on the difference between flowing pressure and static pressure.

Annular blowout preventer   A blowout preventer which uses a rubber plug to seal borehole or tubing diameters. Contrast with a **ram blowout preventer**, which uses a ram to close and seal a wellbore.

Annular capacity   A measure of the total petroleum-carrying capacity of a pipeline.

Annular pressure relief valve (APRV)   A valve which is configured to control and to release the build-up of pressure in an annulus.

Annular production   The production of petroleum from a formation whereby the petroleum flows to the surface through the annulus between a liner and a casing.

Annulus
(1) The spacing between the wall of a wellbore and a casing. *See* Appendix 2C.
(2) The spacing between a liner and a casing. *See* Appendix 2C.

Anthracite   Coal which has a higher calorific value than bituminous coal or lignite.

Anti-bribery and corruption (ABC)   Under a petroleum project contract, a provision to regulate the ongoing behaviour of the contracting parties in the prevention of bribery and corruption activities.

Anti-deprivation rule   A legal principle whereby a transaction could be void if it removes an asset or an interest from the estate of an insolvent person which might otherwise be available to satisfy the claims of the creditors of that person (and which is sometimes mentioned in the context of the effectiveness of the forfeiture remedy under a JOA).

Anti-fouling programme   An alternative term for marine growth prevention system.

Anti-hoarding   A regulatory principle which is intended to prevent a person from developing or maintaining a market position, through the holding of petroleum stocks and/or petroleum-processing, storage, or transportation capacity, which is or which could become anti-competitive.

Anti-knock rating   A quality specification which defines the resistance of a quantity of gasoline to knock. Also called an **octane rating** and *see also* **research octane number**.

Anti-washout provision   A contractual provision whereby a contract will be renewed when an underlying agreement to which the contract relates is also renewed. *See also* **washout provision**.

Anticline   An upwardly convex folded geological structure in which petroleum could be present at the top of the structure. Contrast with a **syncline**, in which petroleum could be present at the base of the structure.

Any delivery   Under a petroleum sales contract, a delivery of petroleum to the buyer which can be made at any time within a defined period of time at the discretion of the seller.

Any safe world port (ASWP)   A ship chartering convention which says that transportation costs will be the same for delivery to any safe port in the world, regardless of distance from the loading port.

API gravity   Originally devised by the American Petroleum Institute (API), a measure of the density of petroleum grades relative to water, measured in degrees (°), which is used to define petroleum fractions and also indicates the relative viscosity of petroleum fractions. Also called **density**. *See* Appendix 6.

Appraisal   The activity of assessing the existence and/or the extent of petroleum in a discovery, including through the drilling of a step-out well.

Appraisal well   An alternative term for a **step-out well**.

Appropriate technology   Technology for petroleum production or processing which is selected to be particularly beneficial to a local population. *See also* **corporate and social responsibility**.

Approved for construction (AFC)   A series of drawings in the design of a petroleum infrastructure item which represent the final iteration before construction begins.

Approved for design (AFD)   A stage in the creation of a petroleum infrastructure item at which a concept is approved for detailed design.

Aquifer   A naturally occurring underground geological structure with sufficient permeability to contain a water reservoir. *See also* **artesian well**.

Aquifer drive   An alternative term for **water drive**.

Aquifer reservoir   An aquifer which is capable of being drained and used as a gas storage facility.

Arbitrage   The simultaneous sale and purchase of petroleum quantities by a person in the same or different markets in order to exploit market price differentials between those quantities for gain.

Arbitration clause   Under a petroleum project contract, a provision whereby disputes between the contracting parties are settled through a formal process of binding arbitration. *See also* **ad hoc arbitration** and **institutional arbitration**.

Arbitration rules   An agreed body of procedural rules which govern the conduct of a formal process of arbitration.

Architectural elements   The various structural elements which together make up the physical composition of a particular formation.

Arctic floater (AF)   A petroleum infrastructure item which can operate in Arctic waters.

Arctic submersible rig   An offshore drilling unit which is adapted for Arctic operations, with a protective housing around all of the underwater elements to protect the unit from damage by moving ice. Also called a **mobile Arctic caisson**. *See also* **concrete island drilling system** and **conical drilling unit**.

Area of mutual interest (AMI) A defined geographical area to which an area of mutual interest agreement applies.

Area of mutual interest agreement (AMIA) A contract which records the commitment of a group of persons to jointly develop petroleum exploration and/or production opportunities in a defined area of mutual interest for a defined period of time, typically with a commitment of exclusivity between them. *See also* **pairing agreement**.

Area rate clause Under a gas sales contract, a provision whereby the contract price can be increased by the seller if a regulator has prescribed a higher gas price for payment by comparable buyers.

Area reduction An alternative term for **relinquishment**.

Area rental Under a concession, a rental payment which is due for payment by the concession-holder to the grantor and is calculated by reference to the extent of the surface area of the concession.

Argillaceous A formation which consists principally of clay or shale.

Arm's length A contractual arrangement which is made between competitive persons, or which is made between associated or affiliated persons and has the commercial rigour of an arrangement made between competitive persons.

Aromatic hydrocarbon A petroleum molecule which is formed by a ring of carbon atoms. Contrast with **aliphatic hydrocarbon**, which is formed by a straight chain of carbon atoms.

Aromatics Principally the benzene, toluene, and xylene isomers which are made by catalytic cracking and reforming and are used to manufacture plastics, synthetic fibres, and detergents.

Aromatisation A process for the conversion of an aliphatic hydrocarbon to an aromatic hydrocarbon.

Array A particular geometrical configuration of seismic sourcing equipment (including hydrophones) which generates seismic data. *See also* **streamer**.

Arrearage Under a petroleum sales contract, the accrued amount of money which is owed by the buyer to the seller at any time on accumulated invoices for the sale of petroleum.

Arrival contract An alternative term for **delivered ex-ship**.

Arrival on location (AOL) A point in time at which a ship or a drilling unit arrives at its intended location.

Artesian well An aquifer from which water flows to the surface under natural pressure.

Articulated loading platform (ALP) An offshore petroleum exploration or production platform whereby the riser is jointed and flexible in order to allow for tidal movements.

Artificial island (AI) A man-made island which is used as a base for petroleum production, processing and storage activities in a sea area.

Artificial lift   The application of manual intervention to assist the flow of petroleum from a formation into a wellbore, where the petroleum would otherwise not flow because of insufficient capillary pressure in the formation. Also called **on-the-pump**, **pressure maintenance**, and **re-pressure**. *See also* **voidage replacement**. Contrast with **open flow**, where petroleum is produced through a wellbore from a formation naturally through inherent capillary pressure in the formation.

As low as reasonably practicable (ALARP)   An objective standard which is applied to define certain activities which are necessary to be undertaken for the reduction of certain defined hazardous risks.

As-billed   A charging methodology by which the owner of a petroleum facility passes a share of the operating costs through to facility users to the same extent that the costs were incurred by the facility owner, without an uplift on those costs in the facility owner's favour.

As-delivered Btu   The number of Btus which exist in a given quantity of raw gas, which is adjusted to reflect water content in the gas.

As-is where-is   Under an asset or share sale and purchase agreement, a formulation which is adopted by the seller of a business, asset, or project whereby the buyer the makes its purchase with full knowledge and appreciation of the current condition of the business, asset, or project and with no warranty, representation, or improvement covenant being given by the seller.

Asphalt   A highly viscous semi-sold form of petroleum, which is a natural element of asphaltic-based crude oil and in which the principal constituent is bitumen.

Asphaltic-base crude oil   One of the principal forms of crude oil, which is based on the presence of naturally occurring asphalt, which produces relatively high concentrations of bottoms when refined.

Assay   A test which is performed on a sample of crude oil in order to determine its chemical composition and other properties. Also called **finger printing**.

Asset management agreement (AMA)   An alternative term for an **operating (operation) and maintenance agreement**.

Asset optimisation (AO)   A predictive maintenance tool which is used to minimise disruption to a petroleum infrastructure item.

Asset swap   An exchange of petroleum project assets or interests between persons, possibly also with a cash balancing element. *See also* **boot**.

Assignment   A transfer of contractual rights by a contracting party to another person. Contrast with a **novation**, whereby rights and obligations are transferred.

Associated free gas   Natural gas which is in immediate contact with crude oil in a formation but which is not in solution with that crude oil. Contrast with **associated gas**, which is in solution with crude oil.

Associated gas (AG)   Natural gas which is in solution with crude oil in a formation and is produced inherently in association with that crude oil. *See also* **associated free gas**, **associated oil**, **non-associated gas**, and **solution oil**. Also called **bradenhead gas**, **casinghead gas**, **dissolved gas**, and **solution gas**.

Associated oil (AO)   Crude oil in a formation which contains associated gas.

Association of British Independent Oil Exploration Companies (BRINDEX)   A United Kingdom-based trade association which promotes British petroleum exploration and production companies (http://www.brindex.co.uk).

Association of International Petroleum Negotiators (AIPN)   An internationally focused organisation which represents the interests of international petroleum industry participants (http://www.aipn.org), known as the **Association of International Energy Negotiators (AIEN)** since 2022.

Association ratio   The proportion of gas to oil in associated gas, or of oil to gas in associated oil.

Assurance   A process by which the extent of a person's compliance with the terms of a contract is confirmed through an independent audit. *See also* **internal due diligence**.

Assured system capacity   A measure of the amount of capacity within a petroleum facility which is capable of being utilised by facility users at any time.

At risk   A contractual condition which allocates certain defined risks or liabilities to a contracting party.

At sea   A point in time at which a ship is released from its moorings at a port and is ready to sail.

Atlantic Basin   A trading area for LNG production and consumption which is bounded by countries having a maritime coastline on the Atlantic Ocean. *See also* **Pacific Basin**.

Atlantic margin   The petroleum-bearing areas of the continental shelf (CS) which are located in the Atlantic Ocean.

Atmosphere (atm)   A unit of pressure. *See* Appendix 1.

Atmospheric distillation   A process of passing heated crude oil through a crude distillation unit in order to produce cuts. Also called **continuous distillation**. *See also* **extractive distillation** and **refluxing**. *See* Appendix 2E.

Atmospheric gasoil   An alternative term for **automotive gas oil**.

Atmospheric pressure   Natural atmospheric pressure, which is commonly measured at sea level. Also called **barometric pressure**. *See also* **absolute pressure** and **gauge pressure**.

Atomisation   The dispersal of a liquid fuel in droplet form, for mixture with air to promote more efficient combustion.

Attenuation   A reduction in the amplitude of a seismic wave as it passes through a formation.

Attic oil   Crude oil deposits which remain in a formation after the permanent cessation of production of crude oil from the formation has taken place.

Attribution   An operational arrangement by which petroleum is loaned and repaid between the members of an identified group of petroleum producers (sometimes to make good a deficiency of a producer identified by the application of an allocation process). *See also* **differential lifting**. Also called **balancing** and **substitution**.

Australian Mining Petroleum Law Association (AMPLA)   An Australian-based organisation which promotes knowledge sharing in the energy and natural resources sector, principally focused on the Australian market (now known as the Energy & Resources Law Association) (http://www.erlaw.org.au).

Australian Petroleum Production & Exploration Association (APPEA)   An Australian-based organisation which represents the interests of the Australian upstream industry (http://www.appea.org.au).

Authigeny   A geological term for the parts of a formation which have been formed in-situ and have not moved over time. Contrast with **allogeny**, whereby parts of a formation have been formed in one location but have moved to another location over time.

Authorisation for expenditure (AFE)   A process under a JOA by which the operator submits a proposal for approving the incurring of certain defined expenditures in the performance of joint operations to the non-operating parties.

Authorised distributor (AD)   A downstream fuels retailing model whereby an independent retailer sells fuel through a chain of branded sites.

Authorised overrun
(1)   Under a petroleum sales contract, a quantity of petroleum which the seller allows the buyer to take delivery of beyond the buyer's contracted offtake rights under the contract.
(2)   Under a JOA, an alternative term for a **permitted overspend**.

Auto-generation   The generation of electric power by a petroleum production or processing facility which is intended primarily to meet the operational needs of the facility and is not intended for commercial sale.

Auto-ignition   The point at which a substance will spontaneously combust in air without the presence of an external ignition source.

Auto-refrigeration   The maintenance of the temperature of LNG at its boiling point, whereby any additional heat which is generated is countered by the energy loss from boil-off such that the LNG remains cold.

Automated meter reading (AMR)   A real-time measurement of a quantity of petroleum as it passes through a specific location.

Automatic custody transfer   A process which automatically measures petroleum streams at a number of defined system points.

Automatic identification system (AIS)   A radar-based system which gives each ship a unique identification code in order to allow the tracking of ship movements.

Automatic shut-down   An alternative term for a **trip**.

Automotive gas oil (AGO)   Fuel oil which is used as a vehicle fuel (akin to diesel). Also called **atmospheric gasoil**.

Autonomous underwater vehicle (AUV)   An underwater survey or repair vehicle which operates without the assistance of a surface operator. *See also* **remote-operated vehicle**.

Auxiliaries
(1) Any items of equipment within a petroleum facility which are not regarded as a direct or essential part of the facility's primary operational purpose.
(2) Secondary operational systems which are intended to back up the operation of a petroleum facility if the primary operational systems fail.

Avenant A documentary rider which formally amends an existing contract or concession and is typically found in Francophone states.

Average freight rate assessment (AFRA) A published assessment of freight rates which are paid for the transportation of crude oil by ship in chartered crude carriers, measured in Worldscale rates and published by the London Tanker Brokers' Panel.

Average sampling In the sampling of petroleum in a tank, a consolidated sample which is drawn from several different layers within the tank.

Average system pressure (ASP) The average of the operating pressures which exist throughout a gas pipeline network.

Avgas
(1) A generic term for any form of aviation fuel. *See also* **aviation turbine gasoline** and **aviation turbine kerosene.**
(2) A particular form of high octane aviation fuel which is used in piston engines.

Aviation turbine gasoline A form of aviation fuel. Also called **ATG, AVTAG,** and **Jet B.**

Aviation turbine kerosene A form of aviation fuel. Also called **ATK, AVTUR, kerosene,** and **Jet A.**

Avoidable costs Petroleum project expenses which can be avoided if a decision is made to alter the course of the project (such as an election not to drill a wellbore). *See also* **but-for costs.**

Avoided deforestation A form of carbon offset which is represented by the reduction of deforestation which was otherwise due to take place. *See also* **afforestation.**

Axial flow meter A form of flow meter.

Azimuth A direction relative to magnetic north, which can be used to define the drilling of a horizontal well.

B

Back cost An alternative term for **embedded cost.**

Back month The last month of a multi-month petroleum lifting or transportation schedule. *See also* **front month.**

Back office   The operational and administrative elements of an energy business (such as accounts and audit). *See also* **front office** and **mid office**.

Back pressure   Resistance in a pipeline to the continuous flow of a gas or a liquid which is caused by constriction, filtration, or surface friction.

Back value pricing   An alternative term for **netback pricing**.

Back-end farm-out   An alternative term for an **earn-in agreement/earn-out agreement**.

Back-flow   The flow of petroleum in a formation from one zone to another which occurs because of inter-zonal pressure differentials. Also called **crossflow**.

Back- in
(1) The right of the non-participant in an exclusive operation under a JOA to later participate in that operation, typically upon the payment of a premium to the participating parties. Also called **reinstatement**.
(2) The right of a person to receive the award of a concession from a state at a future point after certain conditions (relating, for example, to the conduct of certain defined exploration and appraisal activities) have been met.
(3) An alternative term for a **participation right**.
(4) An alternative term for **heads-up**.

Back-in after payout (BIAPO)   A royalty interest which is capable of conversion into an equity interest in the burdened concession (usually at the option of the concession-holder and usually after a certain level of royalty payment has been made or costs have been incurred). *See also* **after payout, before payout**, and **payout point**.

Back-off   The activity of unscrewing the parts of a drill string while it is still in the wellbore.

Back-to-back   Two or more inter-related contracts which effect a seamless pass-through of rights and obligations along a chain of contracting parties.

Back-up   A process whereby one section of drill pipe is held stationary while another section of drill pipe is added to or is removed from a drill string.

Backbone items   Gas or electric power transmission facilities which are key to moving energy from import/production points to consumption points.

Backfill gas   Natural gas from a substitution source which is used as a replacement feedstock for a gas export project.

Background gas   Gas which is entrained in circulating drilling fluid and is produced while drilling a wellbore (and which is possibly also indicative of an impending kick). *See also* **gas-cut mud**.

Background IP   Intellectual property which is already in existence at a point in time. Contrast with **foreground IP**, which has yet to come into existence.

Backhaul
(1) The loading of a cargo into an LNG carrier for transportation on what would otherwise be a ballast voyage.
(2) A pipeline transportation service which requires the movement of petroleum through the pipeline such that the contractual direction of movement (between

input and offtake) is opposite to the physical direction of movement of the petroleum. Contrast with **forward haul**, whereby the contractual direction of movement is the same as the physical direction of movement.

Backload revenues   Revenues from a petroleum project which only begin to accrue towards the end of the project's anticipated lifecycle.

Backloading   The structuring of a petroleum project whereby the incidence of costs and/or the recovery of revenues are deferred to the later stages of the project. Contrast with **frontloading**, whereby costs and revenues arise in the early stages of a project.

Backlog   In the context of the provision of oilfield services, a measure of the accrued amount of work which is due to be completed by an oilfield services company which can be taken as an indicator of the commercial health of the company.

Backstop   An alternative term for a **buy-through**.

Backstop price   The highest price at which petroleum can be sold before it becomes economically uncompetitive with other fuels.

Backward linkage   A situation in which the undertaking of a petroleum project encourages growth in associated economic elements which enable the project to succeed (such as the provision of finance, labour, and raw materials). Contrast with **forward linkage**, whereby the undertaking of a project encourages growth in subsequent economic elements.

Backwardation   A market whereby the price for the delivery of petroleum in the immediate future is higher than the price for delivery of petroleum in the further future. Contrast with **contango**, whereby the immediate future price is lower than the further future price. Also called an **inverted market**.

Bad oil   A quantity of crude oil which contains a sufficient amount of impurities to make it unsuitable for immediate sale or transportation without processing. *See also* **pipeline oil**.

Bailer   A tool which is used to remove cuttings from downhole.

Bailing   The activity of using a bailer to clean a wellbore.

Bailment   The legal aspect of a commercial relationship (found, for example, in petroleum pipeline transportation or petroleum storage) whereby physical possession of property is transferred from one person (the bailor) to another person (the bailee) for safekeeping, with the bailee assuming liability for any loss of or damage to the property but without taking title to the property.

Balance market   An alternative term for a **market of last resort**.

Balancing   An alternative term for **attribution**.

Ball valve   A hollow spherical valve within a pipeline or petroleum transfer system which goes through a gradual 90° turn in order to block the flow of petroleum.

Ballast   Seawater which is carried in the tanks of a crude carrier in order to give improved stability at sea during a ballast voyage, or in the tanks of an offshore drilling unit in order to give improved stability while drilling. *See also* **hydrostatically balanced load**.

Ballast bonus
(1) Under an LNG sale and purchase contract, an additional monetary amount which is payable by the buyer in order to compensate the seller for a long ballast voyage which results from the delivery of the cargo to the buyer.
(2) Under a charterparty, an additional monetary amount which is payable by the charterer in order to compensate the shipowner for the costs of a long ballast voyage which results from the delivery of the cargo.

Ballast control   The activity of keeping an offshore drilling unit stable during drilling operations, regardless of the intervention of prevailing weather and/or tidal conditions.

Ballast passage   An alternative term for a **ballast voyage**.

Ballast voyage   The empty return voyage of a ship after its cargo has unloaded. Contrast with a **laden voyage**, as the cargo-loaded leg of a ship's voyage. Also called a **ballast passage**.

Ballast water   Contaminated water which results from washing a ship's cargo tanks to remove residual crude oil deposits after a cargo has been discharged.

Ballast water exchange at sea (BWEAS)   A methodology for the substitution at sea of ballast water in a manner which is intended to protect the ecological balance of the waters into which the ballast water is discharged.

Ballasting   The refilling of a crude oil tank in a ship with water after crude oil has been discharged from the tank in order to maintain the ship's equilibrium.

Baltic International Maritime Council (BIMCO)   An international shipping organisation which represents the interests of shipowners (http://www.bimco.org).

Bandwidth   A range of volumetric values within which linepack is to be maintained within a pipeline.

Bank gas   An alternative term for **make-up**.

Banking   An arrangement whereby a present entitlement (of energy or money) of a person is deferred to a future receipt. *See also* **gas banking**.

Bar   A unit of pressure. *See* Appendix 1.

Bareboat charter (BBC)   A charterparty whereby a ship is provided to a charterer but without a crew. Also called a **demise charter**. *See also* **full service charter**.

Barefoot completion   An alternative term for **openhole completion**.

Barg   A unit of pressure which is used for the measurement of gauge pressure.

Barge delivery   A delivery of petroleum which is effected by the use of a barge.

Barite   A heavy mineral which is used to add weight to drilling mud.

Barometric pressure   An alternative term for **atmospheric pressure**.

Barratry   Misconduct by a ship's master or crew which damages the ship or its cargo.

Barrel (Bbl/B)   A measure of crude oil or condensate, containing 35 Imperial gallons or 42 US gallons.

Barrel conversion   The conversion of different units of petroleum measurement to give a consistent rate of exchange for determining barrel of oil equivalent. *See also* **converted barrel.**

Barrel of oil equivalent (BOE)   A comparative unit of energy which is based on the approximate amount of energy released by the combustion of one barrel of crude oil. Also called **crude oil equivalent.**

Barrelage   The volumetric measure of the flow of crude oil.

Barrels of liquid per day (BLPD)   A measure of the total daily production of crude oil, liquids, or water from a wellbore.

Barrels of water (BW)   A measure of the volume of water which is used in or which is produced from petroleum project operations.

Barrels per stream day (BPSD)   The number of barrels of oil which are produced through a petroleum production or processing facility on a stream day.

Basalt   A layer of igneous rock which often suggests the existence of a basement within a formation. *See also* **gabbro.**

Base gas   An alternative term for **cushion gas.**

Base insurance   The minimum level of insurance cover which is required by law or by contract in respect of a petroleum project. *See also* **additional insurance.**

Base load   The minimum level of continuous load on an operating electric power-generation system.

Base oil   Lubrication grade oils which are produced from a refinery.

Base period   A defined period of time which serves as a sample for demonstrating incurred costs and/or applied prices, often for use as a benchmark in an adaptation clause.

Base price   Under a petroleum sales contract, the base component of the agreed price to which escalation is then applied. Also called **P zero.**

Baseball arbitration   A colloquial term for a method of dispute resolution whereby the disputing parties each notify the appointed arbiter of their preferred outcome and the arbiter must choose one of those outcomes in making its determination. Also called **pendulum arbitration.** *See also* **golf arbitration** and **shotgun arbitration.**

Basement   A rock layer which does not contain petroleum and, when encountered, will usually lead to the abandonment of drilling.

Basic sediment and water (BS&W)   The proportion of a quantity of petroleum which is made up of water and sediment. Also called **bottom settlings and water.** *See also* **total S&W** and **water bottom.**

Basin   An alternative term for a **formation.**

Basin-centred gas accumulation (BCGA)   An unconventional gas deposit which lies within several low permeability reservoirs across a geographical area.

Basin-opening exploration   Petroleum exploration which is focused in frontier areas. *See also* **infrastructure-led exploration.**

Basing-point pricing   A methodology for pricing petroleum whereby a base price is calculated for a base location, to which standard freight prices are then added to give an overall delivered price which the buyer pays to the seller.

Basis of design (BOD)   A narrative which records the decisions made (and the reasoning behind them) during the design phase of a petroleum project.

Basis point (BPS)   One hundredth (0.01 per cent) of one per cent (so, for example, 100 basis points = 1 per cent), used in the determination of a contractual interest rate.

Basis risk
(1)   A risk that the value of a futures contract will not move in line with the value of the underlying commodity.
(2)   A risk that deliberately structured divergent contractual positions within a portfolio hedging strategy will not move in a divergent manner and so create the intended balanced offset of risk.

Basket price   The global average price of crude oil at any time which is averaged irrespective of the source or the grade of the crude oil.

Basket threshold   An aggregate of liabilities across a number of heads of claim which has to be achieved before a contractual liability provision (such as a claim for a breach of a representation or a warranty or an indemnity right) is triggered.

Batch/batching   A shipment of a grade of petroleum through a multiphase pipeline.

Batch cementing   In cementing, the pumping of cement in separated batches rather than as a single, continuous operation.

Batching sphere   A rubber sphere which is placed between different batches in a multiphase pipeline.

Batholith   An irregular intrusion of an igneous rock into another rock.

Batphone   A dedicated communication line between two places.

Batter   An alternative term for **leg batter**.

Battery   An alternative term for a **tank farm**.

Battery limit (BL)   A defined boundary between different areas of infrastructure and/or responsibility, such as the flange between pipelines having different owners or the fenceline of a petroleum facility. *See also* **inside battery limit** and **outside battery limit**.

Baumé scale   A scale which is used to determine the specific gravity of liquids lighter and heavier than water, which has now been largely supplemented by the API gravity scale.

Beach 2000/Beach 2015   United Kingdom-specific contract terms which are used for the delivery of gas at the point of entry into the NTS (http://www.gasgovernance.co.uk).

Beach gas   Offshore gas production which is landed and sold from a seller to a buyer at a defined onshore delivery point.

Beach price A term, mainly used in the United Kingdom, to define the price at which offshore gas production is sold by a producer at the point of exit from an onshore gas processing facility.

Beach pull The pulling ashore of an offshore petroleum pipeline prior to the connection of the pipeline with onshore facilities.

Beam jack A surface-mounted pump which is used to raise crude oil from a wellbore. Also called a **nodding donkey** and a **pump jack**. *See also* **sucker rod pumping**.

Beam pumping An alternative term for **sucker rod pumping**.

Bear market A market in which commodity prices are declining. Contrast with a **bull market**, in which commodity prices are rising.

Beaufort scale An empirical measure which relates wind speed to observed conditions at sea or on land.

Bed An alternative term for a **bedding plane**.

Bedding plane A surface which separates one layer of stratified rock from another and whereby changes in compositional conditions can usually be observed. Also called a **bed**.

Before payout (BPO) In relation to a back-in after payout provision, the commercial relationship between the parties as it exists before the payout point.

Behaviour-based safety management (BBSM) A safety philosophy which requires adherence by all officers, employees, and contractors.

Behind pipe Petroleum in a formation which is prevented from entering a wellbore by the presence of casing in the wellbore.

Behind pipe reserves Unproduced petroleum within a formation which will require perforation or recompletion of the wellbore to be produced.

Bellmouth gas purchase A gas sales arrangement whereby the buyer has flow control at the delivery point and takes its gas requirements without making a nomination to the seller.

Below rotary table (BRT) True vertical depth as measured from the rotary table on a drilling unit.

Benchmark crude A grade of crude oil (for example, Dated Brent, Tapis, or West Texas Intermediate) which is used in trading as a basis for positive or negative price differentials. Also called a **marker crude**.

Benchmark price A reference price for the sale of petroleum which is based on the aggregation of published trading prices over a given time period.

Benchmark weighted price (BWP) A price for crude oil which is derived from a weighted average of benchmark crudes.

Benchmarking The assessment of a particular position by comparative reference to an analogue position.

Benefication The process of treating a raw commodity in order to improve its chemical or physical properties.

Beneficial cargo owner (BCO)   A person who takes custody of a cargo at a destination port.

Beneficiary   Under a trust, a person who holds an equitable interest in trust property.

Benefit sharing   Under an LNG sale and purchase agreement, a provision whereby the seller and the buyer share the benefits which are achieved from a diversion according to an agreed ratio. Also called **spread splitting** and **upside sharing**. *See also* **net PSM** and **raw PSM**.

Benefits agreement   An alternative term for a **community development agreement**.

Bentonite   A form of clay which expands when wet, making it useful downhole in protecting a formation against invasion. *See also* **gunk slurry**.

Benzene, toluene, xylene (BTX) *See* **aromatics**.

Berm   An alternative term for a **firewall**.

Berth   A designated location in a harbour at which a ship is moored, typically for loading or unloading but also for ship maintenance. *See also* **dry berth** and **wet berth**.

Best alternative to a negotiated agreement (BATNA)   Any commercial solution which is preferable to the execution of a negotiated agreement (such as the continuation of an informally agreed operational arrangement).

Best available technology/technique (BAT)   A performance standard which requires the latest stage of available technology for use in a petroleum project.

Best available technology not entailing excessive cost (BATNEEC)   A performance standard which requires best available technology but with a measure of cost control upon that requirement.

Best environmental practice (BEP)   An objective standard for the conduct of petroleum operations which is determined with regard to a required degree of environmental protection.

Bi-directional pipeline   A pipeline which can take in petroleum for transportation from either end.

Bi-directional terminal   An LNG import terminal with export capacity, or an LNG export terminal with import capacity.

Bid bond   An alternative term for **earnest money**.

Bid round   An alternative term for a **licensing round**.

Bid round circular (BRC)   A teaser which is issued by a state as a precursor to the launch of a licensing round.

Bid round data package (BRDP)   A package of seismic and other data which is purchased by or which is made available to a bidder in a licensing round.

Biddable items   Commercial terms in a concession which are left blank and are open for completion by a bidder in a licensing round, which the state will then take into account when assessing the bidder's bid. Also called **contestable items**.

Bidder's remorse   In a licensing round, the risk to a bidder that a measured and conservatively priced bid proves insufficient to win the auction. *See also* **winner's curse**.

Big knock   A knock-for-knock liability allocation regime which extends to include an indemnified person's other contractors. Contrast with **little knock**, which does not include those other contractors.

Bilateral investment treaty (BIT)   A contract that establishes the terms for the protection of an investment, which is made between a state and an investor. Contrast with a **multilateral investment treaty**, which is made between multiple states for private investor reliance.

Bill of lading   A document by which a master of a ship acknowledges the receipt of a cargo which is consigned to the master for transportation on the ship.

Billion (B)   One thousand million.

Bin   A defined sub-division of a three D seismic survey area. Also called a **binning area**.

Binning area   An alternative term for a **bin**.

Biochemical oxygen demand (BOD)   The amount of oxygen which is consumed by bacterial microorganisms during the aerobic decomposition of organic matter (which is typically determined in order to assess the effectiveness of a waste water management process).

Biocide   A chemical agent which prevents the build-up of microbiological activity in a petroleum infrastructure item. *See also* **sulphate-reducing bacteria**.

Biodiesel   A fuel which is manufactured from a blend of diesel and biological sources such as vegetable oils or tallow. Also called **fatty acid methyl ester**.

Bioenergy   Energy which is produced from the combustion of biomass.

Bioenergy carbon capture and storage (BECCS)   A greenhouse gas mitigation technology which combines bioenergy generation with carbon capture and storage.

Bioethanol   An alcohol-based fuel which is made from the fermentation of crops (principally corn, wheat, and maize).

Biofuel   Any fuel which is generated from a biological process in relation to organic matter (including **bioethanol** and **biogas**).

Biogas   Energy which is produced from the anaerobic digestion of sewage and industrial waste or from the combustion of biomass. *See also* **landfill gas**.

Biogenic gas   Natural gas which is formed within a formation at shallow depths and low temperatures by the anaerobic digestion of sedimentary organic matter. It is typically dry gas found independently of further petroleum shows. Contrast with **thermogenic gas**, which is formed by the thermal cracking of sedimentary organic matter and crude oil deposits.

Biomass   Biologically derived material (principally wood) which can be combusted as a fuel for the generation of electric power.

Biomass to liquid (BTL)   A synthetic fuel which is made from biomass.

Biomethane   A substitute for natural gas which is produced by the removal of carbon dioxide from biogas. Also called **green gas**.

Biostratigraphy   A branch of stratigraphy which uses the existence of fossilised remains to establish the relative age of a formation. *See also* **biozone**.

Biozone   A time interval in a stratigraphic layer which is characterised by the existence of certain fossilised remains. *See also* **biostratigraphy**.

Bird-dog right   A colloquial term for the right of a non-operating party in a JOA to monitor the operator's activities and possibly also to step in to resolve an unremedied failure by the operator.

Bisque clause   Under a loan agreement, a provision whereby the borrower can postpone the repayment of principal and interest to the lender for limited periods in specific circumstances.

Bitumen   A naturally occurring virtually solid form of crude oil, with greater viscosity than ultra-heavy oil.

Bituminous coal   Coal which has a higher calorific value than lignite. Also called **black coal**.

Black coal   An alternative term for **bituminous coal**.

Black oil   Crude oil which contains heavy fractions and has a lower gas–oil ratio than volatile oil (and which is generally black in colour). Also called **low shrinkage oil** and **ordinary oil**. *See also* **volatile oil** and **white oil**.

Black powder   Fine metallic particles which result from the corrosion of the interior surface of a pipeline and can contaminate a petroleum stream flowing through the pipeline.

Black products   Heavy-end refinery products such as residual fuel oil. Contrast with **white products**, as light-end refinery products such as gasoline and kerosene.

Black start   A rapid start-up of an off-line petroleum facility from a standing start.

Black water   Process water which is contaminated with human waste. *See also* **grey water**.

Blackout
(1)   A complete interruption of the provision of energy. Contrast with a **brownout**, as a partial interruption of the provision of energy.
(2)   A provision whereby a party to a JOA which elects not to participate in a joint operation must thereupon withdraw from the JOA and transfer its interests to the participating parties.

Blank pipe   Casing which has not been perforated.

Blanket gas   A gas phase above a liquid phase in a storage vessel which is intended to maintain pressure in the vessel or to protect the liquid phase from contamination.

Blanket indemnity   A wide-ranging indemnity which is given in respect of a general, rather than a specific, liability.

Blanking plate   A heavy steel plate which is imposed at a right angle to the flow of petroleum through a pipeline or a tubular in order to stop or to slow the flow.

Blast furnace gas   Gas which is produced as a by-product from the combustion of coke in a blast furnace, which is typically used as fuel gas in furnace works.

Bleeding The natural tendency of a liquid component to separate from the solid or semi-solid elements of a mixed stream.

Bleeding core A core which is saturated with crude oil or natural gas when it is extracted.

Blending A process of mixing two or more oils or oil products having different properties in order to obtain a new product with intermediate properties.

Blending stock Any petroleum quantities or products which are blended.

Blind drilling A drilling operation in which drilling fluid flows into a formation rather than returns to the surface. *See also* **invasion**.

Blind pool A consortium which has come together to fund petroleum exploration or production activities but without the identification of specific target opportunities at the time of coming together.

Blind ram A blowout preventer which, when closed, forms a seal in a wellbore which does not contain a drill string.

Blind royalty A royalty agreement which contains few or no asset management covenants in favour of the royalty-holder.

Block A fixed geographical description of a marine or terrestrial area which is the subject of the award of a concession, which is sometimes expressed as a part of a wider interest such as a quadrant. On the UKCS a block has an approximate surface area of 10 kilometres × 20 kilometres.

Block image A three-dimensional perspective of a defined sub-surface area.

Block rate A tariff structure for the use of a petroleum facility by which the facility owner charges different rates for different blocks of demand (measured by time or by quantity) to the facility user.

Block valve A valve in a pipeline which impedes or facilitates the flow of petroleum.

Blocky sand A sub-surface sand deposit which takes the form of an angular block with relatively defined edges.

Bloodhound A tool which is used to detect leaks in wellbores, pipelines, or storage tanks.

Bloom A rainbow-like sheen on water which indicates the presence of an oil spill.

Blowdown

(1) A method of producing petroleum from a mixed natural gas/condensate reservoir by letting capillary pressure fall as the natural gas is produced over time without re-injection. Condensate then forms in the reservoir and can be recovered.

(2) The final production of petroleum from a reservoir at the end of a reservoir's life.

(3) Any venting of gas from a reservoir or a petroleum infrastructure item.

Blown oil A method of processing bitumen whereby the viscosity of the bitumen is lowered by heating and injecting it with compressed air. *See also* **toe-to-heel air injection**.

Blowoff line   A safety release device which releases pressure build-ups from a gas pipeline.

Blowout   An alternative term for a **kick**.

Blowout preventer (BOP)   A high-pressure wellhead valve construction which is designed to seal a wellbore quickly in the event of a kick. Also called **blowout preventer equipment**. *See also* **annular blowout preventer** and **ram blowout preventer**.

Blowout preventer equipment (BOPE)   An alternative term for a **blowout preventer**.

Blue hydrogen   Hydrogen which is manufactured from natural gas using a steam methane reforming process, whereby the resultant carbon is captured and stored. *See also* **grey hydrogen**.

Boarding speed   The moving speed of a ship which is adjusted in order to allow a pilot to embark or disembark.

Boarding station   An alternative term for a **pilot boarding station**.

Bodyshopping   An operational model for the conduct of a petroleum project whereby services or staff are contracted in by a person on a short-term contract basis.

Boil-off gas   Regasification which occurs naturally to an LNG cargo in an LNG carrier or a storage tank as an inevitable consequence of imperfect cryogenic insulation. Lighter petroleum fractions (such as methane and ethane) vaporise and the proportionate concentration of heavier petroleum fractions remaining in the cargo increases, so giving the cargo a higher calorific value. *See also* **weathering**.

Boil-off gas combustion system (BOGCS)   A process on an LNG carrier which allows boil-off gas to be rerouted to the LNG carrier's engine room for combustion as fuel.

Boil-off gas recovery system (BOGRS)   A process on an LNG carrier which allows boil-off gas to be re-liquefied and return to the LNG storage vessel.

Boil-off guarantee   Under a charterparty, a defined tolerance, expressed as a percentage of a total cargo of LNG in-ship, which is expected to be lost as boil-off in a voyage.

Boil-off rate (BOR)   A rate, expressed as a percentage of the total volume of a cargo of LNG, at which boil-off will take place over a defined time period.

Boiler fuel gas (BFG)   Gas which is combusted in a boiler for the generation of steam.

Boilerplate   The various administrative and operational provisions which underpin a commercial contract.

Boiling liquid expanding vapour explosion (BLEVE)   An instantaneous vaporisation of a liquid (and corresponding energy release) upon its sudden release from containment under pressure.

Bollinger banding   A pricing premise that high and low bands exist either side of mean prices and when actual prices exceed those bands then actual prices will revert to the mean.

Bomb   A container which is used to hold measurement devices in a wellbore.

Bonded warehouse   A secure customs-controlled warehouse which is used for the storage of dutiable goods (such as gasoline) without payment of duty, whereby those goods are released only if the appropriate duty is paid.

Bonny, Brass, Qua lboe (BBQ)   A composite of crude oil blends from the Bonny, Brass River, and Qua lboe regions (in Nigeria).

Bonus payment   Under a concession, a monetary sum which is payable by the concession-holder to the grantor at certain defined trigger points during the term of the concession. *See also* **production bonus**.

Book   A portfolio of energy trading positions.

Book trading   The trading of energy entitlements within a portfolio for optimisation. Contrast with **prop trading**, which is carried out on a speculative basis. Also called **non-prop trading**.

Bookable reserves   Reserves which are available for booking.

Booked barrel   A barrel of oil which is available for booking.

Booking   The recording of petroleum reserves as an asset in the accounts of a producer which is entitled to produce those reserves.

Book-out   An agreement between all of the participants in a daisy chain to cash out their physical positions.

Boom   A flexible, sea-bed anchored, floating device which demarks an offshore area.

Boomer   The compressed air or electrically generated source of the sound which is applied in an offshore seismic survey.

Booster plant   A compression point on a gas pipeline which is used to increase pipeline pressure.

Boot
(1)   An additional element of cash consideration which is paid by one person to another as part of an asset swap.
(2)   A tubular device which is placed within a wider vessel to effect the separation of petroleum streams.

Bootleg   A sale of petroleum by a seller, or a purchase of petroleum by a buyer, which is made in breach of an exclusive sale or exclusive purchase commitment between the seller and the buyer.

Bootstrapping   A colloquial term for a financial position which is determined by reference to a person's own present internal resources, rather than through access to future or external resources.

Border price   A price at which gas is sold at the border between two countries.

Borderline customer   A buyer of goods or services from a service-provider in one territory which makes payment to the service-provider in another territory.

Bore   The inside diameter of a wellbore.

Borehole   A hole which is drilled by a drill bit as a wellbore.

Borehole ballooning   The flexing and distortion of a wellbore which is caused by drilling fluid temperature and pressure variations which deform the wellbore's walls.

Borehole effect   False influences which are induced in wellbore measurement tools by the physical state of the wellbore.

Borrowing base
(1)   Under a loan agreement, an agreed limitation on the overall amount of money which the borrower can borrow from the lender.
(2)   In reserves-based lending, the amount of the loan facility which is available to the borrower based on the value of the borrower's economically recoverable level of petroleum reserves (as adjusted over time).

Bosman movement   A colloquial term for the transfer by a person of their interests in a petroleum project for value at a time prior to the point at which that person could be required to surrender those interests for no value.

Both to blame   Under a charterparty trading through the United States, a provision for endorsement on a bill of lading whereby in the event of an incident between two ships the charterer (as cargo owner) can recover the value of the lost or damaged cargo from the non-carrying shipowner as part of a circular indemnity by which the non-carrying shipowner then recovers from the carrying shipowner.

Bottle gas   An alternative term for **liquefied petroleum gas**.

Bottleneck   An inherent capacity restriction which impedes the intended proper operation of a petroleum facility. *See also* **debottlenecking**.

Bottom sediment (BS)   An adjustment for a quantity of oil in-tank storage which reflects the presence of sludge.

Bottom settlings and water (BS&W)   An alternative term for **basic sediment and water**.

Bottom stop   An alternative term for a **collar**.

Bottom-up   The circulation of drilling fluid in a wellbore until the point at which the cuttings reach the surface.

Bottom water   An alternative term for **edge water**.

Bottomhole   The lowest point of a wellbore.

Bottomhole agreement   A commitment of a person to make a cash contribution to another person responsible for drilling a wellbore in exchange for the provision of G&G data if the wellbore reaches an agreed target depth and is tested. *See also* **bottomhole money** and **dry hole agreement**.

Bottomhole assembly (BHA)   The lowest-hanging section of a drill string, which includes the drill bit.

Bottomhole choke   A choke in the lower levels of a liner which is used to control flow rates through a wellbore.

Bottomhole location (BHL)   An alternative term for **true vertical depth**.

Bottomhole money   A commitment of a person to make a cash contribution to another person who is responsible for drilling a wellbore in exchange for the provision of G&G data if a defined target depth is reached by the wellbore (whether or not the wellbore produces petroleum). *See also* **bottomhole agreement** and **dry hole agreement**.

Bottomhole pressure   The pressure which is measured at the base of a wellbore.

Bottomhole temperature   The temperature which is measured at the base of a wellbore.

Bottoms   The heaviest petroleum components which remain after crude oil has been refined (such as asphalt and tar). Also called **residuals, residuum**, and *see also* **residual fuel oil**. *See* Appendix 2E.

Bottom-supported offshore drilling unit (BSODU)   Any offshore drilling unit (such as a jack-up) which is partially supported on the sea-bed during drilling operations.

Bottom-up time   The time which is taken for a quantity of drilling fluid to travel in a wellbore from the drill bit to the surface.

Bought-in services (BIS)   Operational services which are procured from a third-party contractor by the operator under a JOA.

Bow tie risk assessment   A risk assessment matrix which, in appearance, assumes the shape of a bow tie.

Boyle's Law   One of the Gas Laws, which states that the pressure of a gas is inversely proportional to its volume.

Box   The hollow (female) end of a section of drill pipe. *See also* **pin**.

Bradenhead gas   An alternative term for **associated gas**.

Braided line   A multiple strand mild steel wire which is used to lower tools into a wellbore. *See also* **slickline**.

Branded LNG   LNG which is sold in a portfolio sale.

Branded reseller   An alternative term for a **dealer-owned service station**.

Break   A sudden and sharp decline in petroleum prices.

Break costs   The transaction costs which are associated with suspending or terminating the performance of a contract or a project. *See also* **unwind costs**.

Breakaway coupling   An alternative term for **emergency release coupling**.

Breakeven analysis   An expense and revenue analysis which is used to determine whether a breakeven point has been achieved.

Breakeven point   The price at which petroleum can be sold so that earned revenues equal or exceed incurred expenses.

Breakout   A sudden departure of petroleum prices from an established pricing pattern.

Breakthrough   *See* **gas breakthrough** and **water breakthrough**.

Breasting dolphin   A dolphin which assists in the berthing of a ship by restricting the longitudinal movement of the ship. *See also* **mooring dolphin**.

Breathing   The emission of vapours from a quantity of petroleum in store. *See also* **evaporation loss**.

Brent Blend   A marker blend of crude oil which is produced from several North Sea sources, from which the Dated Brent crude oil price is derived. *See also* **Brent Forties Oseberg Ekofisk.**

Brent Forties Oseberg Ekofisk (BFOE)   The key crude oil marker blends which make up Brent Blend.

Brent future   A contract for the delivery of Brent Blend crude oil which is to be settled in the future. *See also* **Dated Brent.**

Bridge   An obstruction in a wellbore which is caused by a collapse in the integrity of the wellbore.

Bridge-linked   Elements of an offshore petroleum infrastructure item (such as accommodation, processing, production, and storage modules) which are kept separate and are linked by bridge walkways.

Bridging agreement   A contract which recites and coordinates the interests of a group of persons in relation to a petroleum project.

Bright line test   A clearly defined objective standard which reduces the scope for subjective interpretation and is intended to produce predictable and consistent results.

Bright oil   Crude oil which contains little or no water.

Bright stock   High-viscosity base oil which is produced from a refinery.

Brine   An alternative term for **produced water**.

Bring-down   Under an asset or share sale and purchase agreement, the repetition of the representations and warranties which are made by the parties prior to completion of the sale and purchase.

Bringing in   The act of commencing the production of petroleum from a wellbore. Also called **coming in.**

British Institute Warranties   A set of trading limits.

British National Oil Corporation (BNOC)   A state-owned company which acted as the United Kingdom's national oil company between 1975 and 1982.

British Rig Owners' Association (BROA)   A trade association which represents the interests of companies with United Kingdom interests who operate mobile offshore petroleum infrastructure items (https://www.ukchamberofshipping.com).

British Standards (BS)   A series of standard measures for goods and services which are promulgated by the BSI.

British Standards Institution (BSI)   The agency which is responsible for the promulgation of the British Standards (www.bsigroup.com).

British thermal unit (BTU)   A unit of energy. *See* Appendix 1.

Broached provisions   Provisions which have been opened and partly consumed. Contrast with **unbroached provisions**, which remain unopened.

Brown coal   An alternative term for **lignite.**

Brownfield   The development of a petroleum project with certain pre-existing parameters. Contrast with **greenfield**, as a petroleum project which is developed without pre-existing parameters.

Brownfield production An existing petroleum-producing formation which is not producing but which can be brought back into production because operational or market circumstances make it desirable to do so. Contrast with **greenfield production**, as a newly developed petroleum-producing formation.

Brownout A partial interruption of the provision of energy. Contrast with a **blackout**, as a complete interruption of the provision of energy.

Brucker capsule A particular form of lifeboat which is used in offshore facilities for escape and survival. *See also* **totally enclosed motor propelled survival craft**.

Bubble curve In a phase envelope, the plotted pressure and temperature curve which leads up to the bubble point.

Bubble plot An assessment tool which plots multiple data points in a multi-dimensional format.

Bubble point The point of pressure and temperature drop in a crude oil deposit at which entrained gas will begin to come out of solution (or bubble out). *See also* **volatile oil**.

Bucking the line An increase in the compression of gas at the point of entry into a pipeline in order to overcome a positive pressure differential in the pipeline.

Buckling stress The deformation of a tubular which occurs because of a wellbore's deviation from vertical.

Buffer stock Petroleum stocks which are held as inventory for use during periods of supply interruption. *See also* **strategic petroleum reserve**.

Build own operate (BOO) A petroleum facility development concession model whereby the concession-holder builds, owns, and operates the facility and the state pays periodic fees to meet the facility's construction and operation costs. Model variations can include additional design and finance obligations.

Build own operate transfer (BOOT) A petroleum facility development concession model whereby the concession-holder builds, owns, and operates the facility and the state pays periodic fees to meet the facility's construction and operation costs, with later transfer of ownership of the facility to the state. Model variations can include additional design and finance obligations and operatorship transfer.

Build rate In a high-angle well, a measure of the extent of drift over a certain distance (for example, 6° of incline over 100 feet of drilled depth). *See also* **drift**.

Build-up A positive stock change. Contrast with **drawdown**, as a negative stock change.

Bulge A rapid increase in future petroleum prices.

Bulk rock volume (BRV) A measure of the volume of a solid strata, which is made up of the volume of solid material and the volume of sealed and open pores.

Bulk station A petroleum facility which is used to store and distribute products which typically takes in the bulk delivery of those products by truck or train.

Bulk terminal A petroleum facility which is used to store and distribute products which typically takes in the bulk delivery of those products by ship or pipeline.

Bulkbreak   The disaggregation of a single cargo of petroleum into a number of smaller individual cargoes.

Bull market   A market in which commodity prices are rising. Contrast with a **bear market**, in which commodity prices are declining.

Bullet   A cylindrical steel tank which is used for the storage of LNG or LPG.

Bullet payment   A lump sum payment which is made at a defined point in time.

Bullet perforator   A device which fires bullets through a casing as part of a perforation. Also called a **perforating gun**.

Bullheading   The forcing of natural gas back into a formation by pumping drilling fluids into a wellbore.

Bumper sub   A telescopic joint which is inserted in the top section of a length of drill pipe, to compensate for heave during offshore drilling operations.

Bundled price   Under a petroleum sales contract, a single price for petroleum which is payable by the buyer to the seller and encompasses commodity, transportation, and processing costs.

Bundled service   Under a petroleum sales, transportation, processing, or storage contract, a single package of multiple services which are supplied by a seller or a service-provider.

Bundling   The aggregation of different services and the charges which are associated with them within an integrated business. Contrast with **unbundling**, as the separation and financial disaggregation of the different services and associated charges.

Bundwall   An alternative term for a **firewall**.

Bunker adjustment factor (BAF)   A fluctuating element in the calculation of hire under a charterparty which accounts for fluctuations in the price of fuel. Also called a **fuel adjustment factor**.

Bunker C   An alternative term for **marine gas oil**.

Bunkering   The process of fuelling a ship. Also called **marine fuelling**.

Bunkers   Fuel which is used for the operation of a ship.

Buoyancy-driven accumulation   A conventional petroleum deposit which relies on the migration of petroleum for its creation (in contrast with an unconventional petroleum deposit, whereby petroleum is not free to migrate because of adsorption).

Burial history   An alternative term for **maturity**.

Buried hill   A sub-surface elevation which is covered by younger rock layers.

Buried valley   A sub-surface depression which is concealed by younger rock layers.

Burn rate   The rate at which cash is consumed on a daily or monthly basis over a petroleum project's lifecycle (which is typically expressed in USD/month).

Burner capacity   The maximum heat rate which can be released from the flaring of gas. Also called **burner rating**.

Burner fuel oil   A trade term for fuel oil which is used for heating homes and buildings. *See also* **heating oils.**

Burner rating   An alternative term for **burner capacity.**

Burning point   The lowest temperature at which petroleum will combust when an open flame is held near to it.

Burst pressure   The degree of internal pressure which is required to burst open a pipeline or a tubular from the inside.

Burst rating   A manufacturer's guide to the burst pressure limits and safe operating regime for a pipeline or a tubular.

Bursting disc   A metal disc within a valve arrangement which is intended to burst and so to allow gas to escape within predetermined pressure limits.

Bury barge   A pipelay vessel which lays an offshore pipeline in a scoured trench beneath the sea-bed.

Bury your own dead   A colloquial alternative term for **guilty party pays.**

Bus-stopping   Sequential helicopter movements which take place between offshore platforms. Contrast with **hub-and-spoke,** as bilateral helicopter movements which take place between a shore base and an offshore platform.

Bushing   A pipe fitting which allows two pipe sections of different diameters to be connected.

Business as usual (BAU)   A business continuity planning scenario which assumes the normal execution of standard operational plans.

Business continuity plan (BCP)   An agreed contingency plan which applies to overcome or to mitigate the risks of adverse events which could impact the ongoing integrity of a business.

Business interruption insurance (BII)   A policy of insurance which pays out to the insured person when a defined interruption event affects the insured person's expected revenue flows from a petroleum project.

Business-to-business (B2B)   A transaction which is made between two businesses. *See also* **business-to-consumer.**

Business-to-consumer (B2C)   A transaction which is made between a business and an end-user. *See also* **business-to-business.**

But-for costs   Costs which would not have been incurred in a petroleum project but for a votive decision to incur them. *See also* **avoidable costs.**

Buy-back contract   A concession under which a petroleum company finances exploration costs and the development of a petroleum production project on its own account and in return lifts a defined share of the petroleum produced from the facility over a defined payback period to recover its investment and an agreed element of profit (essentially a risk service contract and principally associated with operations in Iran).

Buy-out payment   A payment which is made by the buyer of a commodity in order to extinguish the associated outstanding liabilities which the buyer has.

Buy-through   Under a petroleum sales contract, a provision for the seller to pro-cure substitute supplies of petroleum when the original source of supply of pet-roleum to the buyer would be interrupted. Also called a **backstop**.

Buy-back   A formulation for the pipeline transportation of gas whereby the trans-porter buys gas from the shipper at a defined point and sells an equivalent quantity of gas back to the shipper at another defined point at a price which additionally re-flects the costs of transporting the gas. Also called **buy/sell transportation**.

Buy/sell transportation   An alternative term for **buy-back**.

Buyer's nomination regime   Under a petroleum sales contract, a provision whereby the buyer nominates the quantities of petroleum for delivery to the buyer. Contrast with a **seller's nomination regime**, whereby the seller's nominations prevail.

Bypass   An auxiliary piping arrangement which is used to carry petroleum around specific equipment or a part of a petroleum facility.

By-product   A secondary or additional product which results from a primary processing operation, which may have value in its own right.

C

Cabinet reservoir   A petroleum project which exists only as a theoretical possi-bility (that is, in a filing cabinet), rather than as an actual project.

Cable tool   A tool which is used to effect percussion drilling.

Cabotage   Petroleum transportation, trading, and logistics management activ-ities, often coastally focused, which take place in a particular state.

Cabotage rules   A state's rules which regulate the rights of non-nationals to par-ticipate freely in cabotage in that state.

Cadastre   A publicly accessible register of the ownership of mining, mineral, and petroleum exploitation rights in a state.

Caisson   A watertight cylindrical steel or concrete chamber which is used for bal-last, buoyancy, or petroleum storage offshore.

Calibration   A process by which a fiscal meter or a check meter is adjusted to make good any deviation from its intended measurement ranges. Also called **adjustment**.

Calibration gas   A gas or a mixture of gases which is used as a comparative standard in the calibration of gas analysis equipment or gas detection equipment. *See also* **zero gas**.

Call-off   A request for services which is made under a master services agreement.

Calorie   A unit of energy. *See* Appendix 1

Calorific value   The amount of heat energy which is released by the combustion of a given quantity of petroleum. Also called **energy content**, **heat content**, and **thermal value**. *See also* **gross calorific value** and **net calorific value**.

Calorific value shrinkage   A lost gas quantity which results from the commingling of gas streams with different calorific values.

Call option   The right, but not the obligation, of a person to buy a particular asset or interest when certain exercise conditions have arisen. *See also* **put option**.

Calvo clause   A provision in a concession whereby a concession-holder expressly renounces any protection which its home government might otherwise give to the protection of its interests under the concession.

Campaign rig   A drilling unit which has been hired in by a petroleum company in order to work on a specific drilling programme.

Canadian Association of Petroleum Landmen (CAPL)   A Canadian-based organisation which represents the interests of landmen in Canada (http://www.land man.ca).

Canted leg jack-up   A jack-up in which the supporting legs are splayed out against the sea-bed for improved support.

Canti-jack   A jack-up upon which the drilling derrick is mounted on an arm which extends outward from the drilling deck. *See also* **slot-type jack-up**.

Cap   Under a petroleum sales contract, a petroleum price ceiling which is agreed between the seller and the buyer as a point beyond which the price will not rise. Also called a **ceiling** and a **top stop**. Contrast with a **collar**, as an agreed petroleum price floor.

Cap and collar   A petroleum pricing construction which applies a cap and a collar to fix a band within which a petroleum price will be set.

Cap and trade   An emissions trading system whereby a regulator caps pollution at a level (the cap) and industry participants holding pollution allowances can trade unused pollution allowances up to the cap.

Cap gas   Associated free gas which is found in a gas cap.

Capacities-based contract   A contract for the pipeline transportation of petroleum which is founded on the principle of the reservation by the transporter, on a shipper's behalf, of a defined amount of capacity in the pipeline (whether or not the shipper uses that capacity and payable for by the shipper through a ship or pay payment). Also called an **advanced reservation of capacity agreement**. Contrast with a **quantities-based contract**, which is founded on the transportation by the transporter of a defined quantity of petroleum between defined points.

Capacity allocation mechanism (CAM)   An operational methodology for a gas pipeline network (or networks) which promotes capacity allocation procedures, operational standards, and access to capacity expansion for the benefit of pipeline users.

Capacity charge   A charge which is payable by the user of a petroleum facility for the reservation of capacity rights in the facility, regardless of whether the capacity is used. Also called an **energy charge**. Contrast with a **commodity charge**, which is payable by the user of a petroleum facility only when the facility is used.

Capacity constraint   A restriction which is imposed or which applies at any time on the useable capacity of a petroleum facility.

Capacity factor   An alternative term for a **load factor**.

Capacity plan   A plan of the spaces which are available for cargo, fuel, fresh water, and water ballast on a ship.

Capacity release   A mechanism for the release of booked but unused capacity in a petroleum facility by the facility user, whereby that released capacity could be made available for use by other persons. *See also* **open access**, **remedy sale**, and **third-party access**.

Capesize   A crude carrier which is too large to pass through the Suez Canal. *See also* **Suezmax**.

Capex uplift   An alternative term for **enhanced cost recovery**.

Capillaries   Minute fissures in a formation through which petroleum or water will flow.

Capillary pressure   Inherent pressure within a formation which leads to the natural flow of petroleum into a wellbore without manual intervention. *See also* **artificial lift** and **solution gas drive**. Also called **natural drive** and **reservoir drive**.

Capital constraint   The aggregate demand for capital expenditure for petroleum project developments which exceeds available capital, so leading to project prioritisation.

Capital expenditure (CAPEX)   The costs which are associated with providing the tangible capital (non-consumable) elements of a petroleum project. Contrast with **operating expenditure**, which relates to intangible item costs.

Capital gains tax   (CGT) take A state take from a petroleum project which is represented by CGT charge, which is levied by the state when a project participant sells an interest in the project for a return greater than the original acquisition cost of the project.

Capital interest   An alternative term **for payment-in-kind interest**.

Capital lease   An alternative term for a **finance lease**.

Capping   The surface closure of a wellbore in order to prevent the continued flow of petroleum through the wellbore.

Caprock   An impermeable layer of sub-surface rock which caps a petroleum-bearing strata and prevents migration to the surface.

Captive arrangement
(1)   Under a petroleum sales contract, a provision whereby the buyer is able to buy petroleum only from the seller or the seller is only able to supply petroleum to the buyer, with no alternative sale or purchase options. *See also* **core customer**.
(2)   A market condition whereby a buyer is able to buy petroleum only from one seller, or a seller is able to supply petroleum only to one buyer, with no alternative sale or purchase options.

Capture   An alternative term for **law of capture**.

Carbon black   A chemical product which is derived from the de-oxygenated combustion of petroleum and is used in the manufacture of rubber-based goods, inks, and paints. *See also* **conversion oil**.

Carbon capture and storage (CCS)   The injection of captured carbon dioxide emissions into a secure formation (such as a depleted gas reservoir) for storage.

Carbon capture and storage unit (CCSU)   A petroleum infrastructure item or process which is used to effect a CCS project.

Carbon credit   An alternative term for a **carbon offset**.

Carbon dioxide equivalent (CO²e) A standardised unit for measuring carbon intensity.

Carbon intensity (CI)   A measure of the amount of carbon by weight emitted per unit of feedstock combusted, which allows a straight-line comparison of carbon emission levels between different feedstocks.

Carbon mitigation/offset agreement (CM/OA)   A contract for the management of carbon reduction and/or carbon offset initiatives in respect of a petroleum project.

Carbon offset   A reduction in emissions of carbon dioxide or GHGs which is effected to compensate for a similar level of emissions which result from a petroleum project process. Also called a **carbon credit**. *See also* **nature-based carbon offset**.

Carbon offset credit (COC)   A record of the amount of carbon which is saved through a carbon offset (and which could be traded).

Carbon reduction   A reduction in emissions of carbon dioxide or GHGs which is made part of a petroleum production process.

Carbon residue   An amount of solid carbonaceous material which remains after the combustion of petroleum.

Carbon sink   A reservoir (whether atmospheric, marine, sedimentary, or terrestrial) which absorbs carbon that has been released from elsewhere.

Carburetted water gas   Synthetic natural gas which is produced by passing water gas through a heated retort.

Care and maintenance (C&M)   The minimum level of investment into a petroleum project which is necessary to be made to keep the project viable. Also called **tickover**.

Care, custody, control (CCC)   A general exclusion under a policy of insurance which disapplies insurance coverage to property not owned by the insured person but in the insured person's care, custody, or control.

Cargo-by-cargo   A petroleum sales contract which assesses the seller's and the buyer's rights and obligations on an individual per-cargo basis.

Cargo containment system (CCS)   The mechanical system by which LNG is stored in a refrigerated state on an LNG carrier.

Cargo handling   The activity of loading or discharging a cargo from a ship.

Cargo manifest   An alternative term for a **manifest**.

Cargo of opportunity   A cargo of petroleum which unexpectedly becomes available for sale.

Carousel
(1)   A storage unit which is used to hold lengths of drill pipe on a drilling unit. *See also* **pipe rack**.
(2)   A reel which is used to store cables or flexible pipes.

Carriage   The transportation of petroleum through a pipeline by a transporter as a service to a shipper and for a fee which is payable by the shipper to the transporter.

Carrier gas   A gas (usually argon, helium, hydrogen, or nitrogen) which is used in a gas chromatograph to take a gas sample through the chromatography process.

Carry/carried interest   The right of a concession-holder (the carried party) to have its proportionate share of the costs of performance of the concession met by another concession-holder (often for a defined period or to a defined amount). Also called a **shared interest**. *See also* **hard carry**, **net spend above carry**, **post-production carry**, **pre-production carry**, **real working interest**, and **soft carry**.

Carry forward
(1)   Under a petroleum sales contract, quantities of petroleum which are taken and paid for by the buyer in excess of its obligations under a periodic take or pay commitment and apply to reduce the next periodic take or pay level. Also called **advance make good**.
(2)   Under a concession, the concession-holder's ability to transfer a minimum work obligation to a later time period or to another concession.

Carrying charge   The total aggregate cost which is incurred by a person in storing petroleum, including storage fees, insurance costs, and administrative charges.

Carry-over   Funds which are unused in a defined financial year and are transferred to a budget for the following financial year.

Carry-over work   An alternative term for **rework**.

Carved-out interest   A petroleum production interest which is created out of a greater interest (such as a royalty agreement which is granted from a concession).

Cas fortuit   An alternative term for **force majeure**.

Cascade   A particular industrial process, pioneered by Conoco Phillips, for the liquefaction of gas to give LNG. Also called **Conoco Phillips optimised cascade**.

Cascade taxation   Taxation which is levied at several different stages of an overall industrial process.

Cased   A wellbore into which casing has been inserted. *See also* a **closed hole**.

Cased and perforated (C&P)   A cased wellbore in which the casing has been perforated.

Cased and suspended (C&S)   A cased wellbore which has been suspended.

Cash accounting   An accounting methodology whereby a revenue item is accounted for when received and an expense item is accounted for when paid out.

Contrast with **accrual accounting**, whereby a revenue item is accounted for when earned and an expense item is accounted for when incurred.

Cash balancing A balancing of interests between persons which is made in cash. *See also* **in-kind balancing**.

Cash call Under a JOA, the payment of estimated cash requirements which are due to be made by each party to the operator in response to a demand made by the operator.

Cash flow at risk (CFAR) A statistical technique which is used to quantify the risk that a cash flow deficiency might be generated by a trading position. *See also* **value at risk**.

Cash-out Under a gas sales contract, the recovery by the buyer of an accrued shortfall price discount balance or make-up balance through a price discount or a cash payment by the seller. Contrast with a **gas-out**, whereby the balance is repayable by the seller in gas or other petroleum quantities.

Cash value A notional cash value which is attributed to non-cash consideration for valuation purposes under a pre-emption provision.

Casing
(1) A steel liner, made up of casing strings, which is inserted into a wellbore in order to secure the integrity of the wellbore's walls. *See* Appendix 2C.
(2) The process of casing a wellbore (also called **running-in pipe**).

Casing hanger The part of a wellhead which supports the top of the first casing string within a wellbore. Also called a **hanger**.

Casing head pressure A measure of casing pressure, which is measured at the wellhead.

Casing pack A method of cementing whereby casing can be later retrieved from a wellbore with minimal difficulty, often so that the casing can be used elsewhere.

Casing point
(1) The depth in a wellbore to which casing is set.
(2) Under a drilling contract, a target depth to which a wellbore is to be drilled and cased.

Casing point election Under a JOA, a decision (which is typically made in the context of approving the associated costs) to run casing into a wellbore and to complete the wellbore.

Casing pressure Pressure which builds up on the casing in a wellbore.

Casing seat The lowest vertical point in a wellbore at which casing is set.

Casing shoe A reinforced section of casing which is placed on the end of a section of casing string in order to protect it from damage.

Casing string A sectional length of casing.

Casinghead Any part of casing which protrudes above ground level or sea-bed level.

Casinghead gas
(1) An alternative term for **associated gas**.
(2) Wet gas which separates naturally at a wellhead, which can be flared or used as fuel.

Casus fortuitus   An alternative term for force majeure.

Cat cracker   A refinery processing unit in which catalytic cracking is carried out.

Catalytic cracking   A process which leads to the production of gasoline and LPG from a refinery. *See also* **cat cracker**.

Catalytic reforming   An alternative term for **reforming**.

Catastrophic release   A major petroleum release which results from an uncontrolled activity.

Catchpot   A filter which removes liquids and solid particles from a gas stream in a pipeline.

Catenary   A measure of the natural curve which occurs when an umbilical is suspended by its ends.

Catenary anchor leg mooring (CALM)   An alternative term for a **single point mooring**.

Catenary mooring   A form of mooring whereby a ship or offshore petroleum infrastructure item is anchored to the sea-bed using multiple free-hanging mooring lines which change configuration according to the movements of the ship or item. Also called **restricted catenary mooring**. *See also* **single point mooring** and **spread mooring**.

Cathodic protection (CP)   A technique which is used to control the corrosion of a metal surface in water by making that surface the cathode of an electrochemical cell.

Catwalk   An elevated narrow walkway.

Caution in advance   An alternative term for the **precautionary principle**.

Cave-in   A collapse of the walls of a wellbore.

Caving   An alternative term for **sloughing**.

Cavings   An alternative term for **cuttings**.

Cavitation   The collapse of a gas-filled cavity in a formation which results from pressure variations which are caused during drilling.

Cavitation damage   The gradual degradation of metal surfaces which occurs when they come into contact with turbulent fluids.

Ceiling   An alternative term for a **cap**.

Cellar deck   The deck which is immediately below the drilling deck of a drilling unit or the working deck of an offshore platform.

Cement job   An alternative term for **cementing**.

Cement log   An evaluation and record of the effectiveness of a cement job.

Cement squeeze   An alternative term for a **squeeze**.

Cementing   The injection of liquid cement into the annulus between a wellbore and casing in order to secure the integrity of the wellbore. Also called a **cement job**.

Cementing head   A device which is attached to the uppermost section of casing and facilitates cementing. Also called a **retainer head**.

Centistoke/centipoise   A unit of measurement for viscosity.

Central estimate   In petroleum exploration, a range of exploration drilling scenarios from which a range of estimated drilling activities can be derived.

Central European Gas Hub (CEGH)   The operator of the virtual trading point (VTP) gas trading platform in Austria.

Central generation   Electric power which is generated in a central facility and is then sent to the point of consumption by transmission and distribution. Contrast with **distributed generation**, whereby electric power is generated directly at the point of consumption.

Central processing platform (CPP)   A petroleum production platform into which a number of satellites are tied in. Also called a **tie-in platform**.

Centraliser   A spacing collar which is used to hold a drill string or casing centrally and away from the sidewall in a wellbore.

Centrifuge   A device which spins up a petroleum sample to high speed, using centrifugal force to separate out the constituent elements of the sample. Also called a **shake-out**.

Centrifuge test   The use of a centrifuge to assess the basic sediment and water content of a crude oil sample.

Certificate of Financial Responsibility (COFR)   A certificate which is required by the US Coast Guard for transporting oil products in US territorial waters, to confirm the shipowner's financial responsibility up to a specified amount for any pollution which is caused.

Certificate of origin (COO)   A certificate which declares the source of production of a cargo of ship-transported crude oil, LNG, or LPG.

Certified emissions profile (CEP)   A certificate which is issued by an independent agency which attests to the carbon intensity of a petroleum delivery or of a petroleum project.

Cess   A tax or levy.

Cessation of production (COP)   The permanent end of production from a petroleum deposit. Also called **permanent cessation of production.**

Cetane   A paraffin-based petroleum which is used as a reference point for determination of the ignition qualities of diesel.

Cetane number (CN)   An index which indicates the ignition quality of a particular fuel (whereby a higher number indicates a shorter time between the injection of a fuel and its ignition).

Chain marketer   A person who retails downstream fuels through a group of owned service stations.

Chance of commerciality (Pc)   A statistical assessment of whether a petroleum project will generate appropriate returns to an investor.

Chance of development (Pd)   A statistical assessment of the likelihood of development if a petroleum discovery is made.

Chance of geologic discovery (Pg)   A statistical assessment of the likelihood of discovery of petroleum.

Chance of success (COS)   An alternative term for **probability of success**.

Chancer   A colloquial term for a petroleum company which holds small interests in multiple joint ventures, often with some indifference to the proper management of the less successful prospects.

Change of control   A change in the ownership structure of a contracting party which allows that party's counterparts to exercise certain protective rights which they might have under a relevant contract (including a termination right or a right to call for the provision of credit support).

Change of state   A physical change of a substance from one phase to another (whether solid, liquid, or gas).

Change order (CO)   An order which is issued under a management of change procedure.

Channelling   The tendency of water or gas which is being produced from a formation to travel towards a wellbore through more permeable rocks, thereby reducing petroleum production from less permeable rocks.

Charge stock   An alternative term for **feedstock**.

Charge system   A combination of source rock, migration, and seal to give a viable petroleum system.

Charles' Law   One of the Gas Laws, which states that the volume of a gas is directly proportional to its temperature.

Chart datum   The water level at which depths on a nautical chart are measured from (which is commonly set as the lowest astronomical tide or as the mean lower low water).

Charterer's option (CHOPT)   An election in respect of the operation of a ship which can be made solely by the charterer under a charterparty.

Charterparty   An arrangement for the hire of a ship which is made between a charterer and a shipowner. *See also* **bareboat charter**, **contract of affreightment**, **employment contract**, **time charter**, and **voyage charter**.

Check meter   Measurement equipment which is relied upon for the verification of fiscal meter readings. Contrast with a **fiscal meter**, as the primary measurement determinant of petroleum quality and quantity.

Check valve   A non-return valve, which allows only a one-way flow of gas or liquids.

Checkerboard leasing   The award of concessions by a state in respect of alternate blocks (with the intention that a discovery will increase the value of an unawarded contiguous block).

Chemical cutter   A tool which uses high-pressure jets of chemicals to release a stuck pipe.

Chiksan   A length of flexible, heavy duty steel hose which is used to transport liquid phase petroleum onto or from a ship.

Child well   An alternative term for an **infill**.

Choke
(1) A reduction orifice used, for example, in an orifice meter. Also called a **venturi**.
(2) The use of smaller-diameter casing to restrict the flow of petroleum through a wellbore.
(3) An identified critical process point within a petroleum project.

Choke line   A high-pressure pipeline which connects a blowout preventer with a choke, through which drilling fluid flows under pressure during drilling and returns to atmospheric pressure.

Christmas tree   A series of valves which are used to control the flow of petroleum from an above-ground or sub-sea wellhead. Also called an **X-tree**. *See* Appendix 2C.

Chromatography   The analysis of gas composition, through the use of a gas chromatograph.

Churn   In gas trading, the number of times which a quantity of gas is traded between initial sale and ultimate consumption. Also called **trading volume**.

Cinderella clause   A colloquial term for a contractual provision whereby a change of position is set to occur at one minute past midnight. Also called a **pumpkin provision**.

Circular indemnity   An indemnity structure between a group of persons whereby a liability works its way back around to the person who was originally intended to bear that liability.

Circulating and conditioning (C&C)   The function of drilling fluid, to circulate through and to condition a wellbore.

City gas   Gas which has been processed for domestic market use and is delivered through a local distribution zone.

City gate   A location at which a local gas distribution entity receives gas from a trunkline, whereby gas pressure is lowered for reticulation.

City gate rate   The rate which is charged to a local gas distribution entity by a gas supplier, referencing the cost of the gas at city gate as the point at which the local gas distribution entity takes title to the gas. Also called **gate rate**.

Clarification   The process of separating liquids and solids (such as oil particles and water).

Clarifier   A separator which is used for clarification.

Classification certificate   A certificate which is issued by a classification society that evidences compliance by a ship or offshore petroleum infrastructure item with the classification society's standards.

Classification society   A non-governmental organisation which sets, oversees, and ensures compliance with technical standards for the construction and operation of ships and offshore petroleum infrastructure items.

Clastic rock   A sedimentary rock which is made up of fragments of other pre-existing rocks.

Clause Paramount   Under a charterparty, a provision for endorsement on a bill of lading which declares that the charterparty is governed by the Hague Rules (1924)/Hague-Visby Rules (1968), relating to liability for loss of or damage to the cargo.

Clawback   A provision in a JOA whereby the transfer of a party's interests to an affiliate is rescinded if the affiliate later ceases to be so. *See also* hive-down/ **hive-up**.

Clay-bound water (CBW)   Water which exists in a clay layer in a formation and does not move when petroleum flows through the formation.

Clean bill of lading   An alternative term for **clean on board**.

Clean break   In the sale of an asset or a business, a provision whereby the seller will have no continuing liability to the buyer or to any third party after the sale has concluded.

Clean cargo   A cargo of crude oil which is made up of white oil. *See also* **dirty cargo**.

Clean LNG   LNG which is produced with the benefit of carbon offset and/or carbon reduction. Also called **advantaged LNG**.

Clean oil   An alternative term for **dry oil**.

Clean on board (COB)   An endorsement which is applied to a bill of lading to record that the carrier received the cargo in good condition and properly packaged (whereupon liability for damage to the cargo or its packaging is assumed by the carrier). Also called a **clean bill of lading**.

Clean petroleum products (CPP)   A collective term for products such as gasoline, jet fuel, naphtha, and condensate, which are carried on clean product tankers. *See also* **dirty petroleum products**.

Clean product tanker   A ship which is used for the carriage of clean petroleum products. *See also* **dirty product tanker**.

Clean rock   A formation which contains no or negligible clay minerals. Contrast with **dirty rock**, which contains clay minerals.

Cleaned bitumen   Bitumen from which all non-petroleum impurities have been removed, usually prior to further processing.

Clear fuel   Gasoline, diesel, or aviation fuel which is not a dyed fuel.

Cleat   A natural fracture in coal which is critical to the existence of coal bed methane.

Clingage   Oil residue which remains on the internal surface of a pipeline or storage tank after the bulk of the oil has been removed.

Clip   A defined parcel of traded natural gas (such as a clip size of 100,000 Btus).

Close of business (COB)   A defined end of a day for petroleum-trading or contractual purposes. Also called **end of day**.

Closed hole   A wellbore which is drilled and cased during or after drilling. Contrast with an **open hole**, whereby a wellbore is drilled without casing.

Closed loop
(1) A process system which automatically regulates a process to a defined outcome without manual intervention. *See also* **open loop**.
(2) An LNG regasification system which uses continuously produced steam as a heat source to regasify LNG. Contrast with **open loop**, which uses sea water as a heat source.

Closed space A space which has been flooded with an inert gas through the operation of an inert gas system.

Closure The effective sealing of a petroleum deposit on the top, base, and vertical surfaces in order to create a trap. Also called a **four-way trap**.

Cloud point The point of temperature of a product such as diesel at which it begins to congeal and take on a cloudy appearance.

Cloud point test A test for fuels which contain dissolved wax, to determine the cloud point.

Coal bed methane (CBM) Methane which is adsorbed onto the surface area of unmined coal deposits. Also called **coal seam gas**.

Coal gas An alternative term for **synthetic natural gas**.

Coal gasification The distillation of bituminous coal to produce synthetic natural gas.

Coal liquefaction A process which converts coal into a liquid fuel.

Coal mine methane (CMM) Residual methane deposits which are found in and are recoverable from active coal mine workings. Contrast with **abandoned mine methane** in disused coal mine workings.

Coal seam gas (CSG) An alternative term for **coal bed methane**.

Coal tar A liquid phase petroleum element which results from the distillation of bituminous coal.

Coal trading agreement (COTA) A contract for the sale and transportation of coal.

Coalescer A filtration vessel which removes water from crude oil.

Co-firing The simultaneous combustion of different fuels in the same combustion system (such as natural gas and landfill gas, or coal and biomass) which is intended to improve overall combustion efficiency.

Co-generation The combined generation of electric power and heat from a single facility. Also called **combined heat and power** and **total energy**.

Co-insurance Insurance whereby a number of insurers share the insured risk. *See also* **re-insurance**.

Co-insured A person who is named as a beneficiary of a policy of insurance, despite not being a principally insured person. Also called an **additional loss payee**.

Coiled tubing (CT) Tubing which is stored on a reel and ready for use.

Coke A solid fuel which is produced from the combustion of coal in the absence of air.

Coke oven gas   Gas which is produced as a by-product from the coking of coal.

Coking   The processing of bottoms into gas oil and naphtha in a refinery. Also called **visbreaking**.

Cold filter plugging point (CFPP)   The lowest point of temperature at which diesel will flow freely.

Cold heavy oil production with sand (CHOPS)   The non-thermal production of heavy oil using sand injection. *See also* **fireflood extraction**.

Cold rig   A drilling unit which is presently not being employed and requires mobilisation. *See also* **hot rig** and **warm rig**.

Cold tap   Coldwork in which a spur is cut into an existing, operational pipeline. Contrast with a **hot tap**, as hotwork in which a spur is cut.

Cold test   The point of temperature at which oil solidifies.

Cold zone   An identified safety area for personnel in a petroleum facility.

Coldstacking   The lay-up of a drilling unit without expectation of near-term use. Contrast with **hotstacking**, whereby a lay-up assumes the possibility of near-term use.

Coldwork   Any work activity which is undertaken on a petroleum facility which is not hotwork.

Collapse pressure   An alternative term for **yield pressure**.

Collapse resistance   A measure of the ability of a tubular to resist yield pressure.

Collar
(1)   A coupling device which joins up two lengths of drill pipe.
(2)   Under a petroleum sales contract, a petroleum price floor which is agreed between the seller and the buyer as a point beyond which the price will not fall. Also called a **bottom stop** and a **floor**. Contrast with a **cap**, as an agreed petroleum price ceiling.

Collar pipe   In a drill string, a section of pipe which is used between the drill bit and the final section of the drill pipe before the drill bit. *See* Appendix 2C.

Collateral support   The provision by a person of any surety or support for another person's contractual payment or performance obligations. *See also* **comfort letter, credit support agreement, letter of credit, parent company guarantee, sleeving**, and **standby letter of credit**.

Collector and separation platform (CSP)   An offshore platform on which petroleum is gathered from multiple sources and separation is carried out.

Collision   The situation in which two moving objects strike each other. Contrast with **allision**, in which a moving object strikes a stationary object.

Collision regulation (colreg)   A series of navigational and operational principles which are intended to mitigate the risk of vessel collisions within a port.

Column height   The vertical measure of a formation which contains petroleum. Also called **reservoir height**.

Column-stabilised drilling unit   A form of semi-submersible.

Combination drive   Any combination of two or more natural energies (such as water drive, gas expansion drive, solution gas release, or capillary pressure) which work together in a formation in order to force water or petroleum into a wellbore.

Combination trap   A mixture of a structural trap and a stratigraphic trap.

Combination utility   A utility supplier of gas and another commodity (such as electric power), often as part of a bundled service.

Combined cycle gas turbine (CCGT)   An electric power-generating facility which incorporates gas and steam turbines to generate electric power, with provision for waste heat recovery. Contrast with a **single cycle gas turbine**, which uses a single turbine to generate electric power, without provision for waste heat recovery.

Combined heat and power (CHP)   An alternative term for **co-generation**.

Combustion energy   A measure of the energy which is released by the process of combustion.

Come out of the hole   An alternative term for a **trip**.

Comfort letter   A written undertaking, which is issued usually by the parent company of a contracting party, whereby the parent company offers a degree of support which is something less than a formal guarantee for the performance of the contracting party's obligations under a contract.

Coming in   An alternative term for **bringing in**.

Comité de gestion   A term which is sometimes used in Francophone states to describe a form of management committee which exists under the terms of a concession.

Comitology   A process by which a group of states formally approve the terms of a proposed multi-state accord.

Commencement of deliveries (COD)   The date upon which a delivery of petroleum begins, according to required contractual standards.

Commercial agent   An agent which acts to secure business for a principal in consideration of the payment to the agent of a defined commission for doing so. Also called a **commission agent** and a **mercantile agent**.

Commercial general liability (CGL)   A policy of insurance which is intended to protect against liability claims for personal injury and property damage arising out of commercial operations.

Commercial grade   Propane which is graded below HD10 and is often used in refineries.

Commercial operations date (COD)   The date upon which a petroleum infrastructure item first starts to operate, according to agreed commercial parameters.

Commercial success ratio   The ratio of the total number of exploration wells which are drilled to the total number of exploration wells which encounter commercially viable petroleum deposits. Also called a **success ratio**.

Commercial well declaration (CWD)   A decision and a declaration which is made by a concession-holder that commerciality has been established in respect of a discovery. Also called a **declaration of commerciality**.

Commerciality   The decision that a discovery should be capable of producing enough net income, at sufficiently low risk and within a reasonable timeframe, to make it worthy of development. *See also* **commercial well declaration**.

Commercially sensitive information (CSI)   Data to which an express or an implied duty of confidentiality applies.

Commingling   A mixture of petroleum from more than one input source within a pipeline or a storage tank.

Commission agent   An alternative term for a **commercial agent**.

Commission piggable   A pipeline which is designed to be pigged only at the point of its initial commissioning and not at any later point during its lifetime.

Commissioning   In the development of a new petroleum facility, the first introduction of petroleum into the facility which takes place after pre-commissioning.

Commissioning and start-up (CSU)   In the development of a new petroleum facility, the combined act of commissioning and the first start of the facility.

Commissioning, handover, and takeover (CHOTO)   In the development of a new petroleum facility, the stage at which the responsibility for the facility passes from the person responsible for its construction to the facility owner.

Commitment shoot   A seismic survey which is required to be carried out by a concession-holder as a compulsory obligation under the terms of a concession, as part of a minimum work obligation.

Commitment well   A wellbore which is required to be drilled by a concession-holder as a compulsory obligation under the terms of a concession, as part of a minimum work obligation. Also called an **obligation well**.

Committed gas   Under a gas sales contract, a provision which commits the seller to deliver gas to the buyer from a specified source.

Committed project   A petroleum production project whereby firm and binding (albeit possibly conditional) commitments have been made by project participants and other stakeholders for the expenditures and activities which are necessary to take the project through to the production stage.

Commodity charge   A charge which is payable by the user of a petroleum facility to the facility owner only when the facility is used. Contrast with a **capacity charge**, which is payable by the user of a petroleum facility for the reservation of capacity rights in the facility, regardless of whether the capacity is used.

Commodity trading advisor (CTA)   A person who is retained to provide advice and services on petroleum trading through the use of derivatives contracts.

Common-carrier facility   A petroleum-processing, transportation, or storage facility which has a contractual or a regulatory obligation to provide access to all potential users. Contrast with an **equity facility** which is owned and/or operated by a person principally for the processing, transportation or storage of that person's own petroleum entitlements.

Common-carrier pipeline   An alternative term for a **multi-shipper pipeline**.

Common data set   An exclusive data set which is agreed between a group of persons which defines the boundaries of the application of a unitisation or a redetermination.

Common depth point   A seismic data acquisition and processing technique which transforms field recordings from seismic surveys into cross-sectional geological images in order to identify optimum drilling locations.

Common property   An alternative term for the **prior appropriation rule**.

Common property resource   Any natural resource which is not naturally owned by a single owner but to which any number of people potentially have open access (such as mobile petroleum to which the law of capture applies).

Common rail injection   A fuel supply system in which two or more high-pressure pumps supply a common manifold.

Common stream   A stream of petroleum in which more than one person has an ownership interest.

Common stream agreement (CSA)   A contract which regulates and administers the interests of a group of persons in respect of a common stream.

Communication   The flow of petroleum which occurs within a formation or between formations.

Community development agreement (CDA)   An agreement which is made between a petroleum project developer, a state, and a local community whereby the developer donates certain financial benefits to the local community as a condition of being permitted to undertake a petroleum project. Also called a **benefits agreement**, an **impacts and benefits agreement**, and a **planning gain agreement**. *See also* **direct distributions**.

Community-driven development (CDD)   An energy project which is developed principally with the needs of a local community in mind, rather than with commercial objectives as the driving force.

Compact LNG (CLNG)   A small-scale LNG production plant.

Company man   The nominated representative of a petroleum company which has hired a drilling unit who is resident on the drilling unit when it is operational.

Company-owned company-operated (COCO)   A downstream fuels retailing model which utilises the company-owned service station model.

Company-owned franchisee-operated (COFO)   A downstream fuel retailing model whereby a franchisee is appointed to run a company-owned service station.

Company-owned service station (COSS)   A downstream fuels retailing model whereby a fuel supply company directly owns and operates a service station in its own name. *See also* **company-owned company-operated**.

Compensator   In respect of an offshore drilling unit, dynamic positioning equipment which offsets heave while drilling is taking place.

Competence assurance programme (CAP)   A process for identifying, assessing, and developing required competence levels for company personnel who are engaged in petroleum operations.

Competent person's report (CPR)   A report on the prospectivity of a company's petroleum assets and interests which is prepared by a recognised independent expert, often as a precursor to a material transaction such as a stock exchange listing, a fundraising, or a merger. Also called a **qualified person's report** and a **technical expert report.**

Competitive bidding   A model for the award of a concession by a state whereby the concession terms are made transparent and are contestable items in a licensing round. Contrast with an **open door award**, whereby concession terms are negotiated directly between the state and an investor.

Competitive drilling   Multiple drilling activities which are conducted on different concessionary or title interests but in respect of a single formation to profit from the law of capture by exploiting mobile petroleum. *See also* **offset well.**

Complete shut-off (CSO)   The total closure of a wellbore during a time when petroleum is flowing through the wellbore.

Completion   The readying of a wellbore (including the installation of permanent wellhead and downhole equipment) for the production of petroleum from an underlying formation.

Completion guarantee   A commitment from a borrower in a project financing to make good certain of the lender's entitlements when the revenues from the project are insufficient to do so. *See also* **debt service undertaking.**

Completion interval   A part of a formation which is open to a wellbore ready for the production of petroleum.

Completion risk (C-risk)   The risk that a petroleum project will not be completed on time, to budget, or to specification.

Complex refinery   Any refinery other than a topping refinery.

Complexity   An alternative term for **refinery complexity.**

Compliant tower   A fixed petroleum production platform with a flexible jacket which moves with tidal forces, typically used in water depths up to 3,000 feet. Also called a **guyed-tower platform**. *See* Appendix 2D.

Compound annual growth rate (CAGR)   A measure of the level of growth of a business or a sector over a defined period of time.

Compressed natural gas (CNG)   Pressurised methane which is used principally as a vehicle fuel (in displacement of diesel or gasoline).

Compressibility   Any change in the overall volume of a given substance which results from a change in the pressure which is applied to that substance.

Compression packer   A packer which is locked in place in a wellbore by the weight of the liner which sits above it.

Compulsory pooling   An alternative term for **forced pooling**.

Concarbon   An alternative term for **Conradson carbon residue**.

Concentrated live load (CLL)   A live load which is distributed unevenly across a petroleum infrastructure item. Contrast with **uniformly distributed live load**, which is an evenly distributed load.

Concentric-tubing workover   A workover which is performed with a smaller-diameter liner within an existing liner.

Concept selection   A stage in a petroleum project when the selection of a core technology item, from a series of possible technology items, is made.

Concession   A written permission which is awarded by a state (as the grantor) to allow a person (as the concession-holder) to explore for and/or to produce the state's petroleum interests in a defined area and for a defined duration. *See also* granting instrument, **licence, lot, production enhancement contract, production sharing contract, revenue sharing contract, service contract, upstream government petroleum contract, usufruct**, and **water lot**.

Concrete island drilling system (CIDS)   An offshore structure which is made principally of concrete and is used for drilling in Arctic waters.

Condeep   A shorthand term for concrete deep water, a form of gravity-based structure which consists of a base of concrete oil storage tanks from which a number of concrete shafts rise and onto which topsides are fitted.

Condensable gas   Gas which is capable of being converted to a liquid (for example, as LNG or as LPG) through changes in temperature or pressure.

Condensate   The heavier (pentane to octane) petroleum fractions which exist as a liquid at ordinary atmospheric temperature and pressure, which typically has an API gravity at or greater than 50°. Also called **gas condensate, light liquid hydrocarbon**, and **pentanes plus**.

Condensate credit   In a gas sales contract, an agreed discount to a dry gas sales price which applies when the seller is also generating separate revenues from the sale of stripped liquids.

Condensate–gas ratio (CGR)   The ratio of condensate to gas in a petroleum stream.

Condensate water   Water vapour which condenses in a fuel tank to give liquid water.

Condensate well   A wellbore which produces condensate from a formation.

Condensation   A transformation of the physical state of a gas to a liquid. Contrast with **vaporisation**, as a transformation of the physical state of a liquid to a gas.

Condition A materials (Con-A)   Under a JOA, materials which are transferred by the operator to the parties for use in joint operations which are as-new in condition.

Condition B materials (Con-B)   Under a JOA, materials which are transferred by the operator to the parties for use in joint operations which have previously been used but which are in sound and serviceable condition and ready for use without further repair or reconditioning.

Condition C materials (Con-C)   Under a JOA, materials which are transferred by the operator to the parties for use in joint operations which have previously been used and are not in sound and serviceable condition and ready for use but which are capable of further repair or reconditioning to make them so.

Condition assessment programme (CAP)   A specialised survey of the physical condition of a ship, a drilling unit, or a petroleum facility.

Condition monitoring (CM)   The ongoing monitoring of the operational state of a petroleum infrastructure item in order to identify a developing fault, as part of a programme of predictive maintenance. Also called **condition-based monitoring**.

Condition monitoring system (CMS)   An operational protocol for the carrying out of condition monitoring.

Condition-based monitoring (CBM)   An alternative term for **condition monitoring**.

Conditioned gas   Raw gas which has undergone any processing after first production.

Conditions of use (COU)   The arrangements which exist between a port owner and a ship which govern the transit of the ship through the port (including the allocation of liability for damage caused).

Conductor casing   The widest diameter of casing, and typically the first run of casing, which is used in a wellbore. Also called a **foundation pile**. *See* Appendix 2C.

Cone penetration test (CPT)   A technique for testing the geotechnical properties of soil and for determining soil stratigraphy. Also called the **Dutch cone test**.

Confidential information memorandum (CIM)   An alternative term for an **information memorandum**.

Confidentiality agreement (CA)   An alternative term for a **non-disclosure agreement**.

Confirmation notice (CN)   A record of key transactional terms which leads to the consummation of a transaction which is made under a master sales agreement.

Confirmation well   An alternative term for a **step-out well**.

Conflict impact assessment (CIA)   An analysis and assessment of local conflict risks which could affect the undertaking of a petroleum project.

Conical drilling unit (CDU)   An offshore drilling unit with a cone-shaped base, which is often used for drilling in Arctic waters. *See also* **Arctic submersible rig**.

Coning   The encroachment of interstitial water or natural gas into an oil column within a producing wellbore.

Connate   Water which is trapped in the pore spaces of a rock since its formation which occurs inherently with petroleum in a formation. Contrast with **edge water**, as sub-surface water deposits which are adjacent to a formation but which are not commingled with petroleum in the formation. Also called **formation water** and **interstitial water**.

Connection gas   Gas quantities which enter a wellbore when the inflow of drilling fluid is interrupted and the bottomhole pressure decreases.

Connectivity   Well logging which measures the degree of connectivity of pore space in a formation (in which high connectivity is conducive to the free flow of petroleum through the formation).

Conoco Phillips optimised cascade (CPOC)   An alternative term for **Cascade**.

Conradson carbon residue (CCR)   A laboratory test which is used to provide an indication of the coke-forming tendencies of a quantity of liquid petroleum. Also called **concarbon**.

Consequential loss   A measure of the loss which is suffered by a person resulting from a breach of contract which represents lost future business and profit opportunities rather than present, actual losses. *See also* **direct loss**. Also called **indirect loss**.

Consolidation
(1)   The restatement of a principal contract and a number of subsequent amendments to that contract into a single consolidated contract which reflects and synthesises all of those amendments, typically with effect from the date of the original contract. *See also* **amendment and restatement**.
(2)   The consolidation of a number of disputes involving the same or related issues within a single set of proceedings. Also called **joinder**. *See also* **umbrella agreement**.

Constant dollars   The time-adjusted measure of a currency whereby monetary values can be compared over time without accounting for inflation.

Constant price case   An alternative term for a **flat price case**.

Construction all risks (CAR)   A single policy of insurance which covers all risks to which a contractor is exposed in the construction of a petroleum project facility.

Construction and tie-in agreement (CTIA)   A contract which is made between the owner of an existing petroleum facility and the owner of a new petroleum facility which regulates rights, obligations, and liabilities between the parties in respect of the construction and the connection of the new facility to the existing facility. *See also* **tie-in agreement**.

Construction environmental management plan (CEMP)   A plan which sets out responsibilities with regards to compliance with environmental legislation and standards during the construction phase of a petroleum project.

Construction management agreement (CMA)   A single point responsibility contract whereby a contractor offers a total project management solution in respect of a series of other contractors (and possibly also assumes project delivery risk and liability). Also called a **wrap**.

Construction support vessel (CSV)   A ship which is used to support the process of constructing and commissioning offshore petroleum facilities.

Constructive total loss   Damage to an insured asset which leads to a determination by an insurer that the cost of repair of the asset exceeds the asset's insured value, such that it will be treated as an actual total loss. *See also* **actual total loss** and **total loss**.

Contact   A defined geological boundary which exists between two different bodies of rock. *See also* **unconformity**.

Contact area   An alternative term for **fluid contact**.

Containment   The storage of LNG within a tank or other vessel.

Containment boom   A flexible boom which is deployed to contain an offshore oil spill.

Contango   A market in which the price for the delivery of petroleum in the immediate future is lower than the price for delivery of petroleum in the further future. Contrast with **backwardation**, whereby the immediate future price is greater than the further future price.

Contestability   A measure of the extent to which a market is open to new entrants.

Contestable items   An alternative term for **biddable items**.

Contiguity   A measure of the extent to which a sub-surface petroleum deposit remains constant and unbroken throughout the entirety of the areas which it traverses.

Continental shelf (CS)   The relatively shallow area of the sea-bed which extends outwards from the shoreline of a landmass.

Contingency   A notional monetary sum which is kept aside to account for unforeseen expenses in the execution of a petroleum project. Also called **unallocated provision**.

Contingent consideration   Under an asset or a share sale and purchase agreement, a payment which is due to be made by the buyer for the purchase of an asset or a business and is deferred in whole or in part and is made dependent on certain conditions being met.

Contingent resources   Estimated quantities of petroleum which are less certain of recovery than reserves but which are more certain of recovery than prospective resources. *See* Appendix 7.

Continuation   Under a concession, a process by which the concession-holder's rights are continued beyond the scheduled date of expiry of the concession, or beyond the scheduled end date of a phase within the concession, for a defined period of time or for a defined purpose, and possibly also subject to the payment of a defined extension fee by the concession-holder. Also called **continuous development**.

Continuous development   An alternative term for **continuation**.

Continuous distillation   An alternative term for **atmospheric distillation**.

Continuous emissions monitoring system (CEMS)   A process for the monitoring of emissions from a petroleum infrastructure item which is carried out on a continuous basis and is used to quantify actual emissions levels. *See also* **predictive emissions monitoring system**.

Continuous pipeline operation   An operational methodology for ongoing corrosion prevention which is intended to allow pipelines and other petroleum infrastructure items to flow petroleum without the need for an interruption for anti-corrosion maintenance.

Continuous sample/sampling   A petroleum sample which is drawn from the flow in a pipeline at a uniform and continuous rate over time. Contrast with **grab sampling**, whereby samples are taken intermittently.

Continuous-type deposit   A petroleum deposit which extends over a large geographical area and is largely unaffected by hydrodynamic influences.

Contra proferentem   A rule of contractual interpretation which provides that in the event of ambiguity of wording a contract will not be construed against the person who had responsibility for drafting that wording.

Contract and collaboration (C&C)   A petroleum project management philosophy which relies on the written word of a contract and close operational collaboration between the contracting parties.

Contract area   Under a concession, a defined geographical area in which the concession-holder's rights can be exercised and obligations must be performed, which could be described two-dimensionally with the use of surface area coordinates or three-dimensionally such that the contract area extends downwards from a defined reference point to a defined stratigraphic depth. *See also* **relinquishment**.

Contract depth/contract distance   Under a drilling contract, a defined depth (vertical) and/or distance (horizontal) which the drilling contractor is engaged to drill a wellbore to.

Contract for differences (CFD)   A bilateral hedging instrument which gives a person a right to receive payment for the difference between a market price and a contract price for an underlying commodity.

Contract market   A methodology for trading crude oil whereby suppliers enter into long-term contracts with buyers, with all terms except price agreed (and where the price is further agreed at the point of sale).

Contract of affreightment (COA)   A contract between a shipowner and a charterer for the carriage of goods by ship (such as a charterparty). Also called **affreightment**.

Contract operator   A person, usually an independent third-party contractor, who is appointed by a contract to operate a particular contract, petroleum project, or petroleum infrastructure item on behalf of the owners of the contract, project, or item.

Contract regarding an interim supplement to tanker liability (CRISTAL)   A supplement to the TOVALOP scheme which existed between 1971 and 1997.

Contracted reserves   Petroleum reserves which are dedicated to the fulfilment of a petroleum sales contract.

Contractor-controlled insurance programme (CCIP)   An insurance package in respect of a petroleum project which is arranged by a project contractor. Contrast with an **owner-controlled insurance programme**, whereby the project owner arranges the insurance package.

Contractual joint venture (CJV)   A joint venture arrangement whereby the parties represent their interests through an unincorporated association (for example, under a JOA). Contrast with an **equity joint venture**, whereby the parties represent their interests through an incorporated association (for example, under a shareholders' agreement in respect of a company). Also called an **unincorporated joint venture**.

Contributed data   Traded petroleum price data which is made available to a price reporting agency.

Control well   An alternative term for a **parametric well**.

Controlled source electromagnetic (CSEM)   A data acquisition geophysical survey which is intended to give a better indication of sub-surface resources locations.

Convention on the International Sale of Goods (CISG)   An alternative term for the **Vienna Convention**.

Conventional/conventionals   Petroleum which can be extracted by traditional (conventional) recovery techniques for exploration and production. Contrast with **unconventionals**, as petroleum which can only be extracted by (unconventional) recovery techniques.

Conventional blendstock for oxygenate blending (CBOB)   A conventional form of gasoline blendstock which is intended for blending with oxygenates downstream of a refinery and produces conventional gasoline. *See also* **reformulated blendstock for oxygenate blending**.

Conventional buoy mooring (CBM)   An offshore mooring point for a ship.

Convergence   When gas is used as a feedstock for electric power generation, the point at which there is parity of attraction in switching the gas supply between use for electric power generation and the direct supply of gas into the market for consumption because realisable electric power and gas prices are the same.

Conversion   Any process which alters the compositional nature of a petroleum stream.

Conversion efficiency   The ratio of useful energy output from a process to the input energy which was required to perform the process.

Conversion factor   An agreed factor which is used to convert quantities of different grades of petroleum to a common measure.

Conversion oil   A crude oil feedstock which is used in the manufacture of carbon black.

Conversion refinery   The most versatile refinery configuration, which incorporates all the functions found in topping and hydroskimming refineries but also features conversion processes such as alkylation, catalytic cracking, coking, hydrocracking, and solvent extraction.

Converted barrel   A barrel of oil equivalent quantity of petroleum to which barrel conversion has been applied.

Convertible farm-out   A form of farm-out agreement whereby a commercial discovery gives the farmee a right to convert the farm-out interests into another form of petroleum project interest (such as a non-possessory pet-roleum royalty).

Conveyance   A transaction for the commercialisation of petroleum which ordin-arily cannot be recognised by a petroleum producer as a bookable reserve.

Cool down   A process whereby the temperature in the tanks and the loading lines of an LNG carrier is reduced from ambient to cargo-ready by the controlled spraying of LNG to prevent the risk of a cargo immediately vaporising if it is pumped into a warm tank. *See also* **spray cooling**.

Cooled with heel   The situation whereby an LNG carrier arrives for the loading of a cargo with its tanks containing heel and the in-tank temperature is such that cool down will not be required and the carrier is ready to load. *See also* **presentation**.

Core   A cylindrical section of rock which has been extracted from a wellbore. *See also* **rotary sidewall core** and **sidewall core**.

Core analysis   The laboratory study of a core which is intended to indicate the possible presence of petroleum in the formation from which the core was extracted.

Core analysis report (CORAN)   The written output from a core analysis.

Core bit   A drill bit which is used for coring (which does not drill out the centre of a borehole but which allows that centre, as a core, to pass through into the core catcher).

Core catcher   A section behind a core bit which retains a core.

Core customer   A person who is party to a captive arrangement.

Coring   The extraction of a core from a wellbore. *See also* **drop coring**.

Coriolis meter   A meter which measures the flow rate of gas.

Corporate and social responsibility (CSR)   A defined set of social responsibility and sustainability guidelines which a petroleum company undertakes to abide by. Also called **environmental, social, and governance** and the **social licence to op-erate**. *See also* **appropriate technology**.

Corporate well   A general commitment of a petroleum company to the effort and expense of drilling a wellbore but which is made without reference to a particular petroleum project.

Correlate   The relation of sub-surface information which is obtained from one wellbore to information which is obtained from other wellbores, so that the charac-teristics of a formation can be better understood.

Correlative rights   The rights of each member of a consortium of petroleum producers to produce, lift, and dispose of their agreed shares of produced petroleum.

Correlative rights doctrine   A corollary to the law of capture that the exercise of capture rights by a person is subject to a corresponding obligation of that person to exercise those rights without negligence or waste.

Corrosion inhibitor   A chemical which is added to the flow of petroleum through a pipeline in order to reduce the level of corrosion of the internal facing of the pipeline.

Corrosion resistant alloy (CRA)   Critical metallurgy which is used in the construction of petroleum facilities.

Cost and freight (C&F)   A petroleum transportation formulation whereby the seller pays the freight costs which are necessary to bring the petroleum to a named port of destination.

Cost bank   An alternative term for a **cost pool**.

Cost hierarchy   An agreed order in which accrued costs are recovered.

Cost insurance freight (CIF)   An Incoterms formulation which is used in ship deliveries whereby the delivery of petroleum from the seller to the buyer takes place at the loading port but the seller bears the transportation obligation (and associated costs) to a nominated unloading port, despite the earlier delivery to the buyer.

Cost of service (COS)
(1)   An economic methodology whereby a service-provider allocates the aggregate costs of providing its services to individual service customers.
(2)   The total costs to a service-provider of providing a service to a service customer, which as a minimum will need to be recovered through revenue payments which are made to the service-provider by the service customer.

Cost oil (cost gas)   Under a PSC, the proportion of produced oil (or gas) which is used to pay back the concession-holder for its investment of capital and operating costs as cost recovery. Contrast with **profit oil (profit gas)**, as produced oil (or gas) which is available for distribution between the grantor and the concession-holder.

Cost plus contract   A time and materials contract whereby the contactor's profit element is set as a percentage of the recovered costs element.

Cost plus contracting   An alternative term for **open book estimate**.

Cost plus pricing   The construction of a price for petroleum which is generated by working forwards from the seller's costs and profit expectations in order to determine the seller's revenue requirements. Contrast with **netback pricing**, whereby the price is determined by working backwards from an ultimate source of revenue, less any operational costs incurred, in order to determine whatever is left for the seller as profit.

Cost pool   The accumulation of incurred capital and operating costs which are eligible for cost recovery under a PSC. Also called a **cost bank**.

Cost recovery   Under a PSC, the stage when the concession-holder recovers its incurred capital and operating costs as cost oil (cost gas). *See also* cost pool, cost saturation, **first tranche petroleum**, and **profit oil (profit gas)**.

Cost reduction in the new era (CRINE)   A 1990s initiative for the reduction of operating costs on the UKCS which was later replaced by LOGIC.

Cost reflectivity   A price which is charged for goods or services by a supplier which must at a minimum reflect the costs incurred by the supplier in making available those goods or services.

Cost saturation   The point in the economic cycle of a PSC when the maximum permissible amount of petroleum production revenues will be applied to recoup recoverable costs.

Cost stack   The aggregation of identified costs which are associated with a particular petroleum project or process.

Cost stop   Under a PSC, a limit on the amount of gross petroleum sales revenues which can be directed by the contractor towards cost recovery.

Council of Petroleum Accountants Societies (COPAS)   A professional trade organisation which provides accounting expertise for the oil and gas industry (http://www.copas.org).

Coupon   An instrument which records the rate of interest which is to be paid periodically by the borrower under a loan agreement.

Cournot profits   Trading profits which are made by a company that offers an identical product to its competitors but where the company seeks to command the greatest market share for its product (assuming that higher output levels will reduce unit costs and increase profits from the product).

Cover standard damages   Under a petroleum sales contract, monetary damages which are payable by the seller to the buyer in respect of a shortfall in the delivery of petroleum by the seller and are intended to cover the incremental cost to the buyer of procuring replacement petroleum quantities.

Crack spread   The price differential which exists between the price of a defined quantity of crude oil put into a refinery and the price which is realised from the sale of the resultant products, which is used as an indication of the gross refining margin which the refinery is achieving. Also called a **distillation margin** and an **indicator margin**.

Cracked gas   Gas which is produced from cracking larger petroleum molecules in a refinery.

Cracked gas cooling   The reduction of the temperature of a gas stream in a catalytic cracking process in order to allow the extraction of different products.

Cracked gas heat recovery   The recovery of waste heat energy from cracked gas, which is used to produce high-pressure steam which in turn is used in a refinery.

Cracking   A process in which a petroleum stream is broken down into separate cuts in a refinery.

Crane barge   A maritime barge which is fitted with a crane for offshore heavy lifting. Also called a **derrick barge**.

Cratering   A collapse of the surface and sub-surface area around a wellbore.

Crawler   A self-propelled machine which X-rays the inside of a pipeline to look for possible welding defects.

Creaming curve   A diagram which presents the relationship between exploration wells drilled in, and the cumulative level of petroleum produced from any discoveries made in, a defined area.

Creaming theory   A statistical technique which theorises that in any exploration province, after an initial period in which the largest fields are found, success rates and average field sizes will decline as more exploration wells are drilled (despite a maturing of the knowledge of the province).

Credit support agreement (CSA)   A contract which provides agreed collateral support for the payment obligations which are owed by a party under a contract for the sale of goods or services.

Creep   A process by which a pipeline gradually increases in length over time because of the high temperatures and/or pressures of the petroleum which it transports.

Creeping expropriation   Expropriation which is effected by a state by indirect means such as changes in prevailing tax or regulatory provisions.

Cricondenbar   The maximum point of pressure at which a gas and a liquid can co-exist as separate phases (and beyond which a gas cannot be formed, regardless of the temperature).

Cricondentherm   The maximum point of temperature at which condensation can occur, and beyond which a liquid cannot be formed, regardless of the pressure.

Crisis, emergency, incident management system (CEIMS)   A predetermined system of responses for a petroleum project incident.

Crisis management team (CMT)   A predetermined multi-disciplinary team which comes together in order to respond to a crisis affecting a petroleum project.

Critical flow   The principle that when gas flows through a choke a disturbance in the pressure downstream of the choke cannot affect the pressure of the flow upstream of the choke.

Critical path analysis   A predictive planning tool which identifies the activities, timings, and anticipated critical risk events which will be inherent in the completion of a petroleum project.

Critical point   A particular point of pressure and temperature at which natural gas and crude oil phases can co-exist.

Critical pressure   The minimum pressure which must be applied to a gas before it can be liquefied to give LNG.

Critical temperature   The point of temperature above which a gas will not liquefy to give LNG, regardless of the pressure which is applied to the gas.

Crooked hole   A wellbore which is unintentionally drilled in a direction other than true vertical.

Cross-block development (CBD)   A single development programme which applies to petroleum deposits underlying several concession areas which have been pooled.

Cross charge   A collateral support mechanism, which is typically used in Australian JOAs, whereby each party creates a registrable charge over its JOA and wider project interests in favour of the other parties as security for a potential payment default.

Cross cutting technology   A thermal energy technology which can be used to support the output of a renewable energy system (such heat recycling or thermal energy storage).

Cross sea   Irregular wave patterns which occur on the sea surface when different wave patterns cross each other at angles.

Crossflow   An alternative term for **back-flow**.

Crosshaul   The situation whereby two ships which are carrying similar cargoes are headed towards each other but in opposite directions, so suggesting the possibility of a swap. *See also* **swap**.

Crossing   The situation whereby a new pipeline or cable crosses another existing pipeline or cable (and the crossing is regulated by a crossing agreement). Also called a **crossover** and an **X-O**.

Crossing agreement   A contract which is made between the owner of an existing pipeline or cable and the owner of a new pipeline or cable and regulates rights, obligations, and liabilities between the parties in respect of the intended crossing.

Crossover
(1)   An alternative term for a **crossing**.
(2)   A development programme which links low and high productivity wellbores to give an averaged production profile from an overall formation.

Crossover valve (X-OV)   A valve arrangement which enables two components with different thread types or sizes to be connected.

Crude by rail (CBR)   A crude oil development project whereby the oil is transported by rail from the point of production.

Crude carrier   A ship which is used for the transportation of crude oil. *See also* **Aframax, Capesize, general purpose, handysize, large range 1, large range 2, medium range, Panamax, Suezmax, ultra-large crude carrier**, and **very large crude carrier**.

Crude dehydration unit (CDU)   A facility which removes entrained water from a crude oil stream.

Crude distillation unit (CDU)   A vertical steel vessel which is used for atmospheric distillation. Also called a **fractionating column**. *See* Appendix 2E.

Crude long   An integrated company whose crude oil refining requirements are less than its crude oil producing capacity. *See also* **crude self-sufficiency**. Contrast with **crude short**, whereby crude oil refining requirements exceed crude oil producing capacity.

Crude oil   Naturally occurring petroleum in a range of semi-solid to liquid phases in sub-surface and surface conditions. *See also* **asphaltic-base crude oil, naphthenic-base crude oil** and **paraffinic-base crude oil**.

Crude oil degassing   The extraction of entrained gas from a crude oil stream. *See also* **degasser**.

Crude oil equivalent   An alternative term for **barrel of oil equivalent**.

Crude oil lifting agreement (COLA)   A contract which is made between the parties to a JOA which provides for the lifting and the disposal of their respective entitlements to the produced crude oil.

Crude oil sale and purchase agreement (COSPA)   A contract for the sale and purchase of crude oil.

Crude oil transportation agreement (COTA)   A contract which is made between a transporter and a shipper for the transportation of crude oil by pipeline.

Crude oil washing (COW)   The removal of slops from a tank after a cargo of crude oil has been discharged. *See also* **loaded on top**.

Crude self-sufficiency   The situation whereby an integrated company has sufficient crude oil production to meet all the feedstock needs of its refining entity. *See also* **crude long**.

Crude short   An integrated company whose crude oil refining requirements exceed its crude oil producing capacity. Contrast with **crude long**, whereby crude oil refining requirements are less than crude oil producing capacity.

Crude slate   A record of the crude oil inputs into a refinery which references the different crude oil blends used. Also called a **product slate** and a **refining slate**.

Cryo-transfer   The offshore transfer of LNG between LNG carriers. *See also* **ship-to-ship**.

Cryogen   A gas which is kept in a liquid state at low temperature and pressure.

Crypto-regulation   The informal and unsanctioned exercise of some or all of the elements of a sector regulatory function by a sector participant rather than by an appointed regulator. Also called **pseudo-regulation**.

Cubes   A colloquial term for cubic metres, typically used to describe the size of a cargo of LNG in-ship.

Cubic foot (ft^3)   The amount of gas or liquid which at reference conditions occupies a volumetric area equivalent to one cubic foot. Also called **scf**, **scuf**, and **standard cubic foot**. *See* Appendix 1.

Cubic metre (m^3)   The amount of gas which at reference conditions occupies a volumetric area equivalent to one cubic metre. *See also* **cubes**. *See* Appendix 1.

Cullen Report   Published in November 1990, the official report into the causes of the Piper Alpha incident (HM Stationery Office ISBN 0101113102).

Cumulative production   At a point in time, the total quantity of petroleum which has been produced from a formation up to then.

Cure period   A contractual right of a person to remedy a breach of contract within a defined period of time, or to face the consequences of the breach.

Curing   The drying of cement after cementing has taken place. Also called **ageing**. *See also* **waiting on cement**.

Curing a title   The remediation of title defects which are apparent from an abstract.

Currency adjustment factor (CAF)   An increase or decrease in costs or charges which is attributable to foreign exchange loss or gain.

Curtailment   The restricted supply of petroleum to a buyer because of an event affecting the seller which is applied proportionately to a group of buyers. Also called **prorationing**.

Cushion   Any quantity of water, drilling fluid, or compressed air which is pumped into a drill string or tubing in order to resist yield pressure.

Cushion gas   Gas which is kept in a gas storage facility as a permanent inventory, for use as a buffer between the facility and the working gas. Also called **base gas** and **pad gas**. *See also* **working gas**.

Custody transfer measurement system (CTMS)   Equipment which is fitted to an LNG carrier in order to determine an LNG cargo's quantity and composition at the point of loading or unloading the cargo.

Custody transfer point   A defined point at which the custody of a quantity of petroleum passes from one person to another (at which point title and/or risk might also transfer).

Customary grounding clause   A provision in an insurance policy whereby the insurers will not cover a ship's grounding, as opposed to a ship's stranding.

Customer collect sale   An alternative term for **free in pipe**.

Cut and cap   A method of abandoning an onshore wellbore by cutting through the diameter of the wellhead below the surface level and capping and then covering the wellhead.

Cut and cover   An onshore pipeline-laying method whereby a trench is dug, the pipeline is laid in the trench and the trench is refilled.

Cut and fill   The reduction of high ground and the infill of low ground to give a uniform height terrain for the laying of a pipeline (onshore or offshore).

Cut oil   Crude oil which contains water.

Cut point   A temperature point at which a product is obtained from the process of atmospheric distillation.

Cutback   Ultra-heavy oil or bitumen to which a diluent has been added to make the ultra-heavy oil or bitumen more readily capable of being transported and used. *See also* **dilbit**.

Cuts   Different petroleum fractions which result from the process of atmospheric distillation, from which products are manufactured. Also called **distillation cuts**.

Cutterstock
(1)   Diesel or another diluting agent which is used to thin heavy oil. Also called **diluent** and **thinner**.
(2)   An intermediate substance which is used to separate different petroleum batches within a multiphase pipeline. Also called **flux stock, interface, slop**, and **transmix**.

Cutting oil   A water-soluble lubricating oil which is used to lubricate metal-cutting tools.

Cuttings   Rock fragments which are broken away from the side of a sub-surface structure during the drilling of a wellbore. Also called **cavings**.

Cuttings sample log   A record of petroleum presence in cuttings which are gathered from circulated drilling fluid on a drilling unit.

Cycle condensation   Condensation which is produced from cycle gas.

Cycle gas   Gas which is produced from a formation which is compressed and re-injected into the formation to maintain reservoir pressure.

Cycle plant   A gas processing plant in which liquids are stripped from a gas stream and the gas is then reinjected downhole for gas lift purposes.

Cycle stock   Unfinished product which is withdrawn from a refining process and is returned to an earlier stage in the refining process for further processing.

Cyclic steam stimulation (CSS)   An alternative term for **steam flood**.

Cycling
(1)   The cycle of redelivering the total quantity of working gas from a gas storage facility and the injection of a replacement quantity (and **cycle rate** is the number of times a gas storage facility's working gas can be cycled during a specific time period). Also called **turnover**.
(2)   An alternative term for **gas lift**.

Cylinder gases   Tonnage gases which are stored in pressurised cylinders.

D

D-value   The largest size of helicopter (measured by the circumference of the rotors when turning) which can be used for the assessment of a helideck's capacity.

Daily average send-out   Under a petroleum sales contract, the total quantity of petroleum which is delivered by the seller during a defined period of time, divided by the total number of days in that period, to define the daily average send-out rate.

Daily contract quantity (DCQ)   Under a petroleum sales contract, the defined quantity of petroleum which is to be delivered by the seller and bought by the buyer on each day.

Daily delivery rate   Under a petroleum sales contract, a defined maximum rate at which the seller's facilities are required to be capable of delivering petroleum to the buyer on each day.

Daily drilling costs (DDC)   The aggregate costs which are associated with hiring a drilling unit and a spread, divided by the number of days for which the drilling unit and the spread are hired.

Daily drilling report (DDR)   Under a drilling contract, a daily report of drilling progress which is submitted by the drilling contractor to the employer company.

Daily operating costs (DOC)   In respect of a petroleum facility, a measure of the daily costs which are associated with operating the facility.

Daily peak   Under a petroleum sales contract, a maximum quantity of petroleum which can be delivered by the seller or which can be required for delivery by the buyer on each day.

Daily production rate (DPR)   The average rate at which petroleum is produced from a formation on a day.

Daisy chain
(1)   A series of linked contracts for the sale and purchase of petroleum, made between a number of intermediaries, before the petroleum is delivered to an end-user.
(2)   A series of sub-sea wellheads which are linked by flowlines.

Damages   A monetary sum which is payable by a contracting party which has breached the terms of a contract, in compensation for the loss which that breach caused to another contracting party. *See also* **cover standard damages, liquidated damages**, and **unliquidated damages**.

Damages at large   An alternative term for **unliquidated damages**.

Damages for detention (DAMFORDET)   An alternative term for **demurrage** or for a **port congestion charge**.

Danforth anchor   A particular form of anchor which has two large flukes that pivot from the main shaft.

Darcy   A unit of measurement for permeability. *See also* **millidarcy**.

Dark spread   A spark spread whereby electric power is generated from the combustion of coal.

Dart   A device which manipulates hydraulically operated tools downhole.

Data acquisition system (DAS)   A system for the measurement of wellbore or formation performance.

Data sale   A sale of petroleum exploration or production data between persons. *See also* **data trade**.

Data trade   An exchange of petroleum exploration or production data between persons. *See also* **data sale**.

Dated Brent   Physical cargoes of Brent Blend crude oil which have been assigned specific delivery dates and result in the publication of the Dated Brent published price benchmark. *See also* **Brent future**.

Day ahead scheduling (DAS)   The scheduling of a petroleum delivery which is made on the day before the intended day of delivery of that petroleum.

Dayrate   Under a drilling contract, a daily monetary rate which is payable by the employer company to the drilling contractor for the daily hire of a drilling unit.

Days all purposes (DAPS)   A measure of the total amount of time (measured in days) which is allotted to the loading, carriage, and unloading of a cargo.

Days since spud (DSS)   A measure of the number of days which have passed since a wellbore was spudded.

Dead cash/dead money   Capital which is locked up in non-productive investments (such as the funding costs of the provision of a parent company guarantee), rather than which is being used actively to generate returns.

Dead oil   Crude oil which contains no or negligible associated gas. Contrast with **live oil**, which contains associated gas.

Dead well   A wellbore from which petroleum will no longer flow without manual intervention such as artificial lift.

Deadfreight   The portion of a ship's cargo-carrying capacity which is not utilised on a laden voyage.

Deadstock   A non-moving petroleum inventory in a pipeline or in a storage tank.

Deadweight tonnage (DWT)   A measure of how much weight (measured in metric tons) a ship can safely carry (including cargo, bunkers, water, and victuals). Also called **deep sea vessel capacity**.

Dealer-owned dealer-operated (DODO)   A downstream fuels retailing model which utilises the dealer-owned service station model.

Dealer-owned service station (DOSS)   A downstream fuels retailing model whereby an independent dealer owns and operates a service station, often subject to a solus tie. Also called **branded reseller**. *See also* **chain marketer** and **dealer-owned dealer-operated**.

Dealer tank wagon (DTW)   A measure of the wholesale price for gasoline which is delivered to a retail outlet which is paid by the retailer to the wholesaler.

Deasphalted oil (DAO)   Crude oil which results from the process of solvent deasphalting.

Death or glory   A colloquial term for an obligation to drill a wellbore to a defined target depth whereby the obligation is regarded as performed if, at an earlier depth, the drilling encounters impenetrable substances (death) or a commercial discovery (glory).

Debottlenecking   An increase of the processing capacity of a petroleum facility through modifications in order to remove inherent capacity restrictions within the facility. *See also* **bottleneck**.

Debrining   The use of water injection for the removal of salt in the creation of an underground salt cavern, as part of the wider process of leaching.

Debt service coverage ratio (DSCR)   A ratio test in project financing and reserves-based lending which assesses the ability of the borrower to repay the debt at a point in time. *See also* **loan life coverage ratio**.

Debt service reserve account (DSRA) In project financing, a bank account that the borrower maintains and which holds a defined number of months of projected project debt service obligations, which can be drawn down by the lender in the event of a borrower default or other interruption in the flow of project debt repayments.

Debt service undertaking (DSU) A commitment from a borrower in the project financing of a petroleum project to make good certain of the lender's entitlements if the revenues from the petroleum project are insufficient to do so. *See also* **completion guarantee**.

Debt to equity (D:E) An alternative term for **leverage**.

Debutaniser A device which separates out butane from a gas stream.

Debute A gas stream which has been stripped of butane.

Decant oil A heavy oil which results from catalytic cracking, which is often used as a feedstock for coke production.

Decentralised energy Energy which is generated and used in dispersed locations, eneralise the need for energy transportation infrastructure, costs, and losses.

Decision gate (DG) A defined point in a petroleum company's internal petroleum project approvals process at which a project direction decision is made. Also called a **define gate/definition gate**.

Declaration of commerciality An alternative term for a **commercial well declaration**.

Decline curve analysis The forecasting of a formation's future petroleum production potential which is achieved by the extrapolation of the formation's past performance profile.

Decline pattern Under a petroleum sales contract, a profile for the delivery of petroleum from the seller to the buyer which is set during a decline period.

Decline period A profile for petroleum production over a given time period which reduces towards the end of a formation's life, which is found typically in a depletion contract.

Decommissioning A process of removing petroleum production and/or transportation facilities following cessation of use, including the plugging of wellbores, the disposal, demolition, removal, and/or clean-up of facilities and any necessary site remediation and restoration. Also called **abandonment, dismantling, removal and restoration**, and **reclamation**. *See also* **abandonment and reclamation obligation**.

Decommissioning security agreement (DSA) An agreement which is made between the parties to a JOA as to how they will fund and provide security for their respective costs of a decommissioning programme.

Decommissioning trigger point Under a concession, a defined point in time in the history of the related petroleum project at which the concession-holders have to start making payments into a decommissioning fund in order to meet the anticipated future costs of decommissioning the project.

Decommvelopment   The simultaneous activities of decommissioning an existing petroleum production field or petroleum infrastructure item and reworking it into a new development proposition.

Deconsolidation point   A defined point at which a single cargo of petroleum is broken down into a number of smaller quantities.

Dedicated design day capacity (DDDC)   The maximum quantity of gas which is dedicated by a supplier to a customer's use, based on the maximum quantity of gas which is used on the most demanding day. Also called the **premise demand factor**.

Deductible   An agreed part of an insured person's loss which will not be covered by an insurance payment which is made by an insurer. Also called an **excess**.

Deed of priority   An arrangement which is made between a number of lenders to determine the order of priority between them as to how they will enforce their respective security interests against a common borrower. *See also* **subordination**.

Deemed contract   A contractual arrangement between persons for the sale and purchase of petroleum where the terms of that contract have not been reduced to writing.

Deep catalytic cracking (DCC)   A form of catalytic cracking.

Deep cut plant   A gas processing facility which is located close to a gas transportation pipeline, removes liquids from a raw gas stream, and has increased processing capacity compared to a shallow cut plant. Also called an **extraction plant**.

Deep draught   The distance from the sea surface to the lowest point on a ship below the sea surface, expressed as a depth.

Deep marine deposit   A sedimentary environment which is found below water depths greater than 200 metres.

Deep pool well (DPW)   A wellbore which is drilled into an already-producing formation but down to a deeper payzone. Contrast with a **shallow pool well**, which is drilled into an already-producing formation but down to a shallower payzone.

Deep sea vessel capacity (DSVC)   An alternative term for **deadweight tonnage**.

Deep water (DW)   Water depths in the range of 400 to 1,500 metres. Contrast with **shallow water**, as water depths up to 400 metres, and **ultra-deep water**, as water depths of greater than 1,500 metres.

Deepening   The further drilling of a wellbore to below its actual depth or its target depth.

De-escalation factor   A methodology for a commodity price decrease which applies when underlying costs have decreased. Contrast with **escalation factor**, where prices increase when costs increase.

Deethan   A gas stream which has been stripped of ethane.

Deethaniser   A device which separates out ethane from a gas stream.

Deferred production agreement   An agreement between concession-holders whereby one or more of the concession-holders' entitlements are deemed to remain in-ground, despite the production of petroleum at full rates from the concession area.

Deferred working interest farm-out   A form of farm-out agreement whereby the farmee commits to obligations which are disproportionately greater than the interests which the farmee would be entitled to, subject to later recovery of the excess element of the farm-out obligation.

Deficiency gas   Under a gas sales contract in which a take and pay provision or a take or pay provision applies, the difference between the quantity of gas which the buyer is obliged to take delivery of and the quantity of gas which the buyer actually takes delivery of.

Define gate/definition gate   An alternative term for a **decision gate**.

Deflection tool   A tool which is used to alter the trajectory of a wellbore as part of a directional drilling exercise. *See also* **stabiliser**.

Deflocculant   A chemical agent which is used in deflocculation.

Deflocculation   The reduction of the viscosity of a petroleum stream though the addition of a diluent. Contrast with **flocculation**, which encourages the coagulation of smaller particles in the stream.

Deg   A shorthand term for **degradation**.

Degasser   A facility which extracts entrained gas from crude oil or drilling fluid.

Degradation   The reduction over time of the efficiency of an item of plant because of wear and tear. Also called **deg**. *See also* **negative degradation**.

Degradation cycle   A forecast of the likely rate of degradation of a petroleum infrastructure item, thereby indicating the need for maintenance.

Dehydration   The removal of water components from a petroleum stream.

Dekatherm (Dth)   One million Btus.

Del credere agent   An agent for a principal which acts as the seller of petroleum which also acts as a guarantor of the buyer's petroleum purchase and payment obligations under the contract between the seller and the buyer.

Delayed coking   A process by which heavier crude oil fractions can be thermally decomposed under high temperatures and pressures to produce a mixture of lighter oils and petroleum coke.

Delineation well   An alternative term for a **step-out well**.

Delinkage   A change in the method of the determination of gas prices from indexation according to a contractually specified indexation package to benchmarking against published crude oil prices.

Deliver or pay (DOP)   Under a petroleum sales contract, a mechanism by which the seller has the option not to deliver petroleum but to make a payment to the buyer but without constituting a shortfall, effectively making it an interruptible contract for the seller.

Deliverability   In a gas storage facility, the total quantity of working gas which can be withdrawn from the storage facility at a given withdrawal rate and in respect of a defined period of time (measured in volume or heat rate) for a facility user. *See also* **injectability**.

Deliverability test   A test which is used to determine a flow rate through a wellbore.

Deliverables master list (DML)   A project management tool which records the progress towards completion of the deliverables which are required for the completion of a petroleum project.

Delivered at place (DAP)   An Incoterms 2010 formulation which is used in ship deliveries whereby the delivery of petroleum from the seller to the buyer takes place at a named destination and the seller bears the transportation obligation (and associated costs) from the loading port.

Delivered at terminal (DAT)   An Incoterms 2010 formulation which is used in ship deliveries whereby the delivery of petroleum from the seller to the buyer takes place at a named terminal and the seller bears the transportation obligation (and associated costs) from the loading port.

Delivered energy   The actual amount of energy which is available for use or consumption at a particular point. This will be less than the amount of generated energy, to account for production, processing, and transportation losses.

Delivered ex-ship (DES)   An Incoterms 2000 formulation which is used in ship deliveries whereby the delivery of petroleum from the seller to the buyer takes place at the unloading port and the seller bears the transportation obligation (and associated costs) from the loading port. Also called **arrival contract, destination contract**, and **ex-ship**. *See also* **free on board**.

Delivered sale   A sale of petroleum whereby the delivery point is the point of connection between the buyer's receiving facilities and a delivery pipeline which the seller uses to transport the petroleum to those facilities. Contrast with **free in pipe**, whereby the delivery point is the point of connection between the seller's production facilities and a pipeline in which the buyer has secured transportation capacity.

Delivery   The delivery of a ship by the owner to the charterer at the start of the charter period. *See also* **redelivery**.

Delivery capacity   Under a petroleum sales contract, the obligation of the seller to maintain petroleum production and transportation infrastructure capacity which is sufficient to meet its petroleum delivery obligations.

Delivery point
(1) Under a petroleum sales contract, a point at which title to the petroleum to be sold (and possibly also custody and risk) transfers from the seller to the buyer and the sales transaction is consummated between the parties. Also called a **sales point**.
(2) Under a pipeline transportation contract, a point at which the shipper delivers petroleum into the pipeline for transportation by the transporter. *See also* **redelivery point**.

Deltaic source   The creation of a formation at the head of a river delta.

Demand guarantee   A bank guarantee which serves as a secondary payment mechanism for a transaction whereby an issuing bank promises payment to an identified beneficiary if a principal payer defaults in payment when due under the transaction. *See also* **performance bond** and **URDG 758**.

Demand-led local content   Local content which is developed in response to the preferences of a petroleum project participant. *See also* **supply led local content**.

De-manning   A reduction in the number of personnel who are engaged in operating a petroleum infrastructure item. *See also* **minimised manning**.

Demise charter   An alternative term for a **bareboat charter**.

Demister   A filter which removes liquid particles from a gas stream.

Demob/demobilisation   Under a drilling contract, the movement of a drilling unit to its original location (or to another agreed location) after the drilling of a wellbore. Contrast with **mobilisation**, as the movement of the drilling unit to a drill site prior to drilling a wellbore.

Demulsibility   The ability of oil to separate from water with which it has been mixed.

Demurrage   In petroleum shipping, a measure of the compensation which is due for payment from the non-shipping party to the shipping party if used laytime exceeds allowed laytime. Contrast with a **port congestion charge**, as payment which is due from the shipping party to the non-shipping party. Also called **damages for detention**.

Dense phase   Gas which is compressed in pipeline transportation to a state where it begins to assume the characteristics of a liquid.

Densitometer/densimeter   An instrument which is used to measure specific gravity.

Density   An alternative term for **API gravity**.

Dépeçage   A jurisprudential theory by which different applicable laws could be applied to different elements of a case or to determine the governing law of a contract.

Depleted field storage (DFS)   A depleted gas field which is used as a gas storage facility.

Depletion contract   A contract for the sale of petroleum whereby the petroleum is sold from a nominated source of supply which is dedicated to the buyer and production from the source by the seller continues through to economic exhaustion of the source. Also called a **depletion-based contract**. Contrast with a **supply contract**, whereby petroleum is sold as a fungible commodity and without an identified source of supply. *See also* **partial purchase contract**.

Depletion drive
(1)   A reservoir in which capillary pressure drops over time as a consequence of the production of petroleum.
(2)   A reservoir from which crude oil is produced with the assistance of associated gas (such that continued crude oil production leads to a drop in reservoir pressure). *See also* **solution gas drive**.
(3)   An alternative term for **gas drive** and **gas-expansion drive**.

Depletion policy   The selected strategic focus of a petroleum-producing state, relating to petroleum production rates and revenue returns.

Depletion programme   A long-term plan for the production of petroleum from a formation to the point of depletion and the COP.

Depletion-based contract (DBC)   An alternative term for a **depletion contract**.

Depocentre   The thickest cross section of a payzone.

Depositional environment   The combined chemical, geological, and physical conditions under which a formation was created.

Depreciation, depletion, and amortisation (DDA)   An accounting principle which recognises the costs that are associated with the development of a petroleum project and allocates them in an agreed order for depreciation over their useful economic lives.

Deprop   A gas stream which has been stripped of propane.

Depropaniser   A device which separates out propane from a gas stream.

Depth map   A contour map which indicates the dimensions of a sub-surface geological feature by depths from a defined surface datum point.

Depth severance   A concessionary interest which is defined by specified stratigraphic depths.

Derating   A reduction of a power-generating unit's dependable capacity, which could be to a point below the unit's original nameplate capacity.

Derecognition   An accounting standard which disapplies disaggregation in respect of an asset when commercial circumstances indicate that a company which previously owned the asset has not effectively disposed of the asset. *See also* **true sale**.

De-risking   A process by which a contracting party's risk profile under a contract is decreased in exchange for that party giving an improved contractual or financial incentive to its contracting counterparty.

Derived data   Petroleum price forecasts and assumptions which are calculated from trade data.

Derived fuel   An alternative term for a **secondary fuel**.

Derrick
(1)   A crane which is used for lifting and lowering goods on a ship.
(2)   A vertical structure which supports the drill string on a drilling unit.

Derrick barge   An alternative term for a **crane barge**.

Desalting   A process for the removal of calcium, magnesium, and sodium chlorides from crude oil.

Desander   An item of equipment which is used to remove solid particles from drilling fluid. Also called a **desilter** and a **hydrocyclone**.

Desiccant   A liquid or solid substance which removes water or water vapour from air.

Design day   A twenty-four hour period of demand which is used as a basis for planning petroleum delivery and/or offtake requirements.

Design still water level (DSWL)   The design of an item of offshore petroleum infrastructure item to match the relevant still water level.

Design water depth   A measure of the deepest point of the sea in which an offshore drilling unit is intended to be capable of operating.

Design wave   The maximum size and frequency of a wave which an offshore petroleum infrastructure item must be able to withstand.

Desilter   An alternative term for a **desander**.

Destination clause   Under a petroleum sales contract, a restrictive provision whereby the buyer cannot require or effect the delivery of the petroleum to a destination other than the originally intended place of delivery without the seller's consent. *See also* **diversion**.

Destination contract   An alternative term for **delivered ex-ship**.

Detailed content   A model for the design of a concession whereby the key provisions are recited in detailed legislation with little room for individual negotiation. Also called **fixed content**. *See also* **hybrid provision** and **individual legislation**.

Deterministic modelling   A method of petroleum project modelling which, from all of the variable inputs, yields a single solution. Contrast with **probabilistic modelling**, which yields a range of possible solutions with weighted likelihoods of outcomes.

Develocat well   The drilling of a wellbore for the exploration and evaluation of a deeper untested formation while also developing a shallower known formation.

Developed/development   The phase of a petroleum project which takes place after a discovery has been appraised, involving the design, construction, installation, and commissioning of petroleum production, processing, transportation, and storage facilities.

Developed in production (DIP)   Developed petroleum reserves which are presently in production.

Developed not producing (DNP)   Developed petroleum reserves which are presently not in production.

Developed reserves (DR)   Petroleum reserves which are expected to be recovered from existing wellbores, which could be developed in production or developed not producing.

Development and production sharing agreement (DPSA)   A PSC which includes the development component but which does not govern exploration. Contrast with an **exploration and production sharing agreement**, which also includes exploration.

Development finance institution (DFI)   A state-backed entity which provides finance to private sector persons for the development of a petroleum project.

Development not viable (DNV)   A discovery which is planned for development but which is presently not viable for development because of external constraints.

Development on hold (DOH)   A discovery in respect of which there is no present plan for development.

Development pending (DP)   A discovery which is planned for development but in respect of which the implementation of the plan of development has yet to be commenced.

Development well   A wellbore which is drilled into a formation to a known productive depth in order to facilitate petroleum production. Also called an **exploitation well**.

Deviated drilling   An alternative term for **directional drilling**.

Deviation   Under a charterparty, a provision which allows a shipowner to deviate from the agreed route (including calling at unscheduled ports) for any reasonable purpose, including to save life or property.

Devil's excrement   A colloquialism which was coined by Juan Alfonzo in the 1970s as a warning about the undesirable consequences of petroleum wealth. *See also* **Dutch disease**.

Dew curve   In a phase envelope, the plotted pressure and temperature curve which leads up to the dew point.

Dew point   The temperature at which a gas will cool sufficiently to condense to a liquid.

Dew point spread   The difference between the dew point and actual air temperature.

Dewatering   The injection of compressed air into a pipeline in order to remove any water which is left over after hydrotesting.

Dewpointing   The removal of heavier petroleum fractions from a gas stream.

Diagenesis   The first stage of the sub-surface metamorphosis of organic matter to form petroleum, representing the chemical, physical, or biological changes which have been undergone by sediments at relatively low temperatures and pressures. *See also* **katagenesis** and **metagenesis**. *See* Appendix 2A.

Diagnostic fracture injection test (DFIT)   A technical analysis of the results of a fracking operation.

Diameters   A measure of a pipeline's length which is expressed as a multiple of the pipeline's diameter.

Diapir   A geological formation in which a core of rock has moved upwards to penetrate an overlying strata.

Diatomite   A sedimentary rock structure which contains petroleum.

Diesel   A lower-grade distillate which is used as a motor vehicle fuel. *See also* diesel-engined road vehicle and **red diesel**.

Diesel-engined road vehicle (DERV)   Diesel which is used as a motor vehicle fuel in the United Kingdom.

Differential   A premium which is added to the price of a grade of petroleum because of the presence of certain compositional advantages in that petroleum (for example, sweet oil or wet gas).

Differential drop   A reduction in gas pressure which occurs when a gas stream flows through a choke. *See also* **orifice meter.**

Differential lifting   The lifting of a quantity of petroleum by a person entitled to do so whereby the quantity so lifted is greater (an overlift) or lesser (an underlift) than the person's lifting entitlement. *See also* **attribution.**

Differential pressure   A measure of the difference between the pressure in a wellbore (due to the drilling fluid column) and the pressure in the surrounding rock.

Differential pressure gauge   A pressure measurement device which calibrates the difference between two or more separate pressure regimes.

Differential range   The extent of the variation between the lowest and the highest pressure readings which a meter can accurately measure.

Differential sticking   An alternative term for **wall sticking.**

Digital oilfield (DOF)   A generic term for a petroleum project process which relies on the use of information technology for improved performance.

Dilbit   Diluted bitumen: the product of bitumen after the application of a diluting agent, to make the bitumen easier to flow and to store. *See also* **cutback.**

Diluent   An alternative term for **cutterstock.**

Dilution provision   An alternative term for a **withering interest provision.**

Dimethyl ether (DME)   A gaseous biofuel which can be used as a vehicle fuel.

Dip   The steepest angle of a tilted bedding plane in relation to a horizontal plane.

Dipping   A manual determination of the depth of oil in a storage tank by the use of a calibrated measuring rod. Also called **tank dipping.**

Direct agreement   In relation to the third-party debt financing of petroleum project costs, an agreement between the lenders and a project-related party relating to the protection of the petroleum project's cashflows.

Direct charges   Under an accounting procedure, reimbursable costs which are directly incurred by the operator in the performance of the joint operations which are charged to the parties. Contrast with **indirect charges**, as reimbursable costs which relate to general intangible support services which are provided by the operator.

Direct distributions   Payments which are made by a petroleum project developer directly to a state or to a group of persons as a condition of being permitted to undertake a petroleum project. *See also* **community development agreement.**

Direct drive   An industrial process which is powered by in-situ fossil-fuelled power-generation facilities. Also called **gas drive.** *See also* **e-drive.**

Direct energy   A measure of the fuel which is used in a petroleum-processing activity. Contrast with **indirect energy**, which is used to prepare intermediate inputs to the activity. *See also* **gross energy requirement.**

Direct hydrocarbon indicator (DHI)   A feature in seismic data which indicates the presence of petroleum in a formation.

Direct loss   A measure of the loss suffered by a person which results from a breach of contract that represents present, actual losses. *See also* **consequential loss**.

Direct negotiation   An alternative term for an **open door award**.

Directional drilling   The drilling of a non-vertical wellbore. Also called **deviated drilling** and **slant drilling**.

Dirty cargo   A cargo of crude oil which is made up of black oil. *See also* **clean cargo**.

Dirty petroleum products (DPP)   A collective term for crude oil, fuel oil, and bitumen, which is carried on dirty product tankers. *See also* **clean petroleum products**.

Dirty product tanker   A ship which is used for the carriage of dirty petroleum products. *See also* **clean product tanker**.

Dirty rock   A formation containing clay minerals which reduce porosity and permeability. Contrast with **clean rock**, which contains no or negligible clay minerals.

Disaggregation   The removal of an asset from a company's balance sheet (including as a consequence of a true sale). *See also* **derecognition**.

Disbonding   In a pipeline, a separation of the internal coating of the pipeline from the pipeline.

Disclosure letter   Under an asset or share sale and purchase agreement, a collateral agreement which qualifies any of the representations and warranties which are set out in the agreement.

Discovered resources opportunities (DRO)   Discovered petroleum resources which presently lack a firm development plan.

Discovery   Potentially economically recoverable quantities of petroleum which have been found in a formation. Also called a **known accumulation**.

Discovery well   The first exploration wellbore which is drilled into a new discovery.

Dismantling, removal, and restoration (DRR)   An alternative term for **decommissioning**.

Dispersant   A chemical which is used to break down a marine crude oil spill into small droplets, prior to dispersal in a sea and the biodegradation of the crude oil.

Displacement gas   In gas pipeline transportation, gas from one input source which is substituted by gas from another input source.

Disponent owner   A person having operational control of a ship but without having legal title to the ship (such as a charterer under a bareboat charter).

Disposal well   A wellbore into which dirty water and waste fluids are injected before being plugged and abandoned. *See also* **service well**.

Dissolved gas   An alternative term for **associated gas**.

Dissolved gas drive   An alternative term for **solution gas drive**.

Distal end   The far end of a pipeline, relative to the input point or a central hub.

Distance tariff Under a pipeline transportation agreement, a tariff which is payable by the shipper to the transporter which reflects the actual distance over which the petroleum is transported. *See also* **postage stamp tariff** and **zonal tariff**.

Distillate A grade of petroleum which results from atmospheric distillation.

Distillation cuts An alternative term for **cuts**.

Distillation margin An alternative term for a **crack spread**.

Distressed cargo A cargo of petroleum which is sold as a distressed sale.

Distressed sale A sale of petroleum which is made by a seller in circumstances of urgency, often with indifference to the quantum of the sale price which is realised by the sale. Also called a **fire sale**.

Distributed generation Electric power which is generated at the point of consumption, avoiding the need for further transmission and distribution. Contrast with **central generation**, whereby electric power is generated in a central facility and is then sent to the point of consumption by transmission and distribution.

Distribution The localised transportation of petroleum through small-diameter pipelines before final delivery to end-users.

Distribution loss Gas which is lost through natural leakage in pipeline deliveries.

District cooling Centrally produced chilled water which is piped to a group of buildings for cooling purposes. *See also* **district heating**.

District energy A composite term for district cooling and district heating.

District heating Centrally produced steam or hot water which is piped to a group of buildings for heating purposes. *See also* **district cooling**.

Diurnal storage A gas storage facility whereby working gas is injected and redelivered on a daily basis. Contrast with **seasonal storage**, whereby injection and redelivery takes place over summer/winter spreads.

Diversified investment fund (DIF) A state-owned fund which promotes national growth in sectors other than the petroleum sector, into which a petroleum company makes payments under the terms of a concession.

Diversion The redirection of an LNG carrier and its cargo from the unloading port to which it was originally destined to an alternative destination, possibly as a modification to the terms of a destination clause.

Diversity factor The ratio of the aggregated demand of a group of petroleum buyers to their individual demands at any point in time.

Diverter/diverter line A device which directs fluids flowing from a wellbore away from a drilling unit (often when a kick is encountered in a shallow wellbore which cannot safely be contained). Also called a **spillway**.

Division order In the United States, an instrument which recites the ownership interests of a party in the produced petroleum from a well or interest which has multiple owners.

Doctor test A test which is used to detect sour gas or sour oil compounds in a petroleum stream.

Documentary letter of credit (DLC)   A letter of credit which serves as a primary payment mechanism for a transaction whereby an issuing bank promises payment to an identified beneficiary, typically in support of the supply of goods or equipment and in import/export situations. *See also* **Uniform Customs and Practices for Documentary Credits 600.**

Dog-leg   A directionally drilled wellbore which reverts to the vertical.

Dog-leg severity (DLS)   An alternative term for **drift.**

Doghouse   On a drilling unit, a small shack which is used for monitoring operations or for storing materials.

Dollar-denominated petroleum production interest (DDPPI)   A royalty whereby the payee (the royalty-holder) is entitled to a defined share of the petroleum which is produced under the concession and is payable in cash. Contrast with a **volumetric production payment interest**, whereby the royalty is payable as petroleum.

Dollar inch kilometre (DIK)   A methodology which is used for calculating pipeline construction costs, reflective of dollars per pipeline diameter (in inches) per pipeline length (in kilometres).

Dolphin   An isolated berthing and mooring point for a ship. *See also* **breasting dolphin** and **mooring dolphin.**

Dome roof   In a petroleum storage tank, a dome-shaped roof which is fixed to the tank. Contrast with a **floating roof**, which floats on the surface of the petroleum in store.

Domestic heating oil (DHO)   Fuel oil which is used for end-user heating purposes.

Domestic market obligation (DMO)   A defined quantity of produced petroleum which is required to be supplied into the producing state's domestic market under the terms of a concession.

Domgas   Gas which is supplied into a domestic market in satisfaction of a domestic market obligation.

Domino decommissioning   The situation whereby the decommissioning of a petroleum facility results in the redundancy and the possible decommissioning of other petroleum facilities which are connected to or reliant upon that facility.

Domino effect   The point at which it becomes uneconomic to produce petroleum from a field where the proportionate share of the operating costs of a processing hub which are shared between a number of user fields are increased for that field because of the decommissioning of other user fields.

Dope   A lubricant which is used to seal and protect the joint threads of a tubular.

Double   Two lengths of drill pipe which have been joined together. *See also* **fourble, single**, and **thribble.**

Double banking operation (DBO)   The location of two ships in parallel alongside a single berth, with the simultaneous loading or unloading of cargoes from both ships.

Double block and bleed valve (DBBV) A manifold combining two block (isolation) valves and a bleed (vent) valve between them, allowing the safe isolation and evacuation of the flow of petroleum from a pipeline.

Double bottom tank (DBT) An alternative term for a **double hull**.

Double containment tank (DCT) An LNG containment tank which is surrounded by an open-top concrete safety container. *See also* **full containment tank** and **single containment tank**.

Double hull A crude carrier which has a double-skinned hull for greater safety. *See also* double bottom tank and **single hull**.

Downdip The relative positioning of different phases within a formation along a dip. For example, a trap could contain (in descending order), gas then oil then water—the water is downdip of the oil and the oil is downdip of the gas. *See also* **updip**.

Downhole Anything which is inserted into a wellbore.

Downhole pressure and temperature (DPT) A measure of the pressure and temperature at a point within a wellbore.

Downhole safety valve (DSV) A device which is inserted in a wellbore in order to isolate petroleum flows. Also called a **sub-surface safety valve**.

Downstream The functions of petroleum processing, refining, and retail marketing. *See also* **refining and marketing**. *See also* the relative functions of **midstream** and **upstream**.

Downtime The proportion of time during which a petroleum facility is, or can be expected to be, not operational. Contrast with **uptime**, as the proportion of time during which a petroleum facility is, or can be expected to be, fully operational.

Downward flexibility quantity (DFQ) Under a petroleum sales contract, the unilateral ability of the seller or the buyer to decrease the quantity of petroleum which is to be supplied during the term of the contract. Contrast with **upward flexibility quantity**, whereby the seller or the buyer can increase the quantity of petroleum.

Drag-along In a joint venture, the right of a selling co-venturer to drag along its fellow co-venturers in a proposed sale. Contrast with a **tag-along**, as the right of a non-selling co-venturer to tag along with a selling co-venturer.

Drag embedment anchor (DEA) A form of mooring a ship whereby an anchor is dragged along the sea-bed until sea-bed penetration uses soil resistance to hold the anchor in place.

Drag reducer An alternative term for a **drag-reducing agent**.

Drag-reducing agent (DRA) A chemical agent which is injected into a petroleum pipeline in order to reduce friction levels between the transported petroleum and the inner wall of the pipeline and so to increase the erosional velocity limit.

Drainage clause Under a gas sales contract, a provision whereby the buyer contracts to take sufficient gas production such that the seller's gas reserves are not kept in-ground and are thereby lost to neighbouring producers through the law of capture.

Drainage time

(1) A measure of the time which is taken for the liquids element of a compound to drain away under standard temperature and pressure.

(2) A measure of the time which is taken for a liquid to drain from a tank under standard temperature and pressure.

Drawdown

(1) The difference in bottomhole pressures which exists between when a wellbore is shut in (with average reservoir pressure) and when the wellbore is flowing (with flowing bottomhole pressure).

(2) A borrower's accessing of the funds which have been made available to it by a lender under a loan agreement.

(3) A negative stock change. Contrast with **build-up**, as a positive stock change.

Drawworks   Hoisting gear on a drilling unit which is used to raise or to lower the drill string from or into a wellbore.

Drift   The degree to which a wellbore is anything other than vertical. Also called **dog-leg severity** and **inclination**. *See also* **build rate**.

Drifting   The deviation of a fiscal meter or a check meter from its intended measurement ranges, thereby requiring calibration.

Drill bit   A tool which is located at the base of a drill string and excavates earth in the drilling of a wellbore.

Drill collar   A thick-walled steel tubular which provides weight to a drill bit during the drilling of a wellbore. *See also* **weight on bit**. *See* Appendix 2C.

Drill-on date (DOD)   The date on which a drill bit begins, or resumes, the drilling of a wellbore. *See also* **spud**.

Drill or drop   Under a concession, a provision whereby (1) the concession-holder is obliged to drill a commitment well within a defined period of time or the concession is subject to relinquishment by the grantor; or (2) the concession-holder has the option to drill a well or to relinquish the concession.

Drill or pay   Under a concession, a provision whereby the concession-holder is obliged to drill a commitment well or alternatively can elect to pay a defined monetary amount to the grantor. *See also* **work obligation payout**.

Drill pipe   Lengths (typically 30 feet) of hollow steel pipe, of various diameters, which connect together to form a drill string. *See also* **double, fourble, single**, and **thribble**.

Drill ready prospect (DRP)   A petroleum discovery which has been indicated by a seismic survey and is ready to be drilled.

Drill ship   A ship which is used to drill a wellbore offshore, capable of drilling in water depths typically up to 12,000 feet. *See* Appendix 2B.

Drill site   A location (whether onshore or offshore) at which a wellbore is drilled.

Drill stem   An alternative term for a **drill string**.

Drill stem safety valve   A shut-down valve which is inserted in a drill string to allow drilling fluid circulation through the wellbore but which can be closed to prevent a kick.

Drill stem test gas   Natural gas which is produced from a formation during drill stem testing.

Drill stem testing (DST)   The testing of the productive capacity of a wellbore with the drill string still in place. Contrast with **well logging**, whereby productive capacity is tested with the drill string removed. Also called **formation evaluation while drilling** and **logging while drilling**.

Drill string   The combination of assembled lengths of drill pipe, drill collars, and a drill bit, which together are used to drill a wellbore. *See* Appendix 2C. Also called a **drill stem**.

Drill to earn   A form of farm-out agreement whereby the buyer performs works directly (or pays the costs of doing the works) and does not earn an interest if it fails to perform its obligation.

Drill to granite   The drilling of a wellbore which is not to a target depth but until a basement is encountered.

Drilled show   An alternative term for a **show**.

Drilling and completion (D&C)   The single act of the drilling and the completion of a wellbore.

Drilling barge   A barge-mounted drilling unit which is used for drilling wellbores in shallow waters.

Drilling break   A sudden increase in the rate of penetration. Contrast with a **reverse break**, which indicates a decrease in the rate of penetration.

Drilling contract   A contract for the drilling of a wellbore, which is entered into between a drilling contractor and an employer company.

Drilling dollar test   A farm-in structure whereby the obligation of the farmee is set as the requirement to incur a defined amount of expenditure rather than to complete a defined schedule of works. *See also* **net well test**.

Drilling draught   The depth to which the lowermost part of an offshore drilling unit rests below the sea surface.

Drilling fluid   A water, oil, or synthetic-based fluid which is used to lubricate a drill bit during the drilling of a wellbore. Also called **mud**. *See also* **normal circulation** and **reverse circulation**.

Drilling fluid cycle time   The measured time which is taken for a given unit of drilling fluid to be pumped into and to be recirculated through a wellbore.

Drilling history (DH)   In respect of a formation, the recorded history of exploration, appraisal, and development drilling which has taken place.

Drilling unit   A portable facility which is used to drill a wellbore onshore or offshore. *See also* **mobile offshore drilling unit**.

Drip gasoline   A naturally occurring form of condensate which is sometimes produced from a gas-producing wellbore and can be used as a vehicle fuel.

Drogue   An instrument which is lowered into the sea in order to indicate deep sea currents through the movement of a connected buoy on the sea surface.

Drop coring   A method of coring in which a core sampler is dropped into a sample, relying on the force of gravity to achieve the required depth of penetration. Also called **gravity coring**.

Drop in fuel   An alternative form of fuel which can be used as is, without modification of the equipment in which it is to be used.

Drop point   An alternative term for a **shot point**.

Dropout   The separation of entrained liquids from natural gas while in a formation, by which the liquids become less easily recoverable. Also called **liquids dropout**.

Dry and abandoned (D&A)   A dry hole which has been plugged and abandoned.

Dry berth   A berth whereby the berthed ship is kept out of the water. *See also* **wet berth**.

Dry but decent   A colloquial term for a dry hole which generates useful data for the assessment of a potentially viable petroleum system.

Dry gas   Natural gas which consists principally of methane. Contrast with **wet gas**, as gas which contains petroleum fractions heavier than methane.

Dry hole   A wellbore which fails to encounter or to indicate economically recoverable quantities of petroleum. Also called a **duster**. *See also* **dry but decent**.

Dry hole agreement   A commitment of a person to make a cash contribution to another person who is responsible for drilling a wellbore in exchange for the provision of G&G data if the wellbore is a dry hole. *See also* **bottomhole agreement** and **bottomhole money**.

Dry hole costs (DHC)   The total costs which are associated with drilling a dry hole.

Dry oil   Crude oil which has been treated to remove basic sediment and water. *See also* **wet oil**. Also called **clean oil**.

Dry pre-commissioning   Pipeline pre-commissioning without the use of hydrotesting. *See also* **wet pre-commissioning**.

Dry shipping   Shipping issues which are associated primarily with contract drafting and negotiation. Contrast with **wet shipping**, which is associated principally with addressing maritime misadventures and incidents.

Dry tow   The transportation of an offshore petroleum infrastructure item to site on a self-propelled barge. *See also* **wet tow**.

Dry tree
(1) In offshore well development, a Christmas tree which is installed above-water. Contrast with a **wet tree**, as a sub-sea Christmas tree.
(2) A sub-sea Christmas tree which is housed in a waterproof cell.

Dual activity   A drilling unit which is capable of drilling two wellbores simultaneously.

Dual completion   The completion of a single wellbore in two separate formations (whereby petroleum from one of the formations is produced within a liner).

Dual-fuel   An electric power-generating facility which is capable of generating electric power from the combustion of more than one feedstock (such as gas and fuel oil) through switching.

Dual-fuel diesel electric (DFDE)   A propulsion system which is used in an LNG carrier. *See also* **tri-fuel diesel electric**.

Dual mixed refrigerant (DMR)   A process for the liquefaction of gas to give LNG which uses a dual phase of mixed refrigerant agents. *See also* **single mixed refrigerant**.

Ductility level blast (DLB)   A low-probability but high-consequence explosion event which by design would not impair the integrity of escape and refuge areas in a petroleum infrastructure item. *See also* **strength level blast**.

Due date   A date upon which an invoice becomes due for payment.

Dull bit   A drill bit which has become blunted or otherwise ineffective.

Dump   A metered volume of oil which is delivered into a pipeline from a dump tank.

Dump flood   An enhanced oil recovery technique whereby water is flooded into a wellbore from an adjacent active aquifer in order to enhance oil production.

Dump tank   A metered oil storage tank.

Dumping   The dumping at sea of ballast water and other waste from a ship.

Duopoly   A commercial situation in which there are only two sellers of a commodity in a given market. *See also* **monopoly** and **oligopoly**.

Duopsony   A commercial situation in which there are only two buyers of a commodity in a given market. *See also* **monopsony** and **oligopsony**.

Duster   An alternative term for a **dry hole**.

Dutch auction   A colloquial term for a joint venture deadlock resolution provision whereby each party submits a sealed bid to indicate the price for which it would be willing to sell its interest in the joint venture, and the party which submitted the highest price must then buy the other party's interest at the lowest submitted price. *See also* **Russian roulette** and **Texas shoot-out**.

Dutch cone test   An alternative term for the **cone penetration test**.

Dutch disease   A colloquial term for the negative impact on a state's economy of a sudden inflow of revenues (such as would result from the discovery of petroleum within the state's territory) which could arise because of the resultant lack of focus on the state's other economic sectors. *See also* **Devil's excrement**, **paradox of plenty**, and **resource curse**.

Duty-holder   On the UKCS, a person who is nominated to prepare, maintain, and implement an operational safety case in respect of a petroleum facility.

Dyed fuel   Fuel or lubricants to which a dye has been added to indicate the intended use of the fuel in a commercial application. Contrast with **clear fuel**, as gasoline, diesel, or aviation fuel which is not a dyed fuel. *See also* **red diesel**.

Dynamic mooring analysis (DMA)   A study of how the mooring position of a ship will change over time.

Dynamic positioning (DYNPOS)   A computer-controlled programme which, through the use of thrusters and propellers, maintains a ship's position on station without the use of anchors. *See also* **position mooring**.

Dynamic tensioning   A method of offshore pipeline laying whereby the pipelay vessel applies constant compensation for horizontal and vertical wave movements in order to prevent excessive stress on the pipeline.

E

E-drive   An industrial process which is powered by electrical energy. *See also* **direct drive** and **re-drive**.

E-line   A multiple strand mild steel armoured cable around an electrical conductor wire.

Early find/early show   A discovery which is made by an exploration well at a shallower depth level than was originally anticipated when the drilling of a wellbore commenced.

Early petroleum (early gas/early oil)   The production of commercially viable quantities of petroleum from a formation before a full-field development programme has been implemented, to give early cash flow to a producer and to secure initial operational experience.

Early production facility (EPF)   A petroleum facility which is developed for the production of early petroleum.

Earn-in agreement/earn-out agreement (EIA/EOA)   A contract for the sale of an interest in a concession whereby the buyer (the earnee) acquires the interest of the seller (the earnor) after completion of the buyer's performance of certain defined work obligations. Also called a **back-end farm-out** and **drill to earn**. Contrast with a **farm-in agreement/farm-out agreement**, whereby the buyer acquires the interest prior to completion of the buyer's performance of certain defined work obligations.

Earnee   The buyer under an earn-in agreement/earn-out agreement.

Earnest money   A non-refundable deposit which is paid by a bidder in a competitive auction situation, as evidence of the bidder's seriousness of intent and willingness to proceed if selected in the auction. Also called a **bid bond**.

Earnings before interest, tax, depreciation, and amortisation (EBITDA)   An accounting measure of the revenues which accrue to a petroleum company from its operations (before deductions from those revenues are made to account for the company's liabilities to pay interest charges, taxes, and to account for depreciation and amortisation costs) as a proxy for the raw operating profitability of the company.

Earnor   The seller under an earn-in agreement/earn-out agreement.

Earthwax   An alternative term for **ozokerite**.

Easement   A real property interest which is granted by a landowner to allow a person to access the landowner's land (for example, to lay and/or to access a pipeline). Also called a **servitude**.

East of Suez (EOS)   A notional boundary point which applies to everything due east of the Suez Canal (viewed from the south with a longitude of 32.3167E). *See also* **west of Suez**.

Eastings   Geographic coordinates which measure distance across an eastwards axis from a fixed point, often using the universal transvereneralitor coordinate system. *See also* **northings**.

EBITDA exploration expenses (EBITDAX)   An accounting measure of a petroleum company's EBITDA which also includes incurred exploration expenses.

Economic basement   In the drilling of a wellbore, a point of depth at which there is no economic sense in continuing to drill because physical conditions indicate that a viable petroleum system will not be found. Contrast with a **technical basement**, as a point at which a wellbore can no longer be drilled because a technical problem has been encountered.

Economic limit   The rate of production of petroleum from a formation below which a producer's net operating cashflows (based on revenues minus expenses) become negative.

Economic rent   A measure of the difference between the revenues which accrue from petroleum production and the costs which were incurred in producing that petroleum.

Economically recoverable petroleum (ERP)   An alternative term for **economically recoverable reserves**.

Economically recoverable reserves (ERR)   The amount of petroleum which a producer reasonably expects will be produced from a formation up to the point at which the continued production of petroleum becomes uneconomic for the producer. Also called **economically recoverable petroleum**.

Edge water   Sub-surface water deposits which are adjacent to a formation but which are not commingled with petroleum in the formation. Also called **bottom water**. Contrast with **connate**, as water which occurs inherently with petroleum in a formation.

Edge well   A wellbore which is drilled on the periphery of a producing formation.

Eductor   A pump which uses the Venturi effect to transport a gas or liquid through an enclosed space.

Effective capacity   The actual level of the operational capacity of a petroleum infrastructure item. Contrast with **nameplate capacity**, as the notional level of operational capacity.

Effective porosity   A measure of the porosity of a rock formation which is available for the flow of mobile petroleum and excludes isolated and occupied pore zones. *See also* **absolute porosity**.

Effective pricing   The real level at which petroleum is priced under a contract or in respect of a market. Contrast with **nameplate pricing**, as the notional, declared pricing level of petroleum which applies under a contract or in respect of a market.

Effective royalty rate (ERR)   The minimum share of monetary revenue or produced petroleum which a state will receive in respect of a given period and a given field or petroleum project. *See also* **rent**.

Egress   Under a drilling contract, the right of the drilling contractor to take a drilling unit away from a drill site. Contrast with **ingress**, as the right of the drilling contractor to bring the drilling unit onto a drill site.

Eight eighths (8/8ths)   An alternative way of describing 100 per cent of a working interest (based on the US customary practice of dividing a lease into eight tracts for subletting).

Electric submersible pump (ESP)   An electrically operated pump which is used downhole for artificial lift. *See also* **hydraulic submersible pump**.

Electro-drill   A drill bit which is powered by an electric downhole motor, without the need to rotate the drill string.

Electromagnetic propagation tool (EPT)   A device which emits microwaves into a formation in order to discover the potential presence of petroleum.

Electronic bulletin board (EBB)
(1) In a petroleum storage facility, a customer-accessed electronic portal which shows available spare storage capacity.
(2) In petroleum sales and trading, a customer-accessed electronic portal which shows available sell-side and buy-side opportunities.

Electronic data interchange (EDI)   Petroleum trading which is effected by screen message exchanges between the trading parties.

Electronic flow measurement (EFM)   The use of electronic devices in order to measure the flow of petroleum in a pipeline.

Elemental capture spectroscopy (ECS)   A spectroscopic technique which analyses the potential yields from different elements within a formation.

Elephant field   A new petroleum discovery which is believed to contain more than 100 million barrels of recoverable oil. *See also* **giant field** and **super-giant field**.

e-methane   Synthetic methane which is produced from the reaction of decarbonised hydrogen with captured carbon dioxide, effectively making it carbon neutral methane.

Embedded cost   At any point in time, the historical, incurred costs of developing a petroleum project which exist up to that point. Also called **back cost** and **sunk cost**.

Emergency disconnect sequence (EDS)   On a drilling unit, a rapid application of shear rams to close a wellbore.

Emergency position-indicating radio beacon (EPIRB)   A distress signal transmitter and tracking beacon which is activated in certain emergency situations.

Emergency release coupling (ERC)   A break point in a petroleum transfer system (such as a loading arm) which actuates valves to interrupt the petroleum flow in an emergency situation. Also called **breakaway coupling** and **powered emergency release coupling**.

Emergency response rescue vessel (ERRV)   A ship or boat which is used for emergency responses in offshore operations. *See also* **fast rescue craft**.

Emergency response training (ERT)   Training in dealing with emergencies for personnel involved in petroleum project operations.

Emergency shut-down (ESD)   A safety mechanism in a petroleum production or processing operation by which a petroleum facility can immediately be taken out of operation. Also called **safety shut-down** and **special shut-down**. Contrast with **routine shut-down**, as a predicted shut-down of a facility.

Emergency systems sustainability analysis (ESSA)   An assessment and analysis of the rigour of a petroleum infrastructure item's emergency systems.

Eminent domain   The right of a state to expropriate private property for public use, subject to the payment of compensation to the property owner.

Emission control area   (ECA) An area of ocean in which controls have been established under **MARPOL** to minimise the sulphur content of marine fuel oil.

Emissions trading scheme (ETS)   A regulatory mechanism which seeks to control harmful environmental emissions. *See also* **cap and trade**.

Employment contract   A term which is used to describe the form of charterparty which is used for the carriage of petroleum by ship.

Empty repo   Empty repositioning: the positioning of a ship without cargo in readiness to load a cargo.

Emulsion   A mixture in which one liquid (the dispersed phase) is uniformly distributed in another liquid (the continuous phase), such as in an oil-in-water mixture.

Emulsion breaker   A chemical agent which breaks down the formation of an oil and water emulsion.

Enabling agreement   An alternative term for a **master sales agreement** or a **master services agreement**.

Enclave project   A petroleum project which is executed principally to meet the needs of the project developer, without regard to the interests of other possible stakeholders, and which remains relatively isolated from local issues.

Encumbrance   A burden on the title to an asset or an interest, such as a charge or a lien.

End of day (EOD)   An alternative term for **close of business**.

End of field life (EOFL)   The point in time at which petroleum can no longer be produced economically from a formation.

End of sea passage (EOSP)   The point at which a ship arrives at its destination port, prior to berthing.

End of well report (EOWR)   Under a drilling contract, the final operational report which is issued by the drilling contractor after the conclusion of drilling the wellbore.

End point   In atmospheric distillation, the highest thermometer reading which is achieved during the distillation process (which usually coincides with the complete vaporisation of the commodity being distilled).

End-to-end plot   A single collection of data which applies throughout a process from start to finish.

End-user   A buyer of petroleum for final consumption. Also called an **ultimate customer**.

Endeavours obligation   A contractual obligation which is qualified from being absolute (such as reasonable endeavours or best endeavours).

Ending stocks   Products which are held in storage at the end of a defined audit or accounting period.

Energy advocacy   The promotion of a particular fuel in the composition of a particular energy market. Also called **advocacy**.

Energy audit   An assessment of energy flows and energy losses in a petroleum process, to identify the points at which efficiencies can be implemented.

Energy bridge   An alternative term for a **floating storage and regasification unit**.

Energy charge   An alternative term for a **capacity charge**.

Energy Charter Secretariat (ECS)   An international organisation which is principally responsible for supporting and sustaining the Energy Charter Treaty (http://www.energycharter.org).

Energy Charter Treaty (ECT)   A multilateral investment treaty which is published by the Energy Charter Secretariat with a focus on investment protection and non-discriminatory energy trade and transit between participating states (http://www.energycharter.org/process/energy-charter-treaty-1994/energy-charter-treaty/).

Energy content   An alternative term for **calorific value**.

Energy density   A measure of the physical footprint of an onshore petroleum project relative to a given level of petroleum output, which allows the comparable assessment of other projects.

Energy Information Administration (EIA)   A US-based data collection agency which provides energy sector statistics and information (http://www.eia.gov).

Energy Institute (EI)   A chartered professional membership body which brings global energy expertise together (http://www.energyinst.org).

Energy trading risk management (ETRM)   A back office business management system which identifies and mitigates commercial and financial risks which are associated with energy trading.

Engler test/Engler distillation   A test which is used for determining the viscosity of petroleum.

Engineering Equipment and Material Users Association (EEMUA)   An international technical association which offers advice to owners and operators of fixed industrial assets (http://www.eemua.org).

Engineering procurement construction installation (EPCI)   A contract for the development of a petroleum facility whereby a single contractor undertakes all the activities which are associated with the design, engineering, procurement, construction, commissioning, installation, and handover of the facility to an owner. Also called **lump sum turnkey** and **turnkey**.

Enhanced cost recovery   Under a PSC, an additional percentage uplift on certain capital costs which are incurred by the contractor which can be cost recovered by the contractor (which is intended to encourage the contractor to incur additional exploration or development expenditure). Also called **capex uplift**.

Enhanced oil production (EOP)   An alternative term for **enhanced oil recovery**.

Enhanced oil recovery (EOR)   A process which improves the flow of petroleum from a formation, such as secondary recovery or tertiary recovery. Also called **enhanced oil production** and **improved oil recovery**.

Enriching/enrichment   A process of modifying a gas stream with certain heavier petroleum fractions (such as propane and butane) in order to increase the gas stream's calorific value. *See also* **spiking** and **Wobbe quality adaptation**.

Enterprise value   A measure of a petroleum company's total value, offering a more comprehensive alternative to equity market capitalisation because the measure includes a company's short-term and long-term debt and the company's cash position.

Entitlement   The portion of current and future petroleum production which is due to accrue to the concession-holder under the terms of a concession.

Entitlement method   A booking methodology for a PSC or a service contract which is based on the contractor's financial entitlements calibrated in money to a number of barrels of oil equivalent at prevailing petroleum prices.

Entrains   Small amounts of crude oil or water which are entrained in natural gas or small amounts of natural gas or water which are entrained in crude oil.

Entrance loss   An irrecoverable energy loss which occurs when a flow of petroleum enters an orifice (such as entry into a storage tank).

Entry/exit charging   A pipeline transportation methodology whereby the shipper pays defined entry point and exit point charges to the transporter, regardless of the distance over which the petroleum is transported within the pipeline.

Environment testing   The testing of an equipment item in the actual conditions in which it is intended to operate.

Environmental and social impact assessment (ESIA)   A process for predicting and assessing the potential environmental and social impacts of a proposed petroleum project.

Environmental impact assessment (EIA)   A baseline environmental status study which is typically carried out prior to the commencement of a petroleum development project and is used to measure the project's subsequent compliance with environmental standards.

Environmental management team (EMT)   A team of persons which is brought together in order to manage an environmental issue which has arisen in respect of a petroleum project.

Environmental, social, and governance (ESG)   An alternative term for **corporate and social responsibility**.

Eon   A sub-division of geological time. *See* Appendix 4. Also called an **eonotherm**.

Eonotherm   An alternative term for an **eon**.

Epeiric sea   A relatively shallow inland sea which is formed by inland marine transgressions from the CS. Also called an **epicontinental sea**.

Epicontinental sea   An alternative term for an **epeiric sea**.

Epoch   A sub-division of geological time. *See* Appendix 4. Also called a **series**.

Equation of state   A description of a substance by reference to a given set of physical conditions such as pressure, volume, and temperature.

Equator Principles   A risk management framework which is applied by financial institutions for determining, assessing, and managing environmental and social risk in energy and infrastructure projects (http://www.equator-principles.com).

Equatorial circumference   The measured circumference of the horizontal equator of a spherical petroleum storage tank.

Equilibrium provision   A stabilisation provision whereby adverse changes to the interests of an investor in a petroleum project are renegotiated in the investor's favour.

Equilibrium vapour pressure   The pressure at which a liquid and a vapour are in equilibrium at a given temperature.

Equity facility   A petroleum-processing, transportation, or storage facility which is owned and/or operated by a person principally for the processing, transportation, or storage of that person's own petroleum entitlements. Contrast with a **common-carrier facility**, which has a contractual or a regulatory obligation to provide access to all potential users.

Equity joint venture (EJV)   A joint venture arrangement whereby the parties represent their interests through an incorporated association (for example, under a shareholders agreement in respect of a company). *See also* **joint venture company**. Contrast with a **contractual joint venture**, whereby the parties represent their interests through an unincorporated association (for example, under a JOA). Also called an **incorporated joint venture**.

Equity lifting   A quantity of equity petroleum which a producer takes delivery of. *See also* **equity petroleum**.

Equity lock up   The commitment of a party to a petroleum project not to divest its project interests for a defined period of time.

Equity oil (equity gas)   An alternative term for **profit oil (profit gas)**.

Equity petroleum   The share of the petroleum which is produced from a petroleum production project, expressed as a proportion of the whole of the produced petroleum, which a producer is entitled to as a result of its equity contributions to that development of the project. *See also* **equity lifting**.

Equity pipeline   A petroleum-transporting pipeline which is operated by a person principally for the transportation of that person's own petroleum entitlements. Contrast with a **multi-shipper pipeline**, which is operated for the transportation of petroleum on behalf of other persons.

Equity rate of return   The rate of return which a debt or equity investor into the costs of developing a petroleum project expects to receive, which is greater than a utility rate of return in recognition of the greater degree of risk which is assumed by the investor.

Equity storage model   A model for the operation of a petroleum storage facility whereby the storage facility owner reserves all of the storage capacity in order to meet its own commercial needs. Contrast with a **storage service model**, whereby a storage facility owner makes all or part of the storage capacity available to third-party users.

Equivalent circulating density (ECD)   An increase in bottomhole temperature which occurs when drilling fluid is being circulated through a wellbore.

Equivalent dip   The depth of petroleum in a storage tank which is determined by reference to the measured ullage in the tank.

Equivalent distillation capacity (EDC)   A further iteration of the Nelson complexity index, which is used to assess refinery complexity.

Era   A sub-division of geological time. *See* Appendix 4. Also called an **eratherm**.

Eratherm   An alternative term for an **era**.

Erosion drilling   The ejection of high velocity drilling fluid (and possibly also sand particles or steel shot) through a drill bit in order to drill a wellbore.

Erosional velocity limit (EVL)   The upper limit of the velocity at which petroleum can travel through a pipeline without starting to erode the pipeline's inner coating.

Escalation factor   A methodology for a commodity price increase whereby underlying costs have increased. Contrast with **de-escalation factor**, where prices decrease when costs decrease.

Escalator/escalation   Under a petroleum sales contract, a provision whereby the agreed contract price is adjusted periodically through indexation.

Escape hatch   Under a JOA, a provision whereby a party becomes relieved of the obligation to make future cost contributions if those costs would exceed an agreed level.

Escrow   A deposit of cash or other securities by a person with another person (the escrow agent) in an identified account (the escrow account), which is to be released by the escrow agent only when certain defined conditions are met.

Essential facility   A petroleum facility which is permitted to operate on a monopoly basis because duplication of the facility would be economically unsustainable (and where, as a condition of such operation, the facility is required to allow third-party access).

Estimated time of arrival (ETA)   A time at which a ship is due to arrive at a loading port or an unloading port.

Estimated time of departure (ETD)   A time at which a ship is due to depart from a loading port or an unloading port.

Estimated ultimate recovery (EUR)   The total quantity of petroleum which is expected to be recovered from a formation over the productive lifetime of the formation. *See also* **actual recovery** and **ultimate recovery**.

Ethane propane mix (EPM)   A mixture of ethane and propane.

Ethyl tertiary butyl ether (ETBE)   An additive which is used in the production of gasoline in order to offer improved air quality benefits. *See also* **methyl tertiary butyl ether**.

European Federation of Energy Traders (EFET)   A pan-European trade organisation which advocates policies, regulatory positions, and commercial contracts which are related to European energy trading (http://www.efet.org).

European Network of Transmission System Operators for Gas (ENTSOG)   An association which promotes cooperation between national gas transmission system operators across Europe (http://www.entsog.eu).

European option   An option agreement which can be exercised by the option-holder only on its expiry date. Contrast with an **American option**, which can be exercised by the option-holder at any time during its currency up to its expiry date.

Eustatic sea level rises   A change in global sea levels which results from a change in the quantity of water in the sea (which could be caused, for example, by melting ice caps).

Evacuation   The transportation of petroleum from the point at which the petroleum is produced to the point at which the commercial value of the petroleum is realised (typically through a sale of that petroleum).

Evaporation loss   The loss of petroleum fractions to the atmosphere through vaporisation. *See also* **breathing**.

Event tree analysis (ETA)   A form of assessment of the risks which are associated with a particular activity.

Evergreen   A right or a contract which subsists without a defined end date, but which is subject to termination by a party at any time.

Ex aequo et bono   A principle whereby an arbitral tribunal will seek to resolve a dispute according to what it thinks is fair and just, rather than according to the rule of law. Also called **amiable compositeur**.

Ex-ship An alternative term for **delivered ex-ship**.

Ex-tank sale A sale of petroleum which takes place at the point of exit from the seller's storage tank. Contrast with an **in-tank sale**, as a sale of petroleum which takes place at the point of entry to the buyer's storage tank.

Exceptions Under a charterparty, a provision which exonerates the shipowner from responsibility to the charterer for damage to a cargo which has been caused by certain named circumstances, where a breach of the charterparty (such as the obligation to carry a cargo safely) occurs.

Excess An alternative term for a **deductible**.

Excess berth occupancy An alternative term for a **port congestion charge**.

Excess gas Under a gas sales contract, gas which is supplied by the seller on terms which are outwith the agreed contractual delivery parameters at the buyer's request (often in consideration of the payment of a price premium by the buyer). Also called **penalty gas**.

Excess LNG Under an LNG sales contract, LNG which is supplied by the seller on terms which are outwith the agreed contractual delivery parameters at the buyer's request (often in consideration of the payment of a price premium by the buyer).

Exchange gas Gas quantities which are swapped between persons at different physical points.

Exchange transportation The notional transportation of petroleum through a pipeline whereby the shipper and the transporter perform matching delivery and offtake commitments but without physically transporting petroleum to meet those commitments.

Exclusive economic zone (EEZ)
(1) A marine area which is defined under the **United Nations Convention on the Law of the Sea** (UNCLOS) as one in which a state has special rights regarding the exploration for and exploitation of resources (stretching out 200 nautical miles from the baseline).
(2) An alternative term for a **joint development zone**.

Exclusive exploitation authorisation (EEA) A form of permit which is awarded in order to allow the production of petroleum from a particular part of a concession area.

Exclusive operation Under a JOA, a petroleum exploration and/or production operation which is carried out by less than all of the parties to the JOA. Contrast with a **joint operation**, as an operation which is carried out by all of the parties to the JOA. Also called a **non-joint operation**. *See also* **non-consent, reinstatement**, and **sole risk**.

Exculpatory clause A contractual provision which limits or excludes a contracting party's liability for default, which is typically used in the context of limiting the operator's liability under a JOA.

Exergy analysis An assessment of the degree of process efficiency which a petroleum facility exhibits, which is typically conducted as part of a wider cost/benefit analysis.

Exhaustibility   The finite and non-renewable nature of a petroleum deposit.

Exit loss   An irrecoverable energy loss which occurs when a flow of petroleum exits an orifice (such as exit from a pipeline).

Expandable sand screen (ESS)   A flexible filter which is used downhole for sand control purposes.

Expansion factor   A multiplying factor which is used to correct pressure, temperature, and velocity changes when calculating the flow rate of gas through a pipeline.

Expansion joint   A device which connects pipeline sections and allows for expansion or contraction in response to temperature changes. Also called a **slack loop**.

Expansion loop   A joint in a pipeline which allows for longitudinal pipeline expansion or contraction in response to pipeline temperature changes.

Expected value (EV)   An economic risk analysis which quantifies the returns which are expected by an investor from a petroleum project. Also called a **hurdle rate**.

Exploitation production sharing agreement (ExpltPSA)   A PSC which governs exploitation but not exploration. *See also* **Exploration production sharing agreement**.

Exploitation well   An alternative term for a **development well**.

Exploration   The activity of exploring for petroleum.

Exploration and appraisal (E&A)   The combined activities of exploration and appraisal.

Exploration and production (E&P)   The combined activities of exploration and production.

Exploration and production sharing agreement (EPSA)   A PSC which includes the exploration component. Contrast with a **development and production sharing agreement**, which does not include exploration.

Exploration licence   A licence which is awarded by a state in order to permit exploration for petroleum in a defined area. Also called an **XL**.

Exploration production sharing agreement (ExplrPSA)   A PSC which governs exploration but not exploitation. *See also* **Exploitation production sharing agreement**.

Exploration well   A wellbore which is drilled to explore for undiscovered petroleum. Also called a **new field wildcat**, a **test well**, and a **wildcat well**. *See also* **rank wildcat**.

Explosimeter   A device which measures the concentration of combustible gases in air.

Explosion protection review (EPR)   An assessment and analysis of the ability of a petroleum infrastructure item to withstand an explosion.

Export credit agency (ECA)   A private or quasi-governmental institution which provides state-backed collateral support for the financing of certain investments. Notable examples include COFACE (France), JBIC (Japan), and KEXIM (Korea).

Exporter of record (EOR)   A declared or official recognised seller of goods which are being exported from a state of origin, with the associated tax and duty liability and responsibility to secure the necessary export documentation. *See also* **importer of record**.

Exposure draft   A contract which is recorded on a public registry (sometimes in a redacted form in order to preserve the confidentiality of certain commercial data).

Expropriation   A direct or indirect state-sponsored act or omission which adversely affects the economic interests of a person who has made an investment in a petroleum project. Also called **taking**. *See also* **creeping expropriation**.

Extended leak-off test (XLOT)   A leak test which is intended to run for a defined extended purpose.

Extended production test (EPT)   An alternative term for an **extended well test**.

Extended reach drilling (ERD)   Directional drilling where the true horizontal distance which is realised by the drilling is greater than the true vertical depth which is realised by the drilling.

Extended reach lateral (XRL)   A horizontal wellbore of approximately 10,000 feet, a distance which is greater than a standard reach lateral.

Extended well test (EWT)   A method for evaluating the characteristics of a formation, often occurring as a precursor to a commercial well declaration. Also called an **extended production test**.

Extensible modelling framework (EMF)   A modelling technique which accounts for future growth potential.

External diameter (ED)   The external diameter of a pipeline section. *See also* **internal diameter**.

External market   A market for petroleum products which is composed of domestic production that is exported for consumption elsewhere. *See also* **internal market**.

Externality   Any environmental, social, or economic factor (which could be positive or negative) which is borne by a producer and so the cost of which is not directly reflected in the sale price of the produced petroleum. Also called a **spillover**.

Extra-heavy oil   An alternative term for **ultra-heavy oil**.

Extraction loss   A reduction in a quantity of wet gas which results from the processes of dehydration and liquids stripping. Also called **shrinkage**.

Extraction plant   A processing plant which removes liquids from a gas stream *See also* a **deep cut plant** and a **shallow cut plant**.

Extractive distillation   Atmospheric distillation in which a solvent is flowed through a CDU in order to separate out different components which have the same vapour pressure.

Extractive Industries Transparency Initiative (EITI)   A set of established global standards which are intended to provide for the good governance of the extractive element of the global petroleum industry and of other extractive industries (http://www.eiti.org).

Extractor   In petroleum measurement, a device which removes small amounts of liquid from a flowing stream and diverts them to a storage container for analysis.

Extreme value analysis (EVA)   A statistical analysis which addresses the possibility of extreme deviations from the median of probability distributions.

Extrinsic value
(1) The valuation of a gas storage facility which is based on the arbitrage of deliverability rates and short-term gas demand/price volatility. *See also* **intrinsic value**.
(2) Under a petroleum sales contract, the use of the contract's flexibility mechanisms in order to respond to external changes in future market price movements. *See also* **intrinsic value**.

F

Facies   The physical characteristics of a rock within a formation.

Facilitation payment   An alternative term for a **grease payment**.

Factor   An agent which is appointed to sell goods on behalf of a principal, has possession of those goods, and charges a commission to the principal for doing so. Also called a **factor agent**.

Factor agent   An alternative term for a **factor**.

Factory acceptance test (FAT)   A process to evaluate whether the manufacture of equipment during and after an assembly process has been conducted in compliance with design specifications.

Fade-out   An alternative term for **nationalisation**.

Fail-safe (FS)   In operational terms, a process response which applies to mitigate and/or to prevent the recurrence of a previously occurring process failure.

Failure mode and effects analysis (FMEA)   An engineering methodology which assesses the reliability of a petroleum infrastructure item or of a process.

Fair market value (FMV)   An estimate of the objective fair market value of an asset or an interest, which is often based on precedent transaction or extrapolated analogue values.

Fair value accounting   An alternative term for **mark-to-market**.

Fairway   A geological orientation along which a parameter occurs.

Fait du prince   The principle that a private contract can be modified by a state if doing so would be in the public interest. *See also* **rebus sic stantibus**.

Fallow acreage   Acreage in respect of which a concession has previously been awarded but where no exploration activity has taken place (whether at all or within a certain period), at which point the concession could be revoked by the grantor and re-awarded to another person.

Fan shooting An acquisition of seismic data which is made using a fan-shaped array.

Farm-in agreement/farm-out agreement (FIA/FOA) A contract for the sale of an interest in a concession whereby the buyer (the farmee) acquires the interest of the seller (the farmor) prior to completion of the buyer's performance of certain defined work obligations. Also called a **front-end farm-out**. Contrast with an **earn-in agreement/earn-out agreement**, whereby the buyer acquires the interest after completion of the buyer's performance of certain defined work obligations.

Farm-in operator Under a farm-out agreement, the appointment of the farmee to act as the operator of the farm-in works.

Farm-in well A well which is drilled as a consequence of the entry into a farm-in agreement/farm-out agreement. *See also* **farm-out well**.

Farm-out well A well which is drilled as a prelude to the entry into a farm-in agreement/farm-out agreement. *See also* **farm-in well**.

Farmee The buyer under a farm-in agreement/farm-out agreement.

Farmer's crude An in-kind royalty which is paid to a landowner in consideration of the development of the landowner's mineral rights by a developer.

Farmor The seller under a farm-in agreement/farm-out agreement.

Fast rescue craft (FRC) A small support vessel which is used for offshore operations in order to rescue personnel from the sea. Also called a **man overboard boat**.

Fast sample loop A secondary pipeline circuit which extracts a sample from a petroleum stream within a pipeline for measurement.

Fast track The results of the initial analysis of a seismic survey which is made without the benefit of full data processing.

Fatal flaw analysis A focused analysis of the key parameters of a petroleum project investment or development opportunity, which is made as a precursor to a decision to drop or to pursue the opportunity.

Fatigue limit state (FLS) The stress level below which an infinite number of loading cycles can be applied to a petroleum infrastructure item without causing structural failure.

Fatty acid methyl ester (FAME) An alternative term for **biodiesel**.

Fault A break in a rock formation along which rocks on one side of the break have been displaced (whether downwards, laterally, or upwards) relative to the other side.

Fault compartmentalisation The natural segregation of a formation into a series of separate petroleum deposits.

Fault plane A surface along which a fault has developed.

Federal contract A contract which is made between an employer company and a service-provider by which the service-provider provides certain services in multiple places and/or for multiple projects at multiple times. *See also* **master services agreement**.

Fédération Internationale Des Ingénieurs-Conseils (FIDIC)   The International Federation of Consulting Engineers: a trade organisation of consulting engineers, best known for the publication of a series of widely used standard form construction contracts (http://www.fidic.org).

FEED EPC (FEPC)   The appointment of a FEED contractor which is expected to move into the subsequent appointment of the same person as an EPC contractor.

Feed gas   Gas which is used as a feedstock for the activities of LNG or LPG production, petrochemical production, or electric power generation.

Feed quality gas (FQG)   Raw gas which has been processed to a standard that is required to give feedstock gas for further processing.

Feed tank   A storage tank from which a petroleum stream is continuously fed for further processing.

Feedstock   Petroleum which is used as a raw material for a further processing activity. Also called **charge stock**.

Feet of pay   The measured vertical thickness of a payzone. Also called **true vertical thickness**. *See also* **net pay**.

Fender   An item of shock-absorbing material which is placed over the side of a ship in order to prevent the risk of damage which is caused by the contact of the ship with port facilities or another ship.

Fiduciary   A person (the fiduciary—such as an agent or a trustee) to whom property or power is entrusted for the benefit of another person, whereupon the fiduciary owes certain duties of care to that other person.

Field   An alternative term for a **formation**.

Field development plan (FDP)   A comprehensive technical and financial proposal which is made by a concession-holder to the state for the development of a petroleum production project when required under the terms of a concession. Also called a **general development and production plan** and a **plan of development**.

Field gas   Natural gas as it is produced from a wellhead, prior to processing.

Field gathering station (FGS)   A central collection point in a gathering system.

Field grade butane   A product which is made up of butane and isobutene and meets a particular specification.

Field life coverage ratio (FLCR)   A ratio test in project financing and reserves-based lending which assesses the ability of the borrower to repay the debt over the anticipated life of the field. *See also* **project life coverage ratio**.

Field mapping   A topographical determination of a formation's possible dimensions by an assessment of above-ground geological features. *See also* **outcrop** and **topography**.

Field pressure   The pressure of a natural gas deposit as it is found in a formation.

Field price   Under a gas sales contract, an agreed price which is payable by the buyer to the seller for field gas at the wellhead.

Field processing   The processing of petroleum which takes place at the point of production and before transportation.

Field superintendent   An alternative term for a **superintendent**.

Filled to spill   A formation which contains petroleum up to the spill point.

Filling the hole   The activity of pumping drilling fluid into a wellbore when a drill string is being removed (to exceed the pressure exerted in the formation and to prevent formation fluids from flowing into the wellbore).

Fill-up rate   The rate at which drilling fluid is pumped into a wellbore during filling the hole.

Filter cake   An alternative term for **mud cake**.

Filtrate   The liquid element of a slurry such as drilling fluid which is recirculated and leaves mud cake behind in a wellbore.

Final investment decision (FID)   The decision which is taken by a concession-holder to finance and develop a petroleum project which, at the time of taking that decision, is still at the design stage.

Finance lease   A leasing arrangement for a petroleum infrastructure item whereby the lessee takes the risks and returns which are associated with ownership of the item, and the lease is treated as an asset on the lessee's balance sheet. *See also* **operating lease**. Also called a **capital lease**.

Finding costs   The aggregate costs of discovering and developing a commercial petroleum discovery.

Finding of no significant impact (FONSI)   A determination that a formation which has been discovered contains no or negligible quantities of petroleum.

Finger printing   An alternative term for an **assay**.

Fingering   The isolation of a pocket of crude oil in a formation by a series of water deposits, thereby making the crude oil difficult to recover.

Finished products   Petroleum products which need no further processing in order to be ready for sale or use. *See also* **semi-finished products**.

Fire and explosion analysis (FEA)   An assessment and analysis of the ability of a petroleum infrastructure item to withstand a fire and/or an explosion.

Fire and gas (F&G)   An operational system for detecting fire and/or gas leaks in a petroleum infrastructure item.

Fire sale   An alternative term for a **distressed sale**.

Fireflood extraction   Tertiary recovery in which a controlled combustion of heavy oil is carried out while it is still in-situ in a formation to allow lighter oil fractions to be produced from the formation. *See also* **cold heavy oil production with sand** and **thermal enhanced oil recovery**. Also called **in-situ combustion** and **thermal recovery**.

Firewall
(1) Earthworks which are built around an oil storage tank to contain the oil in the event of a leak in the tank. Also called a **berm** and a **bundwall**. *See also* **impoundment**.
(2) A distinction which is applied by a person to different classes of that person's contracting counterparties who enjoy different commercial terms.

Firm rights   Arrangements for the sale, transportation, processing, or storage of petroleum which are not interruptible by any of the parties to the underlying transaction. Contrast with **interruptible**, where sales or services can be interrupted.

Firm storage service (FSS)   A gas storage service which is not interruptible by the storage facility owner. Contrast with **interruptible storage service**, which can be interrupted by the storage facility owner.

First aid case (FAC)   A personal injury incident in a petroleum facility which requires a recordable first aid intervention to an injured person.

First-hand price   An alternative term for a **wellhead price**.

First-in first-out (FIFO)   An agreed order in which an accrued series of interests is used, recouped, or disapplied whereby the most mature are applied first. Contrast with **last-in first-out**, whereby the least mature are applied first. Also called **last-in last-out**.

First-in last-out (FILO)   An alternative way of phrasing **last-in first-out**.

First marketable product (FMP)   A methodology for calculating a netback royalty which values the petroleum at the point at which the petroleum has become marketable (after processing and transportation has taken place). *See also* **historical method**.

First oil   The first quantity of petroleum which is identified as having been produced from a formation.

First tranche petroleum (FTP)   Under a PSC, a provision for the state to take a defined early quantity of produced petroleum as profit oil (profit gas) ahead of the incidence of cost recovery in favour of the concession-holder.

Fiscal marksmanship   A measure of the degree to which the designs of the fiscal terms of a concession have succeeded in attracting investors and also in providing a satisfactory level of state take.

Fiscal meter   Measurement equipment which is relied upon as the primary measurement determinant of petroleum quality and quantity. Contrast with a **check meter**, as the measurement of petroleum quantities as a verification of fiscal meter readings.

Fiscal standard   An alternative term for a **metering standard**.

Fiscalisation point   In relation to a concession, the point at which a concession-holder's interest in the produced petroleum (whether represented in cash or in kind) is realised in the concession-holder's favour.

Fischer Tropsch   A process for effecting gas-to-liquids conversion. *See also* **gas-to-liquids**.

Fish   A tool which is lost or stuck in a wellbore.

Fishing   The recovery of a tool (or other debris) which is lost or stuck in a wellbore.

Fishing tool   A tool which is used to recover a fish or other debris from a wellbore.

Fixed choke   A choke with a single-size opening only. Contrast with an **adjustable choke**, which can be opened to different sizes.

Fixed content   An alternative term for **detailed content**.

Fixed installation   An offshore petroleum infrastructure item which is anchored to the sea-bed. Contrast with **floating installation**, which is not anchored.

Fixed interest   Unit interests which are fixed in a unitisation exercise without the prospect of a redetermination.

Fixed platform   A petroleum production platform which is fixed to the sea-bed by its jacket, typically used in water depths up to 1,500 feet. *See* Appendix 2D.

Fixed R/T   A royalty trust which has no power to invest its income in the development of new royalty-paying assets. Contrast with a **growth R/T**, as a royalty trust which has a reserved power to invest its earned income in the development of new royalty-paying assets.

Flag state   The nation in which a ship is registered, and that holds jurisdiction over the ship while it is afloat.

Flange   A flat, circular ring at the end of a pipeline length to which another pipeline length or other equipment can be bolted.

Flanging- up
(1) The completion of the drilling of a wellbore.
(2) The final connection on the installation of a new pipeline.

Flare platform   An offshore platform which is used only for the flaring of natural gas that is generated from petroleum operations.

Flarestack   A vertical or horizontal structure which is used for flaring. Also called a **jumbo burner**.

Flaring   The immediate combustion of natural gas at the point of its production for technical or commercial reasons.

Flash drum   A pressure vessel which is used to reduce the pressure of solution oil in order to encourage the vaporisation of solution gas.

Flash gas   Gas vapours which are released from liquid phase petroleum as a result of a temperature increase or a pressure decrease. Also called **flash vapour**.

Flash point   The lowest temperature at which petroleum will form an ignitable mixture in air when an ignition source is present.

Flash point check   A test which is used to confirm that a liquid phase product vaporises within an anticipated temperature range.

Flash tank   A vessel in which flashing takes place.

Flash vapour   An alternative term for **flash gas**.

Flashing   The vaporisation of liquid phase water by the rapid transformation of water into steam.

Flat gas   Under a gas sales contract, gas which is supplied by the seller to the buyer at a uniform rate over a given period of time.

Flat price case   An economic model which assumes unescalated prices (of commodities and of project components) for the life of a petroleum project. Also called a **constant price case**.

Flexibility stack   The sequence of options for a gas market participant to access flexible gas options.

Flexible energy purchasing (FEP)   An option right whereby a person can buy petroleum from, or can sell petroleum to, another person at any time in order to take advantage of petroleum price fluctuations.

Flexible gas
(1)   Additional sources of gas which can be accessed by a gas market participant when it becomes necessary or attractive to do so.
(2)   A supply of gas which responds positively or negatively to market price signals.

Flexicoking   A process which converts bottoms into lighter petroleum fractions in a refinery.

Flip-flop   A three D seismic survey which is performed by firing two identical air gun arrays from alternate shot points.

Float casing   A method of casing a wellbore whereby a sealed drill string becomes buoyant in drilling fluids in the wellbore, displacing the fluids, and then being cemented into place.

Float collar   A check valve which is inserted at the bottom of a casing string in order to prevent the flowback of cement from the annulus when the cementing process is underway.

Float out   The loading and transportation of jackets or topsides for offshore installation.

Float-over   The mating of one offshore petroleum infrastructure item to another by floating it over the other item when it is submerged, rather than by lifting in place. *See also* **lifting in place**.

Floating drilling production storage and offtake (FDPSO)   An FPSO which also has the capacity to drill wells.

Floating installation   An offshore petroleum infrastructure item which is not anchored to the sea-bed. Contrast with **fixed installation**, which is anchored.

Floating LNG (FLNG)   A marine-based facility which liquefies gas to give LNG or which receives LNG for regasification.

Floating LNG vessel (FLNGV)   A ship-based gas liquefaction plant for the production of LNG.

Floating production facility (FPF)   An alternative term for **floating production, storage, and offtake**.

Floating production, storage, and offtake (FPSO)   A ship which is used to process, store, and export produced petroleum, typically used in water depths up to 6,000 feet. Also called **floating production facility**. *See* Appendix 2D.

Floating roof   In a petroleum storage tank, a flat roof which floats on the surface of the petroleum in store. Contrast with a **dome roof**, which is fixed to a tank.

Floating storage   A crude carrier which is moored offshore and is used as a temporary storage tank for crude oil. Also called **anchored storage**.

Floating storage and regasification unit (FSRU)   A ship which receives LNG from offloading LNG carriers and provides regas send out to the shore. Also called an **energy bridge** and a **liquefied natural gas regasification vessel**.

Floating storage offshore (FSO)   A ship or other vessel (such as a spar) which is used for the offshore storage of produced crude oil.

Floating storage unit (FSU)   A generic term for FDPSOs, FLNGVs, FPSOs, FSOs, and FSRUs.

Flocculant   A chemical agent which is used in flocculation.

Flocculation   The coagulation of solids in a drilling fluid through the addition of a coagulating agent in order to cause the attraction of charged particles. Contrast with **deflocculation**, which encourages the reduction of viscosity.

Flooding   A composite term for **gasflood** and **waterflood**.

Floor   An alternative term for a **collar**.

Flotel   An alternative term for an **accommodation support vessel**.

Flotsam   Floating remains after a ship has foundered or items which are accidentally washed overboard from a ship. Contrast with **jetsam**, as items which are deliberately jettisoned from a ship.

Flow   A moving stream of petroleum.

Flow assurance   An engineering assurance process which relates to the successful transition of petroleum from a reservoir to a point of sale (including effective separation and managing production interruptions, pipeline blockages and pipeline erosion, and temperature and pressure management).

Flow check   A method for assessing whether a kick is occurring in a wellbore or if wellbore conditions are stable by temporarily interrupting or stabilising the flow of drilling fluid into the wellbore.

Flow computer   A computer which computes and analyses all of the operational data from a flowing pipeline.

Flow control   The ability of a person to control the flow of petroleum through a pipeline.

Flow line temperature   The temperature of drilling fluid as it flows out of a wellbore.

Flow meter   A device which measures the flow rate of gas or liquids through a section of equipment. *See also* **axial flow meter, magnetic flow meter,**

multiphase flow meter, rotameter, turbine meter and **virtual flow meter**. Also called a **velocity meter**.

Flow rate   The rate at which petroleum travels out of a formation or through a petroleum facility.

Flow rate stock curve   A measure of how a flow rate is affected by underlying petroleum stock levels.

Flow straightener   A section of modified pipeline which is installed at the point of entry into a flow meter in order to smooth the incoming flow of petroleum and to reduce the chance of measurement errors.

Flow test   The testing of a wellbore by flowing petroleum to the surface in order to determine flow rates, downhole pressure, and the wellbore's capacity to produce petroleum at a commercially viable flow rate.

Flow treater   A device which acts as a combined separator, heater, and processing vessel for a petroleum stream.

Flowback   Water which is generated in petroleum production.

Flowing bottomhole pressure   Bottomhole pressure during normal petroleum production operations.

Flowing pressure   The pressure of a flowing wellbore which is measured at the wellhead.

Flowing well   An alternative term for **open flow**.

Flowline   An alternative term for an **in-field pipeline**.

Flowline bundle   A combined assembly of flowlines and umbilicals.

Flowstream   An alternative term for a **stream contract**.

Flue gas
(1) A gas which is generated from the combustion of petroleum in air, which mostly consists of nitrogen and carbon dioxide and can be injected into a formation for enhanced oil recovery.
(2) Gases which are produced from the combustion of a flammable material in a special chamber. Also called **stack gas**.

Flue gas desulphurisation (FGD)   A process for the removal of nitrogen oxides and sulphur oxides from the post-combustion emissions of an electric power-generating facility before the emissions are discharged to atmosphere.

Fluid catalytic cracking (FCC)   A form of catalytic cracking.

Fluid contact   In a formation, the approximate point of gas–oil contact, gas–water contact, or oil–water contact. Also called a **contact area**. *See also* **hydrocarbon water content**.

Fluid holdup   In a multiphase pipeline, the fraction of a particular fluid which is present in a particular section of the pipeline (where each fraction moves at different speeds due to different densities and gravitational forces). Also called **holdup**.

Fluid saturation   A measure of the volume of the pores in a formation which are filled with petroleum or water.

Fluidity   A measure of a liquid's ability to flow freely.

Fluorescence   The natural property of minerals and fluids to emit light after absorbing light or other electromagnetic radiation, which in certain situations can be indicative of the presence of petroleum in a formation.

Flush phase   An alternative term for **primary recovery**.

Flush production   A high rate of petroleum production which is reached from a new wellbore.

Flushing medium   A catalyst which is used to flush out built-up crude oil deposits from petroleum-processing, transportation, and storage facilities.

Fluvial source   The creation of a petroleum deposit which is influenced by the depositional/erosional flow of rivers and streams.

Flux oil   The blending of residual fuel oil into ultra-heavy oil in order to reduce the ultra-heavy oil's viscosity.

Flux stock   An alternative term for **cutterstock**.

FM rate   Under a drilling contract, the revised dayrate which is payable by the employer company to the drilling contractor during a force majeure event that affects the drilling operations.

Foaming   Foam (bubbles) which becomes apparent at the point of gas and oil separation in a mixed petroleum stream.

Fold
(1)   A geological structure describing a rock formation which has been deformed by compression into bends such as anticlines or syncline.
(2)   An item of seismic data, reflective of the number of seismic receivers which are used to generate a trace. *See also* **full fold** and **single fold**.

Fold and thrust belt   A surface geological feature adjacent to mountainous areas, created as tectonic activity spreads outward, which usually presents as a series of foothills. Also called a **thrust belt**.

Footage   An alternative term for **rate of penetration**.

Footage contract   Under a drilling contract, a provision whereby the drilling contractor is paid an agreed monetary amount by the employer company for each foot of depth or distance drilled.

Force majeure (FM)   A contractual mechanism which relieves a contracting party from a liability for the failure to perform a contractual obligation where the intended performance was prohibited by supervening circumstances. Also called an **Act of God clause, cas fortuit, casus fortuitus**, and **vis major**. Contrast with **all-events**, whereby a contractual obligation is stated as being absolute and not capable of being relieved by force majeure.

Force majeure excepted (FME)   A provision in a contract by which an obligation or liability is expressed to be subject to the application of force majeure relief.

Force majeure pre-emption right (FMPER)   The right of a party to a contract or a petroleum project to exercise a pre-emptive right to purchase a counterparty's interests if the counterparty has declared force majeure in certain circumstances.

Force majeure restoration quantity (FMRQ)   Under a petroleum sales contract, the commitment of a party to make good the non-delivery or the non-offtake of a quantity of petroleum because of an earlier force majeure event which affected that party.

Force of law provision   A stabilisation provision which guarantees the sanctity of the investor's interests in a petroleum project through enacting the concession as a law in its own right, which is only capable of being changed by a new law.

Forced pooling   A US legal mechanism whereby a person can drill a well and exploit a petroleum deposit without voluntary pooling by all the persons having an interest in the deposit (and without a joint operating agreement or unit operating agreement in place) upon securing a regulatory permission to do so. Also called **compulsory pooling** and **statutory pooling**. *See also* **forced pooling JOA**.

Forced pooling JOA   A short form JOA which is entered into between the interested persons when a forced pooling has resulted in a petroleum discovery.

Foreclosure   Strategic behaviour by a person or group of persons which has the objective or the effect of restricting or eliminating competition in a market.

Foreground IP   Intellectual property which has yet to come into existence at a point in time. Contrast with **background IP**, which is already in existence.

Foreshore   The part of the sea-shore which lies between the low and high tide levels.

Forfeiture   Under a JOA, a provision to address payment default by which the defaulting party forfeits to the other parties some or all of its petroleum project interests (under the JOA and the underlying concession). *See also* **absolute forfeiture** and **withering interest provision**.

Formation   A rock formation (which could be igneous, metamorphic, or sedimentary) potentially containing petroleum in one or more reservoirs. Also called a **basin**, a **field**, and an **interval**.

Formation boundary   The horizontal and the vertical limits of a formation.

Formation breakdown   The fracturing of a formation from excessive wellbore pressure. *See also* **fracture pressure**.

Formation damage   A reduction in the permeability of a formation because of an invasion.

Formation evaluation (FE)   The evaluation of an explored and appraised formation as the precursor to the decision to develop the formation.

Formation evaluation while drilling (FEWD)   An alternative term for **drill stem testing**.

Formation integrity test (FIT)   A test of the integrity of a formation after a wellbore has been cased.

Formation pressure   The measured pressure at the base of a wellbore when the wellbore is shut in at the wellhead.

Formation strength   The ability of a formation to resist formation breakdown.

Formation volume factor (FVF) The ratio of a quantity of petroleum at formation temperature and pressure to the same quantity of petroleum at reference conditions.

Formation water An alternative term for **connate**.

Forum shopping The selection by a party to a dispute of what that party considers to be the most favourable forum for the resolution of the dispute.

Forward contract A petroleum trade which is intended for settlement (delivery and payment) at a point during any month later than the month in which the trade is made. *See also* **future contract** and **spot contract**.

Forward curve A predictive tool which estimates future expectations of petroleum supply and/or demand volumes and/or prices over a defined period of time.

Forward haul A pipeline transportation service which requires the movement of petroleum through the pipeline such that the contractual direction of movement (between input and offtake) is the same as the physical direction of movement of the petroleum. Contrast with **backhaul**, whereby the contractual direction of movement is opposite to the physical direction of movement.

Forward linkage A situation in which the undertaking of a petroleum project encourages growth in subsequent economic elements (such as the development of power-generation facilities resulting from the development of a gas discovery). Contrast with **backward linkage**, whereby the undertaking of a project encourages growth in associated economic elements.

Forward price A price that is payable for petroleum in a contract which specifies delivery at a future date and applies a price set in the contract. *See also* **spot price**.

Forward purchase A method of financing the costs of a petroleum project development whereby a producer forward sells defined quantities of petroleum to an investor and later delivers the requisite petroleum quantities to the investor in repayment of the investor's financing. Also called **advance payment financing**. *See also* **stream contract**.

Fossil fuel A carbon-based fuel source, such as coal or petroleum, formed subsurface from decayed organic matter. *See also* **kerogen**. *See* Appendix 2A.

Foundation customer An alternative term for an **anchor buyer**.

Foundation pile An alternative term for **conductor casing**.

Founding The sinking of a ship.

Founding fathers' rights Under pipeline system rules, a rule of priority which applies in the curtailment of pipeline capacity to give priority to certain shippers who first participated in transporting petroleum through the pipeline.

Four corners A colloquial term for the principle that the commercial and legal terms of a relationship between contracting parties will be represented exclusively by the words within the four corners of their contract and not by any extraneous arrangements.

Four D (4D)   A seismic survey which is made up of a collated series of three D seismic surveys which are conducted over a defined time period in order to create a record of the behaviour of the target formation over time.

Four-way trap   An alternative term for **closure**.

Fourble   Four lengths of drill pipe which have been screwed together. *See also* **double, single**, and **thribble**.

Frac hit   The leakage into a wellbore of fluids and proppants from another wellbore which has been used for fracking.

Frac pad   An alternative term for a **frac spread**.

Frac spread   A distinct area upon which fracking is conducted. Also called a **frac pad**.

Frac spread count   The number of frac spreads that are active at any time, which is often used as an indicator of the health of the unconventional petroleum production sector.

Fracking   The mechanical creation of artificial fractures in a formation by pumping in high-pressure fluids and **proppants** to improve permeability and petroleum recovery. Also called **hydrofracking, hydraulic fracture stimulation**, and **hydraulic fracturing**.

Fracking fluids   Water and other fluids which are pumped into a wellbore as part of a fracking operation.

Fracpack   The simultaneous fracking of a wellbore and the insertion of a gravel pack downhole in order to prevent the build-up of sand in the wellbore.

Fraction   Any part of a petroleum stream, which could exist in a natural state at the point of production or which could result from the processing of petroleum.

Fractionating column   An alternative term for a **crude distillation unit**.

Fractionation
(1) The separation of crude oil into its constituent components through a process of atmospheric distillation. *See* Appendix 2.
(2) The process of the separation of NGLs into butane and propane.

Fracture pressure   The point of pressure at which a formation will fracture. *See also* **formation breakdown**.

Fracture zone   A naturally occurring fracture or fissure in a formation which could present difficulties in the intended exploitation of a formation.

Frame agreement   A contract for the provision of goods and/or services whereby outline legal and commercial terms are agreed up-front, with transaction-specific details to follow in a separate appendix.

Free barrels   An alternative term for **headroom**.

Free carry   An alternative term for a **hard carry**.

Free gas   An alternative term for **non-associated gas**.

Free in pipe (FIP)   A sale of petroleum whereby the delivery point is the tailgate of a facility or a plant and the entry point into a pipeline in which the buyer has secured transportation capacity. Also called a **customer collect sale**. Contrast with a **delivered sale**, whereby the delivery point is the point of connection between the buyer's receiving facilities and the seller's transportation facilities.

Free look   A right of a person to examine data relating to an existing or a prospective petroleum project but without an obligation thereafter to make any investment into the project.

Free on board (FOB)   An Incoterms formulation which is used in ship deliveries of petroleum whereby the delivery of petroleum from the seller to the buyer takes place at the loading port and the buyer bears the transportation obligation (and associated costs) from the loading port. *See also* **delivered ex-ship**.

Free point   The lowest point downhole at which a stuck pipe is free to move (and so the section of a wellbore above a freeze point).

Free pratique   The licence which is given to a ship to enter port on an assurance from the captain that the ship is free from contagious disease.

Free ride   The entitlement of a person to enjoy an interest in a petroleum production project but without an obligation to pay the corresponding share of the costs of the production (*see*, for example, a gross overriding royalty).

Free surface effect   The tendency of a liquid in-tank on a ship to move in response to a change in the attitude of the ship. *See also* **pressed up**, **slack tank**, and **sloshing**.

Free time
(1)   An alternative term for allowed laytime.
(2)   A period of time in which a cargo is stored without charge to the cargo owner.

Free water   Water deposits within a formation which are able to flow freely and are not bound to granular surfaces within the formation.

Free water knock-out (FWKO)   An alternative term for a **three-phase separator**.

Free well agreement   An agreement whereby one person undertakes all the costs of drilling a well and in return takes over the interests of certain other persons in the well.

Freespan   An unsupported section of a pipeline (whether by design between pipeline supports or caused by scouring).

Freestone rider   An alternative term for a **Pugh clause**.

Freeze point   The downhole point at which a stuck pipe occurs. Also called a **stuck point**.

Freezing clause   A stabilisation provision whereby the legal and commercial regime which was applicable at the time of making an investment into a petroleum project is thereafter preserved for the benefit of the investor. Also called **stricto sensu**.

Freight   The amounts which are payable by a charterer to a shipowner under a bareboat charter or a voyage charter. *See also* **hire**.

Freight assessment   A determination of the costs which would be associated with the transportation by ship of a petroleum cargo.

Freight manifest   An alternative term for a **manifest**.

Freight rate lifting   A ship lifting of crude oil or condensate which is scheduled to take advantage of a low rate of freight under a charterparty.

Friction costs   The total direct costs (such as money and taxes) and indirect costs (such as time and opportunity costs) which are associated with managing a petroleum project.

Friction-reducing agent   A chemical additive which is inserted downhole in order to reduce the level of friction on tools during the drilling of a wellbore.

Friends and family funding   The raising of early stage finance by a petroleum company, which is done typically in order to cover ongoing operational costs, which is sourced principally from close associates of the company's directors and shareholders.

Front-end engineering design (FEED)   A process step which provides for the detailed technical definition of a petroleum project, after the conceptual design phase has taken place (in contrast with general technical description, whereby a contractor performs against a generalised functional specification). *See also* **pre-FEED**.

Front-end farm-out   An alternative term for a **farm-in agreement/farm-out agreement**.

Front month   The first month of a multi-month petroleum lifting or transportation schedule. *See also* **back month**.

Front office   The customer-facing elements of an energy business (such as marketing, sales and trading). *See also* **back office** and **mid office**.

Front running   A petroleum-trading practice whereby a trader executes a trade for its own account before executing a trade for a client to benefit from an anticipated resultant market movement.

Frontier/frontier area   A formation in which the petroleum prospectivity is unproven and is not supported by any data.

Frontier explorer   A petroleum company that conducts its exploration activities in areas in which petroleum production infrastructure has not yet been put in place. *See also* **tie-back explorer**.

Fronting agreement   A contract for the sale of petroleum whereby a person acts as the principal in buying petroleum from a seller and immediately on-sells the petroleum on back-to-back terms to another person who cannot contract directly with the seller.

Frontloading   The structuring of a petroleum project whereby the incidence of costs and/or the recovery of revenues arise in the early stages of the project. Contrast with **backloading**, where costs and revenues arise in the later stages of a project.

Frost heaving   In relation to an onshore pipeline, movements in the underlying soil which result from alternate freezing and thawing over time which can affect the stability of the pipeline.

Frustration   A common law doctrine which brings a contract to an end if super-vening circumstances have rendered the contract unperformable in accordance with its originally intended manner.

Fuel adjustment factor (FAF)   An alternative term for a **bunker adjustment factor**.

Fuel cell   A facility for the generation of electric power from the reaction of hydrogen and oxygen in the presence of a catalyst but without combustion taking place.

Fuel cost   The cost of the heat content of a given quantity of fuel which is calcu-lated by dividing total fuel costs by the resultant heat energy.

Fuel dye   An additive to a fuel or lubricant in order to create a dyed fuel. Also called a **marker**.

Fuel gas   An alternative term for **own-use gas**.

Fuel gas ratio   The ratio of a given quantity of gas to the part of that gas which is used as fuel gas.

Fuel gas supply system (FGSS)   On a ship or offshore facility, the complete fuel supply system from the fuel storage tank to the point of combustion.

Fuel lost and unaccounted for (FL&U)   An alternative term for **lost gas**.

Fuel oil   A liquid petroleum fraction which is obtained from atmospheric distil-lation and combusted as a fuel in boilers and furnaces. *See also* **automotive gas oil**, **Bunker C, burner fuel oil, fuel oil domestique, heating oils, high-sulphur fuel oil, low pour fuel oil, low sulphur fuel oil, low sulphur marine fuel, low sulphur waxy residue fuel oil, marine fuel oil, Mazut**, and **marine gas oil**. Also called **gas oil**.

Fuel oil domestique (FOD)   Fuel oil which is used in France for end-user heating purposes.

Fugitive emissions   Gas leakages from a pressurised gas facility.

Full and primary   An indemnity provision which allows for a complete recovery of losses by the holder of the indemnity.

Full away on passage (FAOP)   The point in a ship's voyage at which the ship tran-sitions from manoeuvring to departure at full speed.

Full cargo lot   The total quantity of LNG which is expected to be loaded into or unloaded from a single LNG carrier. *See also* **stub quantity**.

Full containment tank (FCT)   An LNG containment tank which is fully sur-rounded by a concrete safety container. *See also* **double containment tank** and **single containment tank**.

Full cost accounting   A petroleum accounting convention by which all operating expenses which are associated with exploraeneralizedpitalised, regardless of the success of the exploration effort. *See also* **successful efforts accounting**.

Full cycle economics   The economic analysis of the full lifecycle of a petroleum project which includes all the costs of project development including exploration, appraisal, development, production, decommissioning, and rent.

Full-field model (FFM)   A predictive computer model which is used to assess the petroleum prospectivity of a formation.

Full fold   A fold which records multiple seismic data points, typically in relation to three D seismic. *See also* **single fold.**

Full lifecycle emissions (FLE)   A measure of the carbon intensity which is associated with the LNG production to final consumption elements of the value chain. *See also* **well to tank.**

Full reach, burthen, and decks   Under a charterparty, the entirety of the space on a ship which is normally available for cargo carrying.

Full service charter   A time charter or a voyage charter, in which a ship is provided to a charterer with a crew. *See also* **bareboat charter**.

Full tensor gravity gradiometry (FTGG)   An airborne or marine survey which measures variations in gravity in order to indicate the presence of a petroleum-bearing sub-surface formation.

Full-time employee (FTE)   A person who is recorded as being a full-time employee of an employing entity, in contrast to a person who acts as an independent contractor.

Full value chain   All of the elements which together make up a value chain.

Full value well cost   A description of the total cost of drilling a wellbore, including the costs of drilling, completion, testing, and plugging.

Full well stream   Petroleum production from a wellbore at the point at which it emerges from a wellhead.

Functional unbundling   The disaggregation of energy generation, supply, and transportation interests at the operational level. Contrast with **ownership unbundling**, whereby disaggregation takes place at the corporate ownership level.

Fundamental analysis   Petroleum price performance analysis which is derived from actual supply and demand factors such as physical inventories, facility operations, physical trades, and market disruptions. Contrast with **technical analysis**, which is derived from studying historical market movements and applying predictive tools

Funded from operations (FFO)   Cashflow from petroleum production operations which is used to fund the ongoing petroleum project development costs.

Fungible/fungibility   A measure of the likeness and interchangeability of certain petroleum products.

Funnelling   An alternative term for **windowing**.

Future contract   A forward contract which is traded on a recognised investment exchange.

Future well   A wellbore which is identified to be drilled at a defined future point.

G

Gabbro   An igneous rock which is similar to basalt.

Gall   Surface damage to metal threads and surfaces which is caused by localised friction.

Galling   The adhesion of two mating metal surfaces which are not protected by a lubricant.

Gallons per minute (GPM)   A unit of measurement of the flow rate of liquid petroleum through a section of equipment.

Gamma ray log   An alternative term for a **radioactivity log**.

Gap-filler   A contractual provision which is implied by statute or by the decision of a court to apply if the contract itself has not provided for a particular matter.

Gas analysis recorder (GAR)   An alternative term for a **gas chromatograph**.

Gas and oil-cut mud (G&OCM)   Gas-cut mud and oil-cut mud which exist together.

Gas absorption   An alternative term for **absorption**.

Gas balance   The degree of balance (or imbalance) which exists in a market in which domestic gas demand is met by domestic gas production.

Gas balancing agreement (GBA)   An attribution agreement which applies to natural gas production between multiple interest owners in a gas stream.

Gas banking   The decision of a gas buyer to make a take or pay payment and to later recover its gas entitlements as make-up.

Gas-blanketed storage   The filling of the vapour space in a petroleum storage tank with a dense mixture of gas and air in order to reduce vaporisation and fire risk.

Gas breakthrough   The situation where gas which has been injected to maintain formation pressure breaks through to a production wellbore. *See also* **water breakthrough**.

Gas bubble   A sustained supply of gas into a market which exceeds aggregate demand, which can also drive down market gas prices.

Gas buster   A device which removes entrained gas from circulated drilling fluid.

Gas cap   A natural gas deposit within a formation which overlies a crude oil deposit (but in which the gas is typically not associated gas).

Gas cap drive   Petroleum production through a wellbore which is driven by the expansion of gas from a gas cap.

Gas carrier cargo handling   A generic term for all of the operational cycle elements of an LNG carrier (including drying, inerting, cool down, loading, discharge, warm-up, and gas-freeing).

Gas chromatograph   A device which performs chromatography in a gas stream. Also called a **gas analysis recorder**.

Gas chromatography mass spectrometry (GCMS)   An analytical process which combines gas chromatography and mass spectrometry in order to identify different substances within a test sample.

Gas condensate   An alternative term for **condensate**.

Gas containment unit (GCU)   An onshore or on-ship vessel which is used for the storage of LNG.

Gas-cut mud (GCM)   Drilling fluid which is recirculated through a wellbore with entrained gas. *See also* **background gas**.

Gas-dangerous place   Any area in a petroleum infrastructure item which is not designated as a gas-safe zone.

Gas day   Under a gas sales contract, a defined twenty-four hour period (which could start at any time).

Gas dehydration unit (GDU)   A device which dewaters a natural gas stream.

Gas disposal project (GDP)   A project for the disposal of natural gas within the context of a wider petroleum project.

Gas down-to (GDT)   A vertical measure of the vertical thickness of a natural gas deposit in a formation, as the deepest point at which gas saturation is measured. *See also* **oil down-to**.

Gas drive
(1) An alternative term for **direct drive**.
(2) An alternative term for **depletion drive**.

Gas expansion   The flow of gas through a wellbore from a freshly drilled formation.

Gas expansion drive   Gas production through a wellbore which is driven by expansion of the remaining gas in a formation as gas is produced. Also called **depletion drive**. *See also* **water drive**.

Gas expansion factor (GEF)   A measure of the volume which is occupied by a substance in a liquid form when compared to the volume which would be occupied by the same substance in a gaseous form.

Gas export regulation manifold (GERM)   A central infrastructure point at which gas export to different destinations is managed and monitored.

Gas Exporting Countries Forum (GECF)   An inter-governmental organisation which represents the interests of the world's major gas-producing and gas-exporting countries (http://www.gecf.org).

Gas-free   A petroleum storage tank to which gas-freeing has been applied.

Gas-freeing   An operation in which the petroleum vapours remaining in-tank after the discharge of a cargo are replaced with an inert gas, and then with air, in order to prevent explosion hazard.

Gas-holder   An underground or surface-mounted steel or concrete tank which is used for the storage of gas. Also called a **gasometer**.

Gas hydrate   An alternative term for **hydrates**.

Gas Industry Standards Board (GISB)   A US-based trade organisation which monitors issues associated with the wholesale and retail gas markets in the United States and that was absorbed into the North American Energy Standards Board in 2002.

Gas initially in place (GIIP)   A measure of the amount of natural gas which is estimated to be contained in a formation before gas production begins. *See also* **oil initially in place**.

Gas Laws   A series of theoretical laws which relate to the behaviour of a gas, by reference to the interrelationships between temperature, pressure, and volume. *See also* **Boyle's Law, Charles' Law**, and **Gay-Lussac's Law**.

Gas lift   Gas which is injected into a formation as part of secondary recovery in order to displace crude oil. Also called **cycling, gasflood**, and **jetting**.

Gas lock
(1)   The emission of gas-cut mud from drilling fluid which interferes with the intended manner of operation of a drilling fluid pump.
(2)   A device which permits that measurement of petroleum in a tank without the loss of gas to the atmosphere.

Gas-mature   Source rock which has been exposed to sufficient temperature and pressure over geological time to generate natural gas. *See also* **oil-mature**.

Gas oil   An alternative term for **fuel oil**.

Gas–oil contact (GOC)   The boundary in a formation above which predominantly natural gas occurs and below which predominantly crude oil occurs (usually a blurred rather than a defined boundary).

Gas–oil ratio (GOR)   The ratio of the quantities of natural gas to the quantities of crude oil which are produced from a quantity of petroleum.

Gas/oil separation plant (GOSP)   An alternative term for a **two-phase separator**.

Gas-out   Under a gas sales contract, the recovery by the buyer of an accrued shortfall price discount balance or make-up balance through the provision of gas or other petroleum quantities by the seller. Contrast with a **cash-out**, whereby the balance is repayable by the seller as a price discount or a cash payment.

Gas plant products   Natural gas liquids which are recovered from a natural gas stream in a gas processing facility.

Gas processing plant (GPP)   A facility through which natural gas streams are processed to remove water, impurities (typically carbon dioxide, hydrogen sulphide, mercury, or helium) and to separate out any gaseous phase petroleum fractions heavier than methane. Also called a **gasoline plant**.

Gas Processors Association (GPA)   A body which promotes industry standard technologies and procedures for natural gas processing, distribution, and measurement.

Gas regulator   A pressure-operated valve which restricts the flow of gas through a pipeline.

Gas-safe zone   A defined area in a petroleum infrastructure item which is engineered such that its atmosphere is at all times free of the presence of gas. *See also* **gas-dangerous place**.

Gas sale and purchase agreement (GSPA)   An alternative term for a **gas sales agreement**.

Gas sales agreement (GSA)   A contract for the sale of raw gas, regas, or processed gas. Also called a **gas sale and purchase agreement**.

Gas sand   A natural gas-bearing sand stratum.

Gas station   An alternative term for a **petrol station**.

Gas storage facility   A facility for the storage of gas, which could be underground or surface-mounted and is operated by reference to the functions of injection and redelivery.

Gas-to-liquids (GTL) conversion   A process by which natural gas is converted into gasoline and diesel in a refinery. Also called **middle distillate synthesis**. *See also* **Fischer Tropsch**.

Gas-to-power (GTP)   A process by which gas is combusted in order to generate electric power.

Gas-to-wire   The immediate consumption of natural gas at the point of its production in order to generate electric power.

Gas transportation agreement (GTA)   A contract which is made between a transporter and a shipper for the transportation of gas by pipeline.

Gas up/gassing up   A process whereby any inert gas in an LNG carrier's tanks (remaining after the LNG carrier has been laid up or dry-docked and the tanks have been purged) is displaced with warm LNG vapours in order to remove accumulated carbon dioxide and to dry the tanks prior to the loading of a cargo of LNG. *See also* **presentation**.

Gas–water contact (GWC)   The boundary in a formation above which predominantly natural gas occurs and below which predominantly water occurs (which is usually a blurred rather than defined boundary).

Gas window   A measure of the sub-surface depth at which gas is formed from kerogen. *See also* **oil window**. *See* Appendix 2A.

Gas year   An annual period of time over which a supply of gas is effected. In the United Kingdom the gas year runs for twelve months from 1 October in each year.

Gas zone   A zone from which natural gas is principally produced.

Gasflood   An alternative term for **gas lift**.

Gasification   A process by which coal or crude oil is converted into gas at standard temperature and pressure.

Gasifier   A facility at which gasification is carried out.

Gasoduct   An alternative term for a gas pipeline.

Gasoline   A lower-grade distillate which is used as a motor vehicle fuel. Also called **mogas**, **petrol**, and **premium motor spirit**.

Gasoline plant   An alternative term for a **gas processing plant**.

Gasometer   An alternative term for a **gas-holder**.

Gasser   A wellbore which principally or exclusively produces natural gas. Contrast with an **oiler**, as a wellbore producing crude oil.

Gate   An alternative term for a valve in a pipeline.

Gate rate   An alternative term for a **city gate rate**.

Gathering   The collection of petroleum from several production sources for transportation to a central point through a network of pipelines.

Gathering and processing (G&P)   The activities of petroleum gathering and processing.

Gathering and transportation (G&T)   The activities of petroleum gathering and transportation.

Gathering line   An alternative term for an **in-field pipeline**.

Gathering system   A network of pipelines which is used for gathering.

Gating decision   A defined milestone in a petroleum project development schedule which leads to the taking of a further decision in respect of the project.

Gauge   The diameter of a drill bit or of a borehole.

Gauge hatch   An alternative term for an **ullage hatch**.

Gauge height   In an oil or liquids storage tank, the distance from a defined reference point at the base of the tank to a defined reference point at the top of the tank.

Gauge pressure   A reading of the pressure of petroleum within a storage tank or a pipeline, which is taken relative to atmospheric pressure.

Gay-Lussac's Law   One of the Gas Laws, which states that the temperature of a gas is directly proportional to its pressure.

Gaz natural liquefié (GNL)   The French language acronym for LNG.

Gazette/gazetting   An official publication which indicates the legal effectiveness of a particular petroleum transaction.

Gearing   An alternative term for **leverage**.

Gel pig   A semi-solid form of pig which can be deployed without an in-built pig launcher and pig receiver facilities on a pipeline.

General & administrative overhead (G&A)   An alternative term for **indirect charges**.

General average   Under a charterparty, a provision by which a party (such as the shipowner) which has suffered loss during a voyage because of an intentional act or sacrifice which was intended to preserve the ship from a peril is compensated by all other related persons (such as the charterer) who benefitted from the loss-generating event.

General development and production plan (GDPP)   An alternative term for a **field development plan**.

General purpose (GP)   A crude carrier with a deadweight tonnage in the range of 10,000 to 25,000 metric tons.

General technical description (GTD)   A petroleum project methodology whereby a contractor performs generalised functional specification for the construction or operation of a petroleum infrastructure item (in contrast with FEED, whereby a contractor performs to a precise requirement as a technical specification).

General terms and conditions (GTCs)   Standard terms which apply to a contract between parties. Contrast with **special terms and conditions**, as bespoke terms which apply between the parties.

Generally accepted accounting principles (GAAP)   A combination of authoritative standards and common practices for recording and reporting accounting information.

Geobody   A description of the geological elements in a formation which results from a geological modelling exercise.

Geographic information system (GIS)   An information system which stores, analyses, and presents spatial geographical data which can be used in petroleum project planning.

Geologic time scale (GTS)   A system which relates the Earth's geological strata to the passage of time since the Earth's formation. *See* Appendix 4.

Geological and geophysical (G&G)   A summary of a formation's essential geological and geophysical attributes.

Geometry pig   A pig which is designed to measure and assess the internal condition of a pipeline.

Geophone   A device which is used in a seismic survey in order to detect sub-surface vibrations which could be indicative of the existence of a petroleum-bearing formation. Also called a **jug**.

Geostatic pressure   The pressure on a formation which is caused by the presence of overburden. Also called **ground pressure** and **lithostatic pressure**.

Geostatics   A process of collecting, analysing, interpreting, and presenting G&G data in order to assess the prospectivity of a formation and to stimulate formation behaviours.

Geothermal gradient   The rate of increasing temperature which corresponds to increasing depth below the surface of the Earth.

Geronimo   A safety slide which is used to facilitate the evacuation of personnel from a ship or an offshore petroleum facility.

Giant field   A sizeable new petroleum discovery which is believed to contain more than 500 million barrels of recoverable oil. *See also* **elephant field** and **supergiant field**.

Gigajoule (GJ)   One billion Joules.

Ginga   A contract for the forward delivery of physical LPG, which is principally used in Asia (http://www.ginga.com.sg).

Girbitol process   A process for the removal of non-petroleum gases from a natural gas stream.

Glacial source   The creation of a petroleum deposit which is influenced by the depositional/erosional action of ice movements.

Global maritime distress safety system (GMDSS)   A communication system which uses satellite and terrestrial radio communication to ensure the sending of the position of a ship in distress.

Global warming potential (GWP)   A measure of how much heat a greenhouse gas traps in the Earth's atmosphere, relative to the amount of heat which is trapped by an equivalent quantity of carbon dioxide, over a defined time period.

Glory hole   A sea-bed excavation in which a sub-sea wellhead sits in order to protect the wellhead from pack ice movements or scouring at the sea-bed level.

Glycol   A liquid desiccant which is used to remove water from a gas stream.

Glycol carriage line (GCL)   A small-diameter pipeline which is used to transport glycol to offshore petroleum production facilities.

Gob hole   A shaft which is drilled into a coal mine in order to allow the exit flow of accumulated methane. *See also* **methane drainage**.

Go-devil   An alternative term for a **pig**.

Goldplating   A colloquial term for the development of a petroleum project in which the developer has an incentive to use the highest-priced available materials (if the project developer can recover the costs which it incurs in doing so). Contrast with **poorboy**, whereby the project developer uses the lowest-priced available materials.

Golf arbitration   A colloquial term for a method of dispute resolution whereby the appointed arbiter decides its preferred outcome, the disputing parties submit their preferred outcomes to the arbiter and the disputing party's preferred outcome which is closest to the arbiter's preferred outcome prevails. *See also* **baseball arbitration** and **shotgun arbitration**.

Good and prudent oilfield practice (G&POFP)   An objective standard of performance which is required of a party in the performance of a contract, measured against the analogue of what another person in similar circumstances would do.

Goodwill agreement   A non-binding undertaking by a petroleum company that it will seek to focus as much procurement expenditure as possible in the state in which a project is located (which is effectively a soft form of local content).

Gooseneck   In drilling, a U-shaped flexible connection which is used for transferring drilling fluid between a rotary and a swivel.

Gosudarstvennyy standart (GOST)   A set of technical standards which are applied to petroleum sales and shipping transactions in Commonwealth of Independent States (CIS) countries.

Government-linked company (GLC)   A company which is directly or indirectly linked to a government.

Government-owned company (GOC)   An alternative term for a **state-owned enterprise**.

Government-sponsored enterprise (GSE)   An alternative term for a **state-owned enterprise**.

Government take   An alternative term for **rent**.

Grab sample/sampling   A petroleum sample which is drawn from the flow in a pipeline on an intermittent basis. Contrast with **continuous sampling**, whereby samples are taken continuously.

Graben   A geological structure which has become displaced from its surrounding features by downward settling. *See also* **horst**.

Grace period   Under a loan agreement, a period of time during which the borrower is relieved of the obligation to make repayments to the lender (whether of principal only or of principal and interest).

Grandfathering   A process by which the benefit of a statutory or contractual provision is maintained in favour of a person entitled to it despite the emergence of subsequent statutory or contractual provisions which would otherwise undermine or remove the benefit.

Granting instrument   An alternative term for a **concession**.

Grass roots refinery   A refinery which has been constructed from the ground up in a single construction phase, in contrast with a refinery which has been created from the accretion of multiple processing facilities over time.

Gravel packing   A downhole filter which prevents the production of unwanted sand from a formation.

Gravi-flow   The flow of a gas or liquid which is compelled by the action of gravity.

Gravimeter   A device which detects changes in the Earth's gravitational pull at a location and could indicate the presence of a sub-surface petroleum-bearing formation.

Gravity-based platform   A floating petroleum production and storage platform with a jacket made of steel or concrete, in which the inherent weight of the platform gives it stability and which is typically used in water depths up to 6,000 feet. *See* Appendix 2B.

Gravity-based structure (GBS)   A steel or concrete jacket which underpins a gravity-based platform.

Gravity coring   An alternative term for **drop coring**.

Gravity differential   In an oil–water emulsion, the degree of difference between the densities of the two substances (where a greater degree of difference indicates a more easily separable emulsion).

Gravity drainage   A basic primary recovery mechanism whereby the movement of petroleum in a formation is naturally effected by the application of gravity.

Gravity segregation   The natural tendency of oil, gas, liquids, and water in a formation to separate into distinct layers according to their respective densities.

Grav-mag   Gravity and magnetic survey: a measurement of the Earth's gravitational and magnetic fields, which is intended to reveal anomalies which could indicate the presence of a sub-surface petroleum-bearing formation. *See also* **aero-mag**.

Grease payment   A colloquial term for a facilitation payment which is made to an official in order to accelerate the performance of an administrative task (which in some circumstances could offend certain anti-bribery and corruption legislation). Also called a **facilitation payment**.

Green gas
(1)   An alternative term for **raw gas**.
(2)   An alternative term for **biomethane**.

Green LNG   LNG which is produced with the benefit of reduced carbon intensity and/or carbon offsets.

Green paradox   A principle (first suggested by Hans-Werner Sinn in 2008) whereby the declaration of increasingly green energy policies could create a rush to produce fossil fuels to capture the remaining value, thereby increasing global warming.

Greenfield   The development of a petroleum project without any pre-existing parameters. Contrast with **brownfield**, as a petroleum project which is developed with pre-existing parameters.

Greenfield production   A newly developed petroleum-producing formation. Contrast with **brownfield production**, as an existing but non-producing petroleum-producing formation which is brought back into production.

Greenhouse gases (GHGs)   Gases (principally carbon dioxide (CO_2), methane (CH_4), and nitrous oxide (N_2O)) which warm the Earth by absorbing heat energy that would otherwise escape into space.

Grey hydrogen   Hydrogen which is manufactured from natural gas using a steam methane reforming process, in which the resultant carbon is not captured and stored. *See also* **blue hydrogen**.

Grey water   Process water which is contaminated but not with human waste. *See also* **black water**.

Grid and raster   An alternative term for a **raster graphic**.

Grid-grade gas (GGG)   Natural gas which has a quality specification which is sufficient to enable that gas to be transported through a pipeline network without further treatment.

Grief stem   An alternative term for a **kelly**.

Gross calorific value   A determination of the calorific value of petroleum which is produced by combustion when water condenses and is combusted and taken into account. Also called **higher heating value**. Contrast with **net calorific value**, whereby the calorific value determination does not account for the combustion of water.

Gross energy requirement   The sum of the direct energy and the indirect energy which are used in a petroleum-processing activity.

Gross gas production   The full extent of natural gas which is produced at a wellhead, including water and liquids.

Gross gas withdrawal   The full measure of compounds, liquids, and non-petroleum gases which are extracted from a natural gas stream through dehydration and liquids stripping.

Gross interest billing   The gross amount which is payable by the JOA parties in respect of a joint operation. *See also* **net interest billing**.

Gross negligence/wilful misconduct (GN/WM)   A widely used test for a certain standard of misfeasance in petroleum contracts, where the term is often defined to mean an act or an omission by a person which was intended to cause, or was in reckless disregard of or was in wanton indifference to, the harmful consequences such person knew or should have known that the act or omission would have on another person's interests.

Gross observed volume (GOV)   In the measurement of oil in-tank, the total volume including sediments and water (other than free water) which exist at prevailing temperature and pressure. *See also* **gross standard volume** and **net standard volume**.

Gross overriding royalty (GOR)   An overriding royalty interest whereby the payee (the royalty-holder) is entitled to a defined share of the petroleum produced under the concession, without paying a corresponding share of the costs of production. Contrast with a **net profit interest**, whereby the royalty-holder's entitlement is net of a share of the production costs. Also called an **overriding royalty interest**.

Gross product worth (GPW)   A method of estimating the value of a barrel of crude oil based on the value of the products derived from it after it has been refined. Contrast with **net product worth**, which also accounts for the associated cost of transportation and refining. *See also* **yield**.

Gross registered tonnage (GRT)   A measure of the total internal volume of a ship which is expressed in register tons. Contrast with **net registered tonnage**, as a measure of the cargo-carrying capacity of a ship. *See also* **register ton**.

Gross rock volume (GRV)   A measure of the volume of rock which is contained between the top of a formation and the first petroleum contact.

Gross split contract   An alternative term for a **revenue sharing contract**.

Gross standard volume (GSV)   In the measurement of oil in-tank, the gross observed volume but with a correction for appropriate temperature and pressure conditions. *See also* **gross observed volume** and **total observed volume**.

Ground floor   A promote level whereby the percentage of the farmor's total required costs which is met by the farmee is equal to the percentage of the net working interest which is being sold down by the farmor. Also called **one-for-one**. *See* Appendix 3.

Ground pressure   An alternative term for **geostatic pressure**.

Grounding
(1)   An alternative term for **stranding**. *See also* **hard aground**.
(2)   A stranding which is predictable and is not covered by marine insurance. *See also* **customary grounding clause**.

Groundwater   Naturally occurring water deposits which are found underground.

Groupe International des Importateurs de Gaz Natural Liquefié (GIIGNL) International Group of Liquefied Natural Gas Importers: a trade organisation which monitors issues associated with LNG sales and transportation (http://www.giignl.org).

Grouped indemnity An indemnity provision in a contract which sits within an expected assembly of indemnities in the contract. Contrast with a **stray indemnity**, which sits outside an assembly of indemnities.

Growth R/T A royalty trust which has a reserved power to invest its earned income in the development of new royalty-paying assets. Contrast with a **fixed R/T**, as a royalty trust which has no power to invest its income in the development of new royalty-paying assets.

Guar A naturally occurring organic bean which is used in fracking as a thickening agent in order to suspend proppants more evenly in the fracking fluid.

Guide fossil A fossil which is found in a rock layer and is used to correlate the geological age of the layer.

Guided expert process An alternative term for **guided owner approach**.

Guided owner approach In a redetermination, the involvement of an expert at all stages of the process to settle disputes between the parties as they arise. Also called **guided expert process**.

Guilty party pays (GPP) A fault-based liability allocation regime in a petroleum project contract whereby a person assumes responsibility for any loss or damage which that person caused. Also called **bury your own dead**. Contrast with **mutual hold harmless**, a non-fault-based liability allocation regime whereby a person assumes responsibility for any loss or damage which is caused to that person's own property and personnel interests.

Gunk Waste deposits which accumulate in a pipeline until it is pigged.

Gunk slurry A mixture of diesel and bentonite which is used to seal a lost circulation zone.

Gusher An alternative term for a **runaway well**.

Guyed-tower platform An alternative term for a **compliant tower**.

H-gas Natural gas which has a relatively high calorific value. *See also* **L-gas**.

Habendum clause Under a concession, a provision which defines the intended duration of the concession.

Hague-Visby Rules (HVR) A set of international rules that are used for the carriage of goods by sea which are incorporated into the terms of a bill of lading.

Halite Salt deposits which occur naturally within a formation.

Hampering   A ship's inability to manoeuvre.

Handling costs   An alternative term for **marketing costs**.

Handover   The transfer of a petroleum-producing field to the grantor of a concession by the concession-holder at the point when the concession comes to an end.

Handysize   A crude carrier with a deadweight tonnage in the range of 15,000 to 35,000 metric tons.

Hanger   An alternative term for a **casing hanger**.

Hard aground   A ship which has run aground and cannot be refloated under its own power. Also called **hard and fast**. *See also* **grounding**.

Hard and fast   An alternative term for **hard aground**.

Hard arm   An alternative term for a **loading arm**.

Hard carry   A carry which is not repayable by the carried party. Contrast with a **soft carry**, which is repayable by the carried party. Also called a **free carry**.

Hard local content   A local content provision which specifies objective minimum percentage requirements which must be met. *See also* **soft local content**.

Hardship clause   Under a petroleum sales contract, a provision for the review of the pricing of petroleum which is triggered by a party claiming it has suffered hardship because of the relationship between the contract price and prevailing market circumstances. *See also* **adaptation clause**.

Harvesting   The reading and analysis of the recorded data from a seismic survey.

Hazard and operability (HAZOP)   A structured examination of a process or operation which identifies and evaluates problems which could present operational risks.

Head   A measure of the height of a column of liquid which is required to produce a specific pressure.

Header   A pipe into which several smaller pipes feed in, or from which several smaller pipes feed out.

Header tank   A liquid petroleum storage tank which maintains gravity pressure in the system, acting as an expansion tank.

Heading   The intermittent flow of crude oil through a wellbore, which typically results from a lack of capillary pressure.

Headroom
(1) For a petroleum company, a measure of the positive difference between its financial resources and its required financial commitments.
(2) The positive productive difference between the nameplate production or processing capacity and the effective production or processing capacity of a petroleum production or processing infrastructure item. Also called **free barrels**.

Heads-up   A colloquial term for the point in the lifecycle of a JOA at which all the parties contribute according to their agreed participating interests (which could be from the inception of the JOA or after a carry period has ended). Also called **back-in** and **straight-up**.

Headstation   The principal point of delivery of petroleum into a pipeline, prior to the transportation of that petroleum through the pipeline.

Health, safety, and environment (HSE)   Defined operational principles in a petroleum project which are focused on the maintenance of an occupational health and safety regime and the safe preservation of the environment. Also called **safety, health, and environment**.

Heat content   An alternative term for **calorific value**.

Heat content pricing   A model for the pricing of gas or LNG by reference to the calorific value of that gas or LNG.

Heat exchanger (HX)   A device in which heat is transferred between two or more separated fluids for fluid cooling and fluid heating processes.

Heat rate   A measure of the efficiency of an electric power-generating facility in converting gas to electric power, whereby 100 per cent of efficiency would imply equality of gas input and electric power output.

Heat recovery   The process of saving waste heat from an industrial process for application to other purposes. *See also* **scavenged heat**.

Heating oils   Fuel oil which is used for heating homes and buildings. *See also* **burner fuel oil**.

Heave   The upwards or downwards motion of a ship or an offshore drilling unit which is caused by sea swells. *See also* **compensator**.

Heavy distillates   Liquids such as residual fuel oil and fuel oil which are produced in a refinery.

Heavy fuel oil (HFO)   A particular form of fuel oil which has particularly high viscosity and density.

Heavy Louisiana Sweet (HLS)   A crude oil blend from the United States.

Heavy oil   Crude oil which has an API gravity in the range of 10° to 22.3°. *See* Appendix 6.

Heavy oil-cut mud (HOCM)   Oil-cut mud which has a high proportion of entrained oil.

Hectare   A metric unit of land measurement which is equal to 2.471 acres or 10,000 square metres.

Hedge/hedging   A derivatives contract (including a forward, a future, a swap, or an option) which is used to reduce exposure to future petroleum price fluctuation risks.

Heel
(1) A minimum quantity of LNG which is retained in-tank in an LNG carrier after unloading, for temperature stabilisation on the ballast voyage.
(2) An alternative term for **sludge**.

(3) The inclination of a ship or an offshore drilling unit to list to one side because of prevailing weather and/or tidal conditions.

Held-by production (HBP)   Under a concession, an extension to the term of the concession which is earned by a commitment of the concession-holder to pay an ongoing production royalty to the grantor.

Helicopter landing officer (HLO)   A person who is designated to act as the manager of incoming and outgoing helicopter movements on an offshore drilling unit or petroleum production facility.

Helicopter rig   An onshore drilling unit which can be moved between drill sites by helicopter. Also called a **heli-rig**.

Heliops   A shorthand term for helicopter operations.

Heli-rig   An alternative term for a **helicopter rig**.

Hell-or-high water   An alternative term for **all-events**.

Henry Hub   A physical meeting point of various gas pipelines in Louisiana (USA) from which collated trading data is used as a price setter for North American gas markets.

Heritage asset   A petroleum facility in respect of which incurred development costs have been fully amortised over time.

High-angle well   A wellbore which is drilled with an inclination of greater than 80° from the vertical, typically leading to a horizontal well. *See also* **angle of deflection**.

High consequence high potential (HCHP)   A risk factor which has a significant impact and a significant risk of occurrence. *See also* **HCLP, LCHP,** and **LCLP**.

High consequence low potential (HCLP)   A risk factor which has a significant impact and a low risk of occurrence. *See also* **HCHP, HCLP,** and **LCLP**.

High-grade lead   A highly prospective opportunity which has been identified within exploration acreage.

High-integrity pressure protection system (HIPPS)   A pipeline operational process which manages the risks associated with over- and under-pressurisation across a pipeline.

High pressure/high temperature (HP/HT)   A wellbore which has a bottomhole temperature greater than 300°F or requires blowout prevention which is rated greater than 10,000 psi.

High seas sale   A sale of ship-transported petroleum whereby the sale is effected at a defined offshore point. *See also* **offshore title transfer point**.

High set
(1) The part of a formation which is nearest to a surface datum point.
(2) The situation in which the rocks of a particular geological period are nearest to a surface datum point over a broad area.

High-speed diesel (HSD)   Diesel with additives which give the diesel lower viscosity and improved combustibility. *See also* **slow-speed diesel**.

High-speed gas   A marketing slogan for gas production in the United Kingdom, which was latterly applied to natural gas production from the UKCS.

High-stability instrumentation (HSI)   Equipment items which do not require frequent maintenance or calibration.

High-sulphur fuel oil (HSFO)   Fuel oil which is relatively high in sulphur compounds. *See also* **low-sulphur fuel oil.**

Higher heating value (HHV)   An alternative term for **gross calorific value.**

Highest astronomical tide (HAT)   The highest tide level which can be predicted to occur under average meteorological conditions and any particular astronomical conditions. *See also* **lowest astronomical tide.**

Hire   The amounts which are payable by a charterer to a shipowner under a time charter. *See also* **freight.**

Histogram   A chart-based display of statistical information which shows the frequency of data items over successive time intervals.

Historical method   A methodology for calculating a netback royalty which values the petroleum at the wellhead, with the deduction of costs incurred beyond that (such as for processing and transportation) from the wellhead valuation. *See also* **first marketable product.**

Hive-down/hive-up   A colloquial term for the method of defeating the application of a pre-emption right to an interest by the transfer by a disposing person of its interest to a subsidiary affiliate and the sale of the ownership of the affiliate (with the interest) to another person. *See also* **clawback**. Also called a **Texas two-step.**

Holding mode   A period in the loading or unloading of an LNG cargo when no LNG is flowing and cryogenic conditions are maintained in the loading or unloading facilities by circulating LNG.

Holdup
(1)   An alternative term for **fluid holdup.**
(2)   A quantity of petroleum which is retained in the tanks and pipework of a petroleum-production or processing facility.

Hole geometry   A description of the dimensions of a wellbore.

Hook-up   A final phase in the construction of a petroleum facility, as individual components within the facility are connected together.

Hook-up, tie-in, and commissioning (HUTIC)   In the development of a new petroleum facility, the combined activities of hook-up, tie-in, and commissioning.

Horizon   An alternative term for a **reservoir.**

Horizontal permeability   A measure of the permeability of a payzone which is referenced according to the payzone's lateral coordinates. Contrast with **vertical permeability**, whereby permeability is assessed from the bottom to the top of the payzone.

Horizontal well   A wellbore which is drilled to enhance reservoir performance by placing the wellbore section within the horizontal axis of a formation. Also called a **lateral well** and *see also* **high-angle well.**

Horst   A geological structure which has become displaced from its surrounding features by upward thrust. *See also* **graben**.

Host government agreement (HGA)   An agreement which is entered into between an investor and a state regarding the investment protection of a petroleum project.

Hosting   The offer of petroleum-processing or storage services to a person by a petroleum-processing or storage infrastructure item owner.

Hosting agreement   An alternative term for a **processing and operating services agreement**.

Hot oil treatment   The treatment of a producing wellbore with heated oil in order to melt accumulated paraffin deposits.

Hot rig   A drilling unit which has been mobilised and is presently being employed. *See also* **cold rig** and **warm rig**.

Hot tap   Hotwork in which a spur is cut into an existing, operational pipeline. Contrast with a **cold tap**, as coldwork in which a spur is cut.

Hot water recovery   A method of enhanced oil recovery whereby boiling water is injected into a wellbore in order to lower the viscosity of crude oil in a formation.

Hotstacking   The lay-up of a drilling unit with an expectation of near-term use. Contrast with **coldstacking**, in which a lay-up assumes no near-term use.

Hotwork   Any work activity which has the potential to produce ignition sources or excess heat (such as welding, cutting, brazing, or grinding). *See also* **coldwork**.

Hourly peak   Under a petroleum sales contract, the maximum quantity of petroleum which is required by the buyer or is delivered by the seller in any one hour during a twenty-four hour period.

Hub   A defined point at which buyers and sellers execute petroleum sale and purchase transactions (and from which a published price index can be derived).

Hub-and-spoke   Bilateral helicopter movements which take place between a shore base and an offshore platform. Contrast with **bus-stopping**, as sequential helicopter movements between offshore platforms.

Hub potential   A development of a petroleum infrastructure item which has the capacity to be the focal point for a number of other satellite developments. *See also* **tie-back reach**.

Hub price   A published price for traded quantities of petroleum at a hub.

Huff and puff   A colloquial alternative term for **steam flood**.

Hull accord   An agreement by which a charterer undertakes to a shipowner to commit to the long-term hire of an as-then unbuilt newbuild.

Hull and machinery   A measure of insurance cover which applies in respect of a ship against physical damage which is caused by a peril of the sea or other covered perils while the ship is in transit.

Hull formula   A requirement in the payment of compensation for an expropriation that such compensation should be prompt, adequate, and effective.

Human capital (HC)   A composite term for employees and contractor personnel who are utilised in petroleum project operations.

Hunting licence   An alternative term for a **reconnaissance permit**.

Hurdle rate   An alternative term for **expected value**.

Hybrid operator   Under a JOA, an incorporated operator or a split operator.

Hybrid platform   A gravity-based structure with a concrete jacket and steel topsides.

Hybrid provision   A model for the design of a concession whereby a brief framework law is put in place which is then supplemented by individually negotiated concessions. *See also* **detailed content** and **individual legislation**.

Hydrate inhibitor   A chemical solution which is used to prevent the formation of hydrates in a gas pipeline.

Hydrate inhibitor pipeline (HIP)   A pipeline which is used to transport a hydrate inhibitor to an injection point into a gas pipeline.

Hydrates
(1)   Water crystals which form on the inner lining of a gas pipeline. *See also* **hydrate inhibitor**.
(2)   Methane deposits which are trapped within ice-like crystalline water structures. Also called **gas hydrate** and **methane clathrate**.

Hydraulic fracture stimulation (HFS)   An alternative term for **fracking**.

Hydraulic fracturing   An alternative term for **fracking**.

Hydraulic power unit (HPU)   A power unit which generates hydraulic energy for use in an operational process.

Hydraulic submersible pump (HSP)   A hydraulically operated pump which is used downhole for artificial lift. *See also* **electric submersible pump**.

Hydrocarbon   An alternative term for **petroleum**.

Hydrocarbon value realisation (HVR)   An allocation and accounting methodology for produced petroleum.

Hydrocarbon water contact (HCWC)   The lowest level within a formation at which mobile petroleum is found. *See also* **fluid contact** and **lowest known petroleum**.

Hydrocracking   A process for the production of aviation fuel, diesel, and LPG in a refinery.

Hydrocyclone
(1)   An alternative term for a **desander**.
(2)   A filtration vessel which removes water from crude oil.

Hydrofining   A process by which unfinished products are treated with hydrogen and certain catalysts to improve their qualities in a refinery.

Hydroforming   A process by which high octane aviation gasoline is produced from aromatics in a refinery.

Hydrofracking   An alternative term for **fracking**.

Hydrometer   An instrument which is used to measure the specific gravity of a liquid.

Hydrophone   A device which is trailed behind a ship as part of an offshore seismic survey array.

Hydrophore   A device which is used to obtain samples of water from different depths.

Hydroskimming refinery   A topping refinery with the addition of hydrotreating and reforming units.

Hydrostatic testing   An alternative term for **hydrotesting**.

Hydrostatically balanced load (HBL)   The balancing of the pressure of water and crude oil in a crude carrier's tanks in order to reduce the risk of outflow pollution if the crude carrier's hull is breached. *See also* **ballast**.

Hydrotesting   The testing of the integrity of a vessel (like a pipeline) through the injection of pressurised water. Also called **hydrostatic testing** and **water testing**.

Hydrotransport   The addition of hot water to oil sands in order to reduce viscosity and to permit transportation by pipeline of the oil sands.

Hydrotreating   The use of a catalyst in order to remove sulphur and hydrogen sulphide from initially refined petroleum products in a refinery.

Hygas   An alternative term for **synthetic natural gas**.

Hyperbaric chamber   A pressurised vessel with variable atmospheric pressures which is used for offshore diver decompression. Also called a **pressure habitat**.

Hyperbaric support vessel (HSV)   A ship which is used to support diving operations.

I

ICC FM clause   A model form force majeure provision which was published by the ICC in 2003 (https://iccwbo.org/publication/icc-force-majeure-clause-2003icc-hardship-clause-2003/).

ICC hardship clause   A model form hardship provision which was published by the ICC in 2003 (https://iccwbo.org/publication/icc-force-majeure-clause-2003icc-hardship-clause-2003/).

Ice class   A ship or offshore petroleum infrastructure item which is certified to be capable of navigating through ice-bound seas.

Ice scour   Abrasive contact which is caused by moving ice formations.

Ideal gas   A hypothetical gas which demonstrates the application of the Gas Laws. Contrast with a **real gas**, which less obviously demonstrates that application.

Idle capacity   A petroleum facility which is neither in operation nor under active repair but which is capable of being placed in operation (usually within thirty days) or which is not in operation but is under active repair and the repair can be completed (usually within ninety days).

IGC Code (IGC)   The International Code for the Construction and Equipment of Ships Carrying Liquefied Gases in Bulk: promulgated by the IMO, an international operational safety standard for ships which carry bulk liquefied gases and certain other substances.

IGF Code (IGF)   The International Code of Safety for Ship Using Gases or Other Low-Flashpoint Fuels: promulgated by the IMO, an international operational safety standard for ships, other than for ships which are covered by the IGC Code, which operate with gas or low-flashpoint liquids as fuel.

Igneous rock   One of the three main rock types (with metamorphic rock and sedimentary rock) which is formed by the solidification of lava (magma) from the Earth's core. *See also* **basalt** and **gabbro**.

Illustrative agreement   A contractual arrangement by which a concession is held by a concession-holder but the costs of the concession, and the revenues which result from the concession, are met by and accrue to an affiliate of that concession-holder.

Imaging   An alternative term for a **seismic survey**.

Imbrication   In a sedimentary rock formation, the arrangement of pebbles in a flat, overlapping pattern.

Impact and benefits agreement (IBA)   An alternative term for a **community development agreement**.

Implementation agreement (IA)   A concession which permits the operation of a petroleum facility by the appointed holder of the implementation agreement.

Importer of record (IOR)   A declared or officially recognised buyer of goods which are being imported into a destination state, with the associated tax and duty liability and responsibility to secure the necessary import documentation. *See also* **exporter of record**.

Impoundment   An arrangement for external spill control around a petroleum storage tank. *See also* **double containment tank**, **firewall**, **full containment tank**, and **single containment tank**.

Improved oil recovery (IOR)   An alternative term for **enhanced oil recovery**.

In-field pipeline   A relatively small-diameter pipeline which is used to connect production wellbores to production facilities, and production facilities to trunklines. Also called a **flowline** and a **gathering line**.

In kind   The provision of petroleum as petroleum, rather than as a cash value of the petroleum.

In-kind balancing   A balancing of interests between persons which is made in units of petroleum. *See also* **cash balancing**.

In personam   A legal right which attaches to a specific person, such as the rights of a party under a contract, and is typically enforceable anywhere in relation to that person. *See also* **in rem**.

In place   A description of the totality of petroleum in place in a formation before production begins. *See also* **GIIP, OIIP, STOOIP, VGIP**, and **VOIP**.

In rem   A legal right which attaches to immovable property and is typically enforceable where the property is situated. *See also* **in personam**.

In-situ combustion   An alternative term for **fireflood extraction**.

In-situ recovery   The recovery of ultra-heavy oil and bitumen through wellbores, without strip mining to remove overburden.

In-situ sale   An alternative term for an **in-tank sale**.

In-tank sale   A sale of petroleum which takes place at the point of entry to the buyer's storage tank. Contrast with an **ex-tank sale**, as a sale of petroleum which takes place at the point of exit from the seller's storage tank. Also called an **in-situ sale**.

In-the-money (ITM)   Under a petroleum sales contract, an assessment that current price performance and future price expectations will combine to make the contract likely to be profit-generative at the point of assessment. Contrast with **out-of-the-money**, whereby the assessment indicates that the contract is likely to be loss-generative.

In-water survey (IWS)   An alternative term for **underwater inspection in lieu of dry-docking**.

Inactive well   A wellbore which has not produced petroleum, typically in respect of a preceding twelve-month period. Contrast with an **active well**, as a wellbore which is currently producing petroleum.

Inboard block   A block which lies in the landward direction from another block. *See also* **outboard block**.

Inclination   An alternative term for **drift**.

Inclinometer   A device which measures the degree of drift in the drilling of a wellbore.

Incompetent formation   A formation which is composed of loose, uncompacted materials.

Incorporated joint venture   An alternative term for an **equity joint venture**.

Incorporated operator   Under a JOA, the appointment of an incorporated entity to act as the operator.

Incoterms   A series of standard terms which apply to regulate the sale and transportation of various commodities (including petroleum), first published by the International Chamber of Commerce (ICC) in 1923 and most recently issued in 2020 (http://www.iccwbo.org).

Independent power project (IPP)   A privately owned entity which owns and operates a facility which is used to generate electric power.

Independent system operator (ISO)   An operator of a petroleum transportation infrastructure item with management which is functionally separated from the affiliated activities of petroleum production or trading. *See also* **independent transmission operator**.

Independent transmission operator (ITO)   An operator of a petroleum transportation infrastructure item which is an entity legally separated from the affiliated activities of petroleum production or trading. *See also* **independent system operator**.

Indexation   The reference of a price for petroleum to movements over time in the value of a defined index. Also called **linkage**. *See also* **weighting**.

Index/hub switch   Under a petroleum sales contract, a provision whereby the method of determining the price migrates from reliance on indexation to reliance on a hub price.

Indicated throughput   The difference between an opening and a closing meter reading, which indicates a metered throughput during a metered phase.

Indicator margin   An alternative term for a **crack spread**.

Indirect charges   Under an accounting procedure, reimbursable costs which relate to general intangible support services which are provided by the operator in the performance of the joint operations. Also called **general & administrative overhead**. Contrast with **direct charges**, which cover reimbursable costs which are directly incurred by the operator.

Indirect energy   A measure of the fuel which is used in the preparation of intermediate inputs to a petroleum-processing activity. Contrast with **direct energy**, which is used in the activity. *See also* **gross energy requirement**.

Indirect loss   An alternative term for **consequential loss**.

Individual legislation   A model for the design of a concession whereby the key provisions are recited in an individually negotiated contract and are then given force of law by legislative ratification. *See also* **detailed content** and **hybrid provision**.

Indivisible interests   A series of contractual obligations which cannot be separated for differentiated performance.

Induced seismicity   Seismic or tectonic activity which is attributable to human activities.

Industrial and cultural noise (I&CN)   Interruptions to a petroleum process which are caused by local environmental issues.

Industry-led solution (ILS)   A solution for the production of petroleum which is based on existing industry-accepted technology rather than on new or untried technology.

Industry mutual hold harmless (IMHH)   An inter-contractor liability allocation mechanism (based on mutual hold harmless) which is managed by OGUK and applies to contractors working on the UKCS (http://www.logic-oil.com).

Inert gas system (IGS)   A system in a petroleum infrastructure item for the flooding of spaces with an inert gas and the reduction of oxygen content as a means of fire suppression. *See also* **closed space**.

Inerting   An alternative term for **padding**.

Inertinite   A form of kerogen which is generally not prospective for the presence of petroleum. *See also* **vitrinite**.

Inerts   Chemically unreactive gases (such as nitrogen) which are blended into a gas stream to reduce high calorific values. *See also* **nitrogen injection**.

Infidelity clause   A colloquial term for a provision in a petroleum sales contract whereby the level of the buyer's take or pay commitment is increased if the buyer elects to buy additional quantities of petroleum from another seller.

Infill   A wellbore which is drilled in addition to a principal producing wellbore in order to increase the rate of petroleum production from an underlying formation. Also called a **child well**.

Inflatable packer   A flexible packer which can be inflated by gas or liquid from the surface, which is often deployed as a temporary measure in order to strengthen weakened casing in a wellbore.

Inflexion point   An alternative term for a **kink point**.

Inflow control valve (ICV)   A downhole valve which partially or completely restricts the flow of water or petroleum into a wellbore.

Information memorandum (IM)   A relatively detailed documentary summary of the key commercial elements of an asset, project, or business which is intended to be sold which is prepared by the proposed seller, often following on from the issue of a teaser. Also called a **confidential information memorandum**.

Infrastructure code of practice (ICOP)   A regulatory framework by which petroleum infrastructure item owners will declare the availability of spare capacity for third-party access.

Infrastructure-led exploration (ILX/ILEx)   Petroleum exploration which is focused in frontier areas with existing midstream infrastructure in order to reduce petroleum project development costs. *See also* **basin-opening exploration** and **tie-back explorer**.

Ingress   Under a drilling contract, the right of the drilling contractor to bring a drilling unit onto a drill site. Contrast with **egress**, as the right of the drilling contractor to take the drilling unit away from a drill site.

Inherently safer design (ISD)   An equipment design methodology which reduces and avoids hazards to the greatest possible extent, recognising that perfect design with the total elimination of hazards might be unachievable.

Inhibited mud   Drilling fluid which contains an inhibitor, which is intended to prevent both water loss into a formation and the build-up of mud cake.

Inhibitor   A chemical additive which is used to retard a potential and undesirable chemical reaction in a petroleum product or process.

Initial boiling point (IBP)   The temperature at which the first drop of a distillation product is obtained.

Initial fill   An alternative term for **linefill**.

Initial potential   A measure of a wellbore's early production levels which can be used to indicate the future maximum production potential of the wellbore.

Initial production rate (IPR)   The rate at which petroleum flows freely from a wellbore over a defined period when petroleum is first produced.

Initial reservoir pressure   The point of pressure that must exist within a formation before petroleum production from the formation begins.

Injectability   In a gas storage facility, the total quantity of working gas which can be injected into the storage facility at a given injection rate and in respect of a defined period of time (which is measured in volume or heat rate) for a facility user. *See also* **deliverability**.

Injection   The function of putting gas into a gas storage facility. Contrast with **redelivery**, as the withdrawal of gas from a storage facility.

Injection string   An alternative term for a **kill string**.

Injection water   Water which is injected into a wellbore through an injection well.

Injection well   A wellbore which is drilled into a formation for gas or liquids injection in order to maintain capillary pressure in the formation. Also called an **input well**. *See also* **make-up gas** and **recycling**.

Inland barge rig   An alternative term for a **swamp barge**.

Innage
(1)   A measure of the quantity of petroleum in a storage tank, which is measured from a datum point at the bottom of the tank to the surface of the petroleum.
(2)   A measure of the space which is occupied in a product container. Contrast with **outage**, as a measure of unoccupied space.

Input well   An alternative term for an **injection well**.

Inside battery limit (ISBL)   Any operational activities which take place inside a battery limit. Contrast with **outside battery limit,** whereby operational activities take place outside the battery limit.

Inside the fence   A measure of an oilfield service company's equity participation in a petroleum project which it provides services to.

Insourcing   The sourcing of resources for the performance of a petroleum project by a petroleum company from resources which exist within the company. *See also* **outsourcing**.

Inspection, maintenance, and repair (IMR)   A process for maintaining and repairing petroleum project facilities and equipment.

Installation, hook-up, and commissioning (IHUC)   A phase in the construction of a petroleum facility at which the facility is installed and is then hooked-up and commissioned.

Institute Warranty Limits (IWL)   Under a charterparty, defined geographical limits within which a ship may safely trade without incurring additional insurance costs. These were replaced in 2003 by the International Navigating Limits but are still frequently referred to. *See also* **trading limits**.

Institutional arbitration   An arbitration which is conducted according to the procedural rules of a particular arbitral body. Contrast with an **ad hoc arbitration**, as an arbitration which is conducted without applying the procedural rules of a particular arbitral body. Also called an **administered arbitration**.

Intangible drilling costs   In drilling, costs which do not have a recoverable salvage value associated with them (such as ground survey costs or the costs of consumables). Contrast with **tangible drilling costs**, which relate to identifiable assets or interests.

Intangibility provision   A stabilisation provision which guarantees the sanctity of the investor's interests in a petroleum project by a promise of the state not to renegotiate the terms of the project unilaterally.

Integrated company   A petroleum company which is actively represented in each of the upstream, midstream, and downstream sectors. *See also* **vertical integration**.

Integrated control and safety system (ICSS)   An operational methodology which combines process controls and functional safely within a single discipline.

Integrated energy market (IEM)   A multi-state or regional market in which energy flows between countries and sector participants without legal, commercial, or technical barriers.

Integrated gasification combined cycle (IGCC)   The generation of electric power from coal by coal gasification.

Integrated manufacturing   The shared use of common manufacturing plants and/or process elements by different companies in the manufacture of petrochemicals.

Integrated operations (IO)   The collaboration of people, processes, information, and communications in order to increase rates of petroleum production.

Integrated project management team (IPMT)   Under a JOA, a team of personnel which is drawn from representatives of all the JOA parties in order to undertake a joint operation.

Integrity pact   A commitment which is made between the parties to a contract that they have not breached any relevant anti-bribery and corruption provisions prior to entry into the contract.

Intellectual property rights (IPR)   The ascribed ownership of the collection of registered and unregistered intellectual property which is generated from a petroleum project activity.

Intelligent completion   A completion of an intelligent well. Also called a **smart completion**.

Intelligent well   A wellbore which is equipped with downhole monitoring and management equipment in order to allow ongoing adjustments for the optimisation of petroleum production. Also called a **smart well**.

Interbedded sand   A sub-surface sand deposit which lies in a defined layer between other sedimentary layers.

Interfacial tension   The surface tension which occurs at the interface between two unmixed liquids.

Inter-fuel substitution   The ability of a petroleum facility to use alternative forms of fuel interchangeably. *See also* **switching**.

Inter-government agreement (IGA)   An agreement which is made between two or more states regarding the investment protection of a petroleum project. *See also* **multilateral investment treaty**.

Interburden   Rock and soil deposits which lie between formations at different stratigraphic depths. *See also* **overburden**.

Interconnector   A pipeline which is used to transport petroleum between two defined points, in which the flow of petroleum in the pipeline can be reversed either way.

Intercontinental Exchange (ICE)   An electronic trading platform and collection of clearing houses which have a focus on petroleum products trading (http://www.intercontinentalexchange.com).

Interface
(1)   An alternative term for **cutterstock**.
(2)   The contact point between the boundaries of two unmixed liquids (such as oil and water).

Interim period   Under an asset or share sale and purchase agreement, the period of time which elapses between executing and completing the agreement.

Intermediate fuel oil (IFO)   A particular form of marine gas oil which has a relatively high proportion of heavy fuel oil.

Intermittency   Periods of time when renewable energy sources are not capable of generating electric power (and when non-renewable energy sources could be required to make good the deficiency).

Intermodal transport   An overall method of petroleum transportation which uses different transportation forms. *See also* **modal share**.

Internal diameter (ID)   The internal diameter of a pipeline section. *See also* **external diameter**.

Internal due diligence (IDD)   A person's internal assurance process which is conducted in respect of a petroleum contract or project to which that person is party. *See also* **assurance**.

Internal gas consumption (IGC)   Lost gas in, and fuel gas which is used in respect of, a gas pipeline.

Internal market   A market for petroleum products which is composed of domestic production and imports which are consumed in-country. *See also* **external
market**.

Internal rate of return (IRR)   A method of calculating the level of return that is
made by an investor on its investment into a petroleum project which is based on
financial returns referenced against time.

International Association of Classification Societies (IACS)   An international
organisation which represents the interests of classification societies (http://www.
iacs.org.uk).

International Association of Drilling Contractors (IADC)   A US-based trade
organisation which monitors issues associated with onshore and offshore drilling
(http://www.iadc.org).

International Association of Oil and Gas Producers (IOGP)   An international
trade organisation which represents the interests of upstream industry participants
(http://www.iogp.org).

International Centre for Settlement of Investment Disputes (ICSID)   An independent arbitral body which provides a forum for resolving energy project investment protection disputes (http://icsid.worldbank.org/en/).

International Chamber of Commerce (ICC)   An international organisation
which is committed to promoting common trade standards and regulation (https://
iccwbo.org/).

International Commission on Stratigraphy (ICS)   An international scientific
body which defines the standards that are used for the determination of geologic
time (www.stratigraphy.org). *See* Appendix 4.

**International Convention for the Prevention of Pollution from Ships
(MARPOL)**   The principal international convention covering the prevention of
pollution of the marine environment by ships from operational or accidental causes
(http://www.imo.org).

International Energy Agency (IEA)   A member state-based international organisation which monitors the full spectrum of international energy issues (http://
www.iea.org).

International Gas Union (IGU)   An international trade organisation which promotes the political, technical, and economic progress of the international gas industry (http://www.igu.org).

International Maritime Organisation (IMO)   An agency of the United Nations
which monitors the safety and security of shipping and the prevention of marine
pollution (http://www.imo.org).

International national oil company (INOC)   A national oil company (NOC)
which has at least some of the characteristics of an international oil company (IOC).

International Navigating Limits (INL)   Since 2003, a replacement for the
Institute Warranty Limits as trading limits for a ship.

International oil company (IOC)   An internationally owned/focused oil company. *See also* **national oil company**.

International Organization for Standardization (ISO)   An international body which is responsible for the promotion of common technical standards (http://iso.org).

International Petroleum Industry Environmental Conservation Association (IPIECA)   An international advisory agency which advises the petroleum industry on social and environmental issues (http://www.ipieca.org).

International safety management certificate (ISM)   A certificate which is issued to confirm the compliance of a ship with the ISM Code.

International Ship and Port Facility Security Code (ISPS)   An international code which is issued by the IMO and defines minimum security levels which will apply between ships and loading ports and unloading ports.

International Standby Practices 98 (ISP 98)   A set of standard rules and best practices which is published by the International Chamber of Commerce (ICC) in 1998 and relating to the use of standby letters of credit.

International Swaps and Derivatives Association (ISDA)   An international trade organisation which monitors issues associated with over-the-counter derivatives (http://www2.isda.org).

International Tanker Owners Pollution Federation (ITOPF)   An international non-profit marine pollution advisory group, which was previously responsible for the administration of TOVALOP (http://www.itopf.com).

Interoperability   A mechanism which is intended to facilitate the interconnection of different systems and features within an overall network.

Interpellation   Under a charterparty, a provision which is designed to relieve the shipowner of the obligation to proceed on a potentially long positioning voyage to a port in the face of uncertainty as to whether the charterer will elect to cancel the charter if the ship fails to arrive by a certain date.

Interruptible
(1)   Under a petroleum sales contract, the right of the seller to interrupt the sale of petroleum to the buyer, or of the buyer to interrupt the delivery of petroleum from the seller, in accordance with defined parameters. *See also* **storage interruptible** and **strategic interruptible**.
(2)   Under a pipeline transportation agreement, the right of the transporter to interrupt the provision of transportation services to the shipper, or of the shipper to interrupt the provision of the transportation services by the transporter, in accordance with defined parameters.

Contrast with **firm rights**, which are not ordinarily interruptible.

Interruptible storage service (ISS)   A gas storage service which can be interrupted (within defined limits) by the storage facility owner. Contrast with **firm storage service**, which is not ordinarily interruptible by the storage facility owner.

Interstice   A pore space within a formation.

Interstitial gas   An alternative term for **pore gas**.

Interstitial water   An alternative term for **connate**.

Intertanko   An international organisation which represents the interests of the maritime oil, gas, and chemical products shipping industry (http://www.intertanko.com).

Interval
(1)   An alternative term for a **formation**.
(2)   A stage in running casing into a wellbore.

Intervention   An alternative term for **well intervention**.

Inter-well   The movement of a drilling unit between identified drill sites.

Intra-day delivery   Traded petroleum which is intended for delivery on the same day as the day upon which the trade is made.

Intrinsic value
(1)   The valuation of a gas storage facility which is based on seasonal storage opportunities and forward gas price differentials over longer term spreads (without taking account of extrinsic value).
(2)   Under a petroleum sales contract, the inherent value of the contract which can be used to hedge against future market price movements. *See also* **extrinsic value**.

Invaded zone   The part of a formation within which an invasion has occurred.

Invasion   The undesirable penetration of drilling fluid into a formation from a wellbore. *See also* **blind drilling, leak-off point, leak-off rate, seal off**, and **skin damage**.

Inventory
(1)   A quantity of stored petroleum.
(2)   A record of stored materials.
(3)   Under an accounting procedure, the operator's periodic audit and assessment of joint property at any time.

Inventory conflict   The risk that a particular shipper's ship-transported cargo of petroleum cannot be loaded or unloaded when scheduled because of a conflict with the demands of other shippers or because of storage facility constraints.

Inventory conflict rules   An agreed set of multi-shipper rules which establish the priorities which are necessary to manage an actual or an anticipated inventory conflict.

Inventory financing   A loan which is secured in the lender's favour over certain petroleum stocks in store which the borrower owns.

Inversion   A process by which seismic data is transformed in order to give a quantitative description of a formation.

Inversion data   Seismic data which is reprocessed in order to better assess the petroleum prospectivity of a formation.

Inverted market   An alternative term for **backwardation**.

Investment multiple   In respect of a petroleum project, the ratio of petroleum project revenues to petroleum project costs (where a greater ratio indicates a more profitable project).

Invitation to tender (ITT)   As part of a procurement exercise, a formal invitation to potentially interested persons to bid for the provision of goods or services to a petroleum project.

Irish Balancing Point (IBP)   A notional point on the Irish gas pipeline network at which shippers trade gas quantities.

Irreversibility   An investment principle whereby a petroleum project investment decision which has been made, or the installation of petroleum facilities which has been commenced or completed, cannot later be undone.

ISM Code   An international standard for the safe operation of ships and for the prevention of maritime pollution, which is created by the IMO.

Isobath   A sea-bed contour line on a map which connects together all points which have the same level of depth below the sea's surface.

Isochore   A contour line on a thickness map that joins up the points of a stratum which have the same vertical thickness.

Isochore map   A geological tool which indicates the isochoric contours of a stratum.

Isocontainer   A transportable vessel which contains refrigerated LNG or pressurised CNG or LPG.

Isolation philosophy   The design of a petroleum-processing system such that an individual equipment item can be removed for maintenance, repair, or calibration with no or minimal interruption to operational integrity.

Isolithic map   A map of a formation upon which points of similar lithology are indicated by a connected series of contours.

Isomorphic mimicry   A failed form of regulatory reform whereby a regulatory structure is replicated but without the successful replication of the underlying functionality of that structure.

Isopach   A contour line on a thickness map that joins up the points in a stratum which have the same true thickness.

Isopach map   A geological tool which indicates the isopachic contours of a stratum.

J

J-lay   A method of offshore pipeline installation, so-called because of the pipeline's J-shaped profile as it comes off the pipelay vessel. *See also* reel pipelay and **S-lay**.

J-tube   A method for the connection of a sub-sea pipeline to a marine riser, whereby the pipeline assumes a J-shaped configuration.

Jack board   A device which supports the end of a length of drill pipe while another length is being joined to it.

Jack rabbit   A device which is run through a tubular in order to check sizing and freedom from impediments prior to use.

Jack-knife rig   A drilling unit which has a folding rather than a fixed derrick.

Jack-up   A drilling unit which is characterised by support legs which sit on the sea-bed and against which the drilling unit is jacked up, capable of drilling in water depths typically up to 600 feet. Also called a **self-elevating drilling unit**. *See* Appendix 2B. *See also* **cantilevered jack-up** and **slot-type jack-up**.

Jack-up barge (J-UB)   A barge-based jack-up which is used for shallow-water drilling operations.

Jacket   The supporting base (made of concrete or steel) for a topside on an off-shore platform. Also called a **platform jacket**.

Japan Customs-cleared Crude (JCC)   An LNG pricing index which is based on the reported values for a basket of crude oil import prices into Japan which are published each month by the Japanese government (http://www.paj.gr.jp). Also called **Japanese Crude Cocktail**.

Japan Korea Marker (JKM)   An LNG benchmark price assessment which is used for spot physical cargoes DES into Japan and South Korea (http://www.platts.com/price-assessments/natural-gas/jkm-japan-korea-marker).

Japanese Crude Cocktail   An alternative term for **Japan Customs-cleared Crude**.

Jar   A percussion tool which is used downhole to deliver a heavy downward blow to a fish.

Jet A   An alternative term for **aviation turbine kerosene**.

Jet B   An alternative term for **aviation turbine gasoline**.

Jet cutoff   A procedure for severing a stuck pipe in a wellbore by the downhole detonation of a ring of explosive charges.

Jet cutter   A tool which is used downhole to effect a jet cutoff.

Jet perforation   Perforation which is effected by the downhole detonation of a ring of explosive charges rather than by the use of a perforating gun.

Jet sledging   A method of offshore pipeline laying from a bury barge whereby high-pressure hoses are used to scour a trench for the pipeline to sit in.

Jetsam   Items which are deliberately jettisoned from a ship. Contrast with **flotsam**, as items which are accidentally lost from a ship.

Jetting   An alternative term for **gas lift**.

Jettison and washed overboard (J&WO)   The loss of a cargo from a ship through deliberate (jettison) or accidental (washed overboard) intervention.

Jetty   A land-based pier which extends into the sea for ships to berth at for the loading or for the unloading of petroleum.

Jobber   A wholesale distributor of gasoline (and other products) to retail outlets.

Joinder   An alternative term for **consolidation**.

Joint   An alternative term for a **single**.

Joint account   A bank account which is maintained by the operator under a JOA and through which debits and credits associated with the conduct of the joint operations pass.

Joint and several (J&S)   The liability of a group of persons which, as between them, is joint and several. *See also* **joint liability** and **several liability**.

Joint application agreement   An alternative term for a **joint study and bidding agreement**.

Joint development zone (JDZ)   A petroleum-bearing area which overlaps disputed international boundaries and is agreed to be developed jointly between the interested states for their mutual benefit (but without applying the principles of unitisation and often without the relevant states conceding sovereignty over the disputed boundary). Also called an **exclusive economic zone**.

Joint industry project (JIP)   A research and development project which is funded by collaborative effort between a number of persons.

Joint interest billings (JIB)   The charges which are made to the parties by the operator under a JOA in order to recoup the costs of the joint operations.

Joint liability   The liability of a group of persons which, as between them, is joint to another person (such that any one person within that group could assume the entire liability). *See also* **several liability**, which creates individual liabilities between the group members to another person.

Joint lifting   A lifting which is made by a number of persons acting in unison as a single lifting, rather than through a series of individual liftings.

Joint movement   The shipment of a quantity of petroleum by a shipper through the facilities of more than one person.

Joint oil data initiative (JODI)   A global initiative which is intended to promote greater transparency in relation to oil (and latterly gas) market data (https://www.jodidata.org/).

Joint operating agreement (JOA)   An agreement which apportions the rights, benefits, obligations, and liabilities that are derived from a concession between multiple parties in proportion to their defined respective percentage interests, and which typically appoints one of those parties to act as the operator.

Joint operating body   An alternative term for a **joint operating committee**.

Joint operating committee   A committee which is representative of the interests of the state and the concession-holders in the operation of a concession. Also called a **joint operating body**.

Joint operation   A petroleum exploration and/or production operation which is carried out by all of the parties to a JOA. Contrast with an **exclusive operation**, as an operation which is carried out by less than all of the parties to a JOA.

Joint production   The simultaneous production of crude oil and natural gas from a formation (but not necessarily in fixed proportions).

Joint property   Property which is acquired or generated by, and belongs collectively to, the parties to a JOA.

Joint study and bidding agreement (JSBA)   A contract which regulates the relationship between a group of persons who apply jointly for the award of a concession and that acts as a precursor to a fully termed JOA coming into place between those persons. Also called a **joint application agreement**. *See also* **pairing agreement**.

Joint tariff   A rate sheet which is issued to a shipper by the facility owners which are associated with a joint movement.

Joint venture company (JVCo)   In relation to an incorporated joint venture, the project company which is incorporated and owned by the shareholders in order to develop a petroleum project.

Joint well   A wellbore which is drilled as a joint operation.

Joule   A unit of energy. *See* Appendix 1.

Joule Thompson low temperature separator (JLTLS)   An alternative term for a JT valve.

JT valve   A valve in a gas pipeline which forces a flow of gas into a temperature drop zone and allows the resultant condensation of liquids for dropout and collection (through a process known as **throttling**). Also called a **Joule Thompson low temperature separator**.

Judicial restraint   A legal doctrine which states that the courts of a state will decline to exercise jurisdiction over a contractual dispute which relates to the sovereign acts of that state.

Jug   An alternative term for a **geophone**.

Jug hustler   A member of a seismic survey crew who has responsibility for the management of geophones.

Jumbo burner   An alternative term for a **flarestack**.

Jumper   A short length of rigid or flexible pipeline which is used to connect flowlines, surface facilities, and sub-sea facilities. *See also* **SURF**.

Junk   Items and debris which are lost downhole.

Junk basket   A device which is used to retrieve junk from a wellbore. Also called a **junk retriever**.

Junk retriever   An alternative term for a **junk basket**.

Junk shot   An explosive charge which is detonated in a wellbore in order to break up junk prior to retrieval.

Junked and abandoned (J&A)   A wellbore which has been plugged and abandoned after tools have been lost in it and have not been retrieved.

Justified for development (JFD)   The determination that a petroleum discovery is commercially and technically ready to become a petroleum development project.

K

Katagenesis The second stage of the sub-surface metamorphosis of organic matter to form petroleum, representing significant changes at which kerogens begin to form petroleum (principally as oil) at increased temperatures and pressures. *See also* **diagenesis** and **metagenesis**. *See* Appendix 2A.

Keeper An exploration well which is intended for further development and completion.

Keepwhole contract A contract for the processing of gas whereby the facility owner returns dry gas to the facility user but keeps any extracted liquids as security for payment of the processing fee by the facility user.

Kelly A square or hexagonal steel tube which is connected between a rotary table and the uppermost section of drill pipe in a drill string. Also called a **grief stem**.

Kelly bushing (KB) A device placed around a kelly, which connects the kelly with a rotary table.

Kelly bypass A system of valves which permit the circulation of drilling fluid through a wellbore without the use of a kelly.

Kelly drilling An alternative term for **rotary drilling**.

Kerogen Organic material fused in sedimentary rocks which, over time and when subjected to the right combination of temperature and pressure, will convert into petroleum. *See* Appendix 2A.

Kerosene A liquid fuel which is widely used as an aviation fuel, similar in composition to paraffin. *See also* **aviation turbine kerosene**.

Key performance indicator (KPI) A measurable objective which is used to determine progress towards the fulfilment of a specific goal.

Keyseat barge A barge from which drilling operations are performed through a keyway.

Keyway A slot which is cut into the hull of a barge, through which a drilling unit is mounted and from which a wellbore is drilled.

Keyway jack-up An alternative term for a **slot-type jack-up**.

Kick An increase in the pressure of the petroleum or water in a formation which exceeds the downward pressure of drilling fluid in the wellbore, which leads to the uncontrolled escape of petroleum or water upwards from the formation and through the wellbore. Also called a **blowout**. *See also* **runaway well**.

Kick fluids Fluids which enter a wellbore as part of a kick.

Kick off
(1) The intentional deviation of a wellbore.
(2) The bringing of a wellbore into production.

Kicker A bonus payment which is paid upon the occurrence of a defined event.

Kickoff point   The point in the drilling of an intended horizontal well at which the wellbore begins to deviate from the vertical. *See also* **angle of deflection**.

Kill   The injection of drilling fluid into a wellbore in order to achieve overbalance and to prevent or to control a kick. Also called a **well kill**.

Kill line   A high-pressure pipeline which leads into a wellbore, through which drilling fluid is pumped in order to control a threatened blowout.

Kill rate   The velocity (which is expressed as a unit of volume per unit of time) of the drilling fluid which is pumped into a wellbore as part of a kill.

Kill string   Small-diameter tubing which is used inside production tubing for the continuous injection of chemicals (such as corrosion inhibitors) downhole. Also called an **injection string**.

Killer well   An alternative term for a **relief well**.

Kilojoule (KJ)   One thousand Joules.

King wire   The central core of a stranded wire rope.

Kink point   A defined point on an S-curve at which a level of petroleum pricing will apply. Also called an **inflexion point** and a **pivot point**.

Kitchen   Source rock in which the conditions of content, temperature, and pressure have combined to create petroleum. Also called **organic-rich rock** and **thermally mature rock**. *See* Appendix 2A.

Knock   The undesirable tendency of gasoline to ignite spontaneously. *See also* **anti-knock rating**.

Knock-for-knock (K4K)   An alternative term for **mutual hold harmless**.

Knocked down (KD)   An item of equipment which is supplied in a completely unassembled state and requires assembly to be fully functional. *See also* **partly knocked down**.

Knock-out   The removal and separation of oil, water, and impurities from a gas stream by a scrubber.

Knock-out drum   A device whereby suspended liquids are separated from a gas stream.

Known accumulation   An alternative term for a **discovery**.

L

L-gas   Natural gas which has a relatively low calorific value. *See also* **H-gas**.

Laden voyage   The cargo-laden leg of a ship's voyage. Contrast with a **ballast voyage**, as the empty return voyage of a ship after its cargo has been unloaded. Also called a **loaded leg**.

Lady days   A set of dates which mark the start of each quarter of a calendar year: 25 March, 24 June, 29 September, and 25 December. *See also* **quarter days**.

Lag time   The time which is taken for cuttings to travel the distance in a wellbore from the drill bit to the surface.

Lagging indicator   An economic prediction tool which relies on a retrospective assessment of historical data. Contrast with a **leading indicator**, which relies on prospective assessments.

Lagniappe   A colloquial term for an extra quantity of petroleum which is made available by a seller to a buyer as a gesture of goodwill.

Lahee classification   A classification tool which was devised in 1944 and is used widely in North America to categorise the prospects for success for a wellbore before drilling, based on the anticipated levels of risk and return from the wellbore.

Laid-up tonnage   Shipping capacity which has been withdrawn from active service and has been put into lay-up.

Landed price
(1)   The actual delivered cost of crude oil to a refinery (including production and transportation costs and any import tariffs).
(2)   The actual delivered cost of petroleum or products to a defined point (including production and transportation costs and any import tariffs).

Landfill gas   A combination of methane and carbon dioxide which results from aerobic digestion which is often used as a feedstock for distributed generation. *See also* **biogas**.

Landing a wellhead   The activity of attaching wellhead equipment to a wellbore where that equipment was not originally installed at the time of completion of the wellbore.

Landman   A negotiator of petroleum exploration leases with landowners, typically found in Canada and the United States.

Landowner royalty   The entitlement of a person to all or part of the gross production of petroleum from a formation but without incurring the deduction of any of the associated production costs (effectively a gross overriding royalty).

Large range 1 (LR1)   A crude carrier which has a deadweight tonnage in the range of 45,000 to 80,000 metric tons.

Large range 2 (LR2)   A crude carrier which has a deadweight tonnage in the range of 80,000 to 160,000 metric tons.

Last mile   The final element of a gas supply chain, connecting supply to intended customers. In LNG projects this term is used to describe regasification facilities.

Last twelve months (LTM)   The measurement of a company's financial condition which is measured on a lagging indicator of the preceding twelve-month basis. Contrast with **next twelve months**, which is a leading indicator.

Last twelve months EBITDA (LTMEBITDA)   An accounting measure of a company's EBITDA which is calculated over a twelve-month period immediately preceding the date on which the assessment is made.

Last twelve months EBITDA exploration expenses (LTMEBITDAX)   An accounting measure of a company's last twelve months' EBITDA which includes incurred exploration expenses.

Last-in first-out (LIFO)   An agreed order in which an accrued series of interests is used, recouped, or disapplied whereby the least mature are applied first. Contrast with **first-in first-out**, whereby the most mature are applied first. Also called **first-in last-out**.

Last-in last-out (LILO)   An alternative way of phrasing **first-in first-out**.

Late fail   A production well which, after it has gone into commercial production, sees a collapse in petroleum production rates to a level which, if that level had been apparent at the outset of developing the well, would have led to an initial decision not to develop the well.

Late life   A formation or petroleum infrastructure item which is coming towards the end of its productive life.

Late life compression (LLC)   The decision to impose artificial lift processes as late as possible in the life of a formation in order to maintain production pressures (and also to reduce a petroleum project's carbon intensity).

Lateral   A small-diameter pipeline which ties into a trunkline, often as part of a wider gathering system.

Lateral well   An alternative term for a **horizontal well**.

Law of capture   A principle by which a landowner will not be liable to an adjacent landowner for exploiting any quantities of mobile petroleum which have freely migrated to the exploiting landowner's lands from the adjacent landowner's lands. Also called **capture**. *See also* **competitive drilling, correlative rights doctrine, prior appropriation rule, res nullius,** and **robber well**.

Law of one price   An economic theory whereby arbitrage opportunities which exist between different countries are immediately exploited by market participants, leading to rapid convergence to a single market price.

Lay barge construction   A pipeline-laying technique which is used in shallow-water areas, in which the slow forward motion of a barge allows a pipeline to be laid from the stern of the barge.

Laycan   A defined range of dates during which a ship is scheduled to load or to unload a cargo at a port.

Layday   A day within a laycan.

Laydown area   An onshore area at which materials are stored during the construction of a petroleum facility.

Layer of protection analysis (LOPA)   A methodology for determining the safety integrity level which a safety instrumented system is required to provide.

Layering   The natural separation of petroleum fractions in a quantity of petroleum which is stored in-tank, whereby the lighter fractions rise higher up the tank. Also called **stratification**.

Laytime   The period of time during which a ship is berthed at a port. *See also* **allowed laytime**, **running hours**, and **used laytime**.

Lay-up   The temporary or long-term cessation of trading of a drilling unit or a ship which is left in dormancy for the lay-up period.

Leaching   A process by which an underground salt cavern is created and is made ready for use as a gas storage facility. Also called **solution mining**.

Lead   An identified formation which is not yet sufficiently defined to represent a viable drilling target.

Leading indicator   An economic prediction tool which relies on a prospective assessment of anticipated data. Contrast with a **lagging indicator**, which relies on retrospective assessment.

Leading Oil and Gas Industry Competitiveness (LOGIC)   An affiliate of OGUK which is responsible for improving the efficiency of working practices on the UKCS (http://www.logic-oil.com).

Leak test   A test which is used to determine the structural integrity of a formation, which is conducted immediately after a wellbore has been drilled below an existing casing level. Also called a **leak-off test** and a **pressure integrity test**. *See also* **negative pressure test** and **positive pressure test**.

Leak-off test   An alternative term for a **leak test**.

Leak-off point   In a leak test, the point of pressure at which drilling fluid begins to enter the formation from the wellbore. *See also* **invasion**.

Leak-off rate   In a leak test, the rate at which drilling fluid enters the formation from the wellbore after the leak-off point.

Lean gas   Gas which is composed principally of methane and has a relatively lower calorific value. Contrast with **rich gas**, as gas which is composed principally of the heavier petroleum fractions and has a relatively higher calorific value.

Learning needs assessment (LNA)   An assessment of the capacity-building needs of a petroleum-producing state, often as a precursor to the provision of training and capacity-building assistance under the terms of a concession.

Lease and operate (L&O)   A model for the deployment of an FSU, by which a contactor leases the FSU to an operator and also provides ongoing operating (operation) and maintenance services. Also called **production services**.

Lease automatic custody transfer (LACT)   A device which facilitates the accurate measurement of the quantity and the quality of petroleum as it is transferred by automated means from the custody of one person to another (which is typically at the point of entry from a gathering system into a pipeline).

Lease condensates   Natural gas liquids which are recovered from a gas stream in a separator.

Ledge   An irregularity in the diameter of a wellbore which is caused by drilling into alternately soft and hard stratas.

Leg batter   The degree of slope of the legs of a steel jacket, such that the base of the jacket is wider than the base of the topsides. Also called **batter**.

Legacy data   Historical petroleum production data which accompanies an asset when it is sold.

Legacy local content   Local content investments which survive the lifetime of a petroleum project over the long term.

Legal minimum stock (LMS)   In relation to petroleum storage, a legislative requirement that a defined minimum stock level of petroleum is held in-tank at all times.

Legitimate expectation   The expectation of a petroleum project investor that it will be allowed by a state to develop and operate the project profitably (often as part of a claim for alleged expropriation).

Length overall (LOA)   The maximum length of a ship's hull which is measured parallel to the waterline.

Lens   A lenticular sedimentary rock formation which is surrounded by an impervious rock formation.

Letter of credit (LC)   A composite term for a **documentary letter of credit** and a **standby letter of credit**.

Letter of quiet enjoyment (LQE)   A letter which is issued by a lender to recite an undertaking of quiet enjoyment.

Levelised tariff   The average value of a tariff (taking account of fixed and variable components) which is payable over the lifetime of a contract which charges the tariff.

Leverage   The relationship between a company's debt and its equity capital (so that a company is highly leveraged if debt greatly exceeds equity). Also called **debt to equity** and **gearing**.

Levered   Petroleum project costs which are met through debt rather than through equity reserves. *See also* **unlevered**.

Levered partner   A petroleum project participant which pays a share of the costs of developing a petroleum project that are proportionately greater than the project equity which that party has.

Levered return   The amount of cashflow which a petroleum project generates after the costs of financing have been met. *See also* **unlevered return**.

Lex petrolea   A suggested body of law which is derived from and relates specifically to the activities of petroleum exploration, development, and production.

Licence   A concession for petroleum E&P which is characterised by the payment of a royalty by the licensee to the state and by the state's taxation of the licensee's licence-related business profits. Also called a **royalty/tax regime**.

Licensing round   A process by which a state awards a concession to a person for the conduct of petroleum E&P activities as part of a multi-party competitive process within a defined time period. Contrast with an **open award**, as an ongoing concession award process. Also called a **bid round**. *See also* bid round circular, bid round data package, **limited procedure**, **open procedure**, and **selective procedure**.

Lien   A right of a person (a lienholder) to keep possession of property belonging to another person until a debt which is owed by that other person to the lienholder is discharged.

Lifecycle assessment (LCA)   An assessment of the environmental effects of a petroleum product over its entire lifecycle (including production, use, and disposal).

Lifter   A person who undertakes a lifting.

Lifting   The taking of delivery of a producer's petroleum entitlements (usually at the wellhead), which is typically conducted in accordance with a defined lifting schedule, which could also reflect differential lifting. *See also* **joint lifting** and **lifting limits**.

Lifting costs   The costs (expressed in a currency per barrel of oil equivalent) which are associated with producing a defined quantity of petroleum, which then allows a comparative analysis of the economic efficiency of different petroleum production projects to be undertaken.

Lifting in place (LIP)   The mating of one offshore petroleum infrastructure item to another by the use of a heavy-lift crane. *See also* **float-over**.

Lifting limits   The defined lower and upper limits on the quantity of petroleum which can be the subject of a lifting.

Light distillates   Gasoline, naphtha, methane, ethane, propane, and butane which are produced in a refinery. Also called **light ends**. *See* Appendix 6.

Light ends   An alternative term for **light distillates**.

Light liquid hydrocarbon   An alternative term for **condensate**.

Light loaded   A crude carrier which is carrying anything less than its full cargo capacity.

Light Louisiana Sweet (LLS)   A crude oil blend from the United States.

Light oil   Crude oil which has an API gravity greater than 31.1°, up to 50°. *See* Appendix 6.

Light tight oil (LTO)   An alternative term for **tight oil**.

Lightering   The offshore transfer of a quantity of petroleum from a large ship to a smaller ship. *See also* **ship-to-ship**.

Lignite   Coal which has a lower calorific value than anthracite or bituminous coal. Also called **brown coal** and **sub-bituminous coal**.

Limit state   A condition of a petroleum infrastructure item which, if exceeded, means that the item no longer meets its original design criteria (such that the limit state indicates the conditions for a possible failure). *See also* **accidental limit state**.

Limited access   A gas pipeline distribution system in which a transporter transports its own gas entitlements under rules which are different to those that apply to other shippers in the pipeline.

Limited entry technique   A method of fracturing a formation whereby fluid is injected into the formation through a limited number of, and not all of the, perforations in a casing.

Limited procedure   A model for the award of a concession by a state whereby the state approaches specific bidders and invites them to negotiate directly. *See also* **open procedure** and **selective procedure**.

Limited recourse finance   The use of third-party financing to meet the costs of developing a petroleum project (such as in project finance) whereby the lenders have limited direct recourse to the sponsor for repayment of the financing. *See also* **non-recourse finance**.

Line drive   In waterflooding, a straight-line pattern of water injection wells.

Line handling   The process of retrieving, handling, and securing ship mooring lines during the mooring and unmooring of a ship.

Line item   A unit of information which appears as a line on its own in a larger data item, such as a single element in a budget or a specific task in an overall work programme.

Line kilometre   A linear kilometre which is used as a unit of measurement for the acquisition of two D seismic data. Contrast with **square kilometre**, as the unit of measurement which is used for the acquisition of three D seismic data.

Line loss   An alternative term for **lost gas**.

Line of sight (LOS)   A data transmission system (such as microwave transmission) which depends for its operation on an unbroken line of visual communication between the transmitting and the receiving facilities.

Line press   The recorded difference in the measurement gauges of a petroleum storage tank when the tank's valves are open and closed.

Lineament   An apparent linear feature on the Earth's surface (such as a fault) which could be indicative of a sub-surface petroleum deposit.

Linear programme (LP)   A form of computer model which is used to optimise process inputs and outputs.

Lined pit   An excavation which is lined with an impermeable coating for the storage of drilling fluid and other contaminants from onshore petroleum production operations.

Linefill   The minimum quantity of petroleum which is needed within a pipeline for the first time of operation so as to allow the pipeline to operate. Also called **initial fill**.

Lineheater   A device which heats petroleum as it flows through a pipeline.

Linepack   A quantity of petroleum in a pipeline at any time which can supply output demand without increased input. Also called **residence petroleum**.

Liner   A casing set which is suspended within the principal casing set within a wellbore. Also called **tubing**. *See* Appendix 2C.

Liner completion   A completion of a wellbore in which a liner is used to effect communication between a reservoir and the wellbore.

Liner patch   Steel tubing which is lowered into existing casing in a wellbore in order to repair a leak in the casing.

Linkage   An alternative term for **indexation**.

Liquefaction   The transformation of gas into a liquid form (as LNG) through refrigeration. Contrast with **regasification**, as the transformation of LNG into the gaseous state.

Liquefied biogas (LBG)   Biogas which is enhanced with methane and is liquefied for use as a vehicle fuel.

Liquefied ethylene gas (LEG)   Ethylene which is refrigerated to minus 104°C in order to transform it to a liquid for easier transportation.

Liquefied natural gas (LNG)   Gas (principally methane) which is refrigerated to minus 160°C in order to transform it to a liquid for easier transportation.

Liquefied natural gas regasification vessel (LRV)   An alternative term for a floating storage and regasification unit.

Liquefied natural gas unit shipping cost (LUSC)   A per-unit cost of shipping a defined unit of LNG to a defined point which is derived from a division of the total costs of shipping by the delivered cargo content (usually measured in USD/mmBtu).

Liquefied petroleum gas (LPG)   Any mixture of propane and butane which is extracted from a gas stream and stored under pressure. Also called **bottle gas** and **tank gas**.

Liquefied refinery gas (LRG)   Propane or butane which is produced as part of a refinery process.

Liquid market   A petroleum-trading market which is characterised by the existence of multiple sellers and buyers with the freedom to contract with each other and multiple opportunities to trade petroleum.

Liquidated damages (LDs)   Monetary damages which are payable for a breach of contract and are agreed between the contracting parties at the time of entry into the contract. *See also* **penalty clause**. Contrast with **unliquidated damages**, which are not pre-agreed at the time of entry into the contract.

Liquids   A composite term for condensate and natural gas liquids.

Liquids and solids remaining on board (LASROB)   A measure of the liquids and the solids which remain on board a ship after a cargo has been discharged. Also called **on-board quantity**. *See also* **retention on board**.

Liquids dropout   An alternative term for **dropout**.

Liquids stripping   A process by which petroleum fractions which are heavier than methane are extracted from a gas stream.

Lithology   A summary of the physical, sedimentary, and mineralogical characteristics of a formation.

Lithostatic pressure   An alternative term for **geostatic pressure**.

Little knock   A knock-for-knock liability allocation regime which does not extend to include an indemnified person's other contractors. Contrast with **big knock**, which does include those other contractors.

Littoral deposit   An alternative term for a **shoreline deposit**.

Live load   The measure of a load which is determined by and varies in accordance with the use of a petroleum infrastructure item.

Live oil   Crude oil which contains associated gas. Contrast with **dead oil**, which contains no or negligible associated gas.

Liveboating   An offshore diving operation which is conducted from a moving boat.

Living quarters (LQ)   The part of an offshore petroleum infrastructure item which houses crew living quarters.

LNG carrier (LNGC)   A ship which is used for the transportation of LNG.

Load   Water which is pumped into a formation during a fracking operation.

Load balancing   A process by which demand (offtake) and supply (input) is balanced between buyers and sellers in respect of a gas pipeline which connects supply and demand.

Load factor
(1) The ratio of the actual output of a petroleum-processing facility to the maximum possible output of the facility.
(2) The ratio of the average usage of a pipeline to the peak usage of that pipeline over a specific period of time.

Also called a **capacity factor**.

Load following   Under a gas sales contract, the situation in which the seller's delivery profile matches the buyer's demand profile.

Load loss   A part of a crude oil cargo which is lost in transit or storage (usually through the vaporisation of the lighter petroleum fractions).

Load oil
(1) Oil which is produced from and then pumped back into a formation for any reason (for example, to fracture the formation). *See also* **recoverable oil**.
(2) Oil which is pumped into a wellbore in preparation for, or as part of, a downhole process.

Load reduction   An alternative term for **load shedding**.

Load shedding   The planned reduction of the output of a petroleum-processing facility. Also called **load reduction**.

Loaded leg   An alternative term for a **laden voyage**.

Loaded on top (LOT)   A method of cleaning an oil storage tank on a ship whereby the residual water/oil mix is collected in a tank and is allowed to separate, clean water is pumped out and the next cargo is loaded in-tank on top of the slops. *See also* **crude oil washing**.

Loading arm   A flexible connection which is used to load or unload petroleum into or from a ship at a port. Also called a **hard arm**.

Loading conflict   A conflict which arises between two or more ships that arrive at the same time at a loading port in order to load a cargo. *See also* **unloading conflict**.

Loan life coverage ratio (LLCR) A ratio test in project financing and reserves-based lending which assesses the ability of the borrower to repay the debt over the period of the loan. *See also* **debt service coverage ratio**.

Loan-to-own Under a loan agreement, debt which is convertible to an equity interest in the borrower at the lender's option in the event of a repayment default by the borrower.

Local content (LC) A regulatory provision that preference must be given by petroleum project investors to locally sourced (rather than imported) goods, services, and personnel. *See also* **hard local content** and **soft local content**.

Local distribution company (LDC) A person who takes gas from a defined point and distributes it to end-users, often within a local distribution zone.

Local distribution zone (LDZ) A defined local area in which gas is distributed to end-users.

Local oil company (LOC) A company which holds an effective monopoly in a geographical area.

Localisation An alternative term for **nationalisation**.

Location damages Monetary damages which are payable to a landowner as compensation for damage caused to the surface and any growing interests as a consequence of the conduct of an onshore petroleum operation.

Locked box accounting Under a contract for the sale of a business, a provision whereby a fixed price is agreed for the business by reference to a defined target date and parameters without post-completion adjustments. Contrast with **post-closing price adjustment accounting**, whereby the price for the business is subject to post-completion adjustments.

Lock-in agreement A commitment which is made between negotiating persons that they will only negotiate with each other for a defined period of time.

Lock-out agreement A commitment which is made between negotiating persons that they will not negotiate with any other persons for a defined period of time.

Lock-out tag-out (LOTO) A health and safety methodology for suspending the operation of an equipment item while repairs are undertaken.

Lock-out tag-out try-out (LOTOTO) A health and safety methodology for suspending the operation of an equipment item while repairs are undertaken, including testing the equipment item. Also abbreviated to **LTT**.

Lock up A contractual provision whereby a person is committed to a contract or a project for a defined period of time, without the right to withdraw or to transfer its interests.

Logging while drilling (LWD) An alternative term for **drill stem testing**.

Logs Records of data which are drawn from a wellbore through well logging.

Loll A measure of the angle of heel of a ship.

London Tanker Brokers' Panel (LTBP) An independent authority which publishes the average freight rate assessment (http://www.ltbp.com).

Long-lead item (LLI)   An item which is needed for a petroleum project and is procured significantly ahead of the time when it is needed because of the anticipated length of time which will be taken to deliver it.

Long position   Under a petroleum sales contract, the position of the seller of petroleum which already owns or has access to the petroleum which it has contracted to sell. Contrast with a **short position**, whereby the seller does not own or have access to the petroleum.

Long-range storage (LRS)   A gas storage facility from which the total quantity of working gas can be fully delivered to the storage facility users over a relatively long timeframe. *See also* **medium-range storage** and **short-range storage**.

Long string   An alternative term for **production casing**.

Long term contract (LTC)   A contract which is intended to subsist for what the contracting parties regard to be a long term.

Long-term fixed quantity (LTFQ)   A petroleum sales contract which indicates a relatively fixed and rigid supply profile.

Long-term interruptible (LTI)   An interruptible gas sales contract.

Long-term sales agreement (LTSA)   A long-term contract for the sale and purchase of petroleum.

Long-term services agreement (LTSA)   A long-term contract for the provision of services.

Long ton (LT)   An imperial weight measure of 2,240 pounds. *See also* **metric ton** and **short ton**.

Longshore current (LSC)   An ocean current which runs parallel to a beach. *See also* **tidal current**.

Longtail   A point in time at which the level of petroleum production from a formation begins a slow but constant period of decline.

Lookback
(1) Under an asset or share sale and purchase agreement, an agreement between the seller and the buyer whereby the agreed sale price is adjusted retrospectively to take account of certain conditions after the sale has completed.
(2) A retrospective accounting adjustment.

Looping   The addition of a secondary, parallel section of pipeline over a defined distance. Also called **twinning**. *See also* **pipeline string**.

Loss damage delay (LDD)   The principal causes of action against a carrier of goods for breach of contract.

Loss of production income insurance (LOPI)   A policy of insurance which covers a producer's lost income resulting from an interruption to petroleum production.

Lost circulation   Drilling fluid which is lost into a formation during the drilling of a wellbore rather than which flows back to the surface through the wellbore. Also called **lost return**. *See also* **thief zone**.

Lost circulation material   A substance which is added to drilling fluid in order to prevent lost circulation.

Lost circulation zone (LCZ)   An alternative term for a **zone of lost circulation**.

Lost gas   Gas which is lost within a pipeline system and is not delivered to the shipper by the transporter when due. Also called **fuel lost and unaccounted-for, line loss, shrinkage**, and **unaccounted-for gas**.

Lost hole   A wellbore which has collapsed internally.

Lost return   An alternative term for **lost circulation**.

Lost time injury (LTI)   An alternative term for a **time losing injury**.

Lot   A form of concession which is typically granted in respect of an onshore area. *See also* **water lot**.

Low consequence high potential (LCHP)   A risk factor which has a low impact and a significant risk of occurrence. *See also* **HCHP, HCLP**, and **LCLP**.

Low consequence low potential (LCLP)   A risk factor which has a low impact and a low risk of occurrence. *See also* **HCHP, HCLP**, and **LCHP**.

Low line gas   An alternative term for **low pressure gas**.

Low pour fuel oil (LPFO)   A form of fuel oil which is used particularly in Nigeria.

Low pressure gas   Gas which is recovered from atmospheric distillation and is compressed to higher pressures for further industrial uses. Also called **low line gas**.

Low shrinkage oil   An alternative term for **black oil**.

Low specific activity (LSA)   Radioactive material which has limited activity per unit of mass.

Low sulphur fuel oil (LSFO)   Fuel oil which is relatively low in sulphur compounds. *See also* **high-sulphur fuel oil**.

Low sulphur marine fuel (LSMF)   A fuel oil which is used as bunker fuel for ship engines.

Low sulphur waxy residue fuel oil (LSWR) A form of fuel oil.

Low temperature separator (LTX)   A separator which uses refrigeration in order to improve the rate of recovery of gas-entrained liquids.

Low water mark   The lowest point of a tidal range, which is taken to denote the boundary point between landward and seaward.

Lower explosive limit (LEL)   The lowest concentration (by percentage) of a gas in air which is capable of combusting with an ignition source, whereby concentrations of gas below the LEL are too lean to combust. *See also* **upper explosive limit**.

Lower heating value (LHV)   An alternative term for **net calorific value**.

Lower marine riser package (LMRP)   A flexible joint which attaches the lowest section of a riser to a sub-sea blowout preventer.

Lowest achievable emissions rate (LAER)   A target standard for noxious emissions to air which result from the conduct of a petroleum project activity.

Lowest astronomical tide (LAT)   The lowest tide level which can be predicted to occur under average meteorological conditions and any particular astronomical conditions. *See also* **highest astronomical tide**.

Lowest carbon intensity project solution (LCIPS)   A solution for the production of petroleum which demonstrates the lowest carbon intensity of a range of potential solutions.

Lowest known petroleum   The deepest observed occurrence of a petroleum deposit within a formation. *See also* **hydrocarbon water contact**.

Loxodrome   A notional latitudinal arc on the mapped surface of the Earth which crosses all meridians at the same angle relative to north. Also called a **rhumb line**.

Lump sum turnkey (LSTK)   An alternative term for **engineering procurement construction installation**.

M-100   An alternative term for **mazut**.

Magnetic flow meter   A form of flow meter which uses a magnetic field to assess flow rates.

Magnetometer   A device which indicates sub-surface magnetic fields, which can be indicative of the existence of petroleum in a formation. *See also* **aero-mag** and **grav-mag**.

Main cryogenic heat exchanger (MCHE)   An essential process element in gas liquefaction which reduces the gas temperature to minus 160°C to create LNG.

Maintenance of uniform interest (MUI)   An alternative term for a **uniform interest clause**.

Make good   The obligation of a person to deliver, or take delivery of, a quantity of petroleum which that person has previously failed to deliver, or to take delivery of. Also called **restoration quantity**.

Make rate   The rate at which processed petroleum or petrochemicals are produced.

Make-up
(1)   Under a petroleum sales contract, a quantity of petroleum which the buyer has a right to take delivery of and that equates to the quantity of petroleum which the buyer has previously paid the seller for under a take or pay payment. Also called **bank gas** and *see also* **recoupment**.
(2)   To assemble a drill string prior to or during the drilling of a wellbore.
(3)   Under a gas transportation agreement, the transporter's remediation of lost gas in a pipeline in the shipper's favour.

Make-up gas   Gas which is injected into a formation which contains a mixed stream of gas and condensate in order to maintain inherent capillary pressure within the formation and also to prevent dropout. *See also* **injection well** and **recycling**.

Makers list   A schedule which identifies the name and the nationality of all the fabrication companies and equipment suppliers which will be involved in the construction of a new build.

Man overboard boat (MOB)   An alternative term for a **fast rescue craft**.

Managed pressure drilling (MPD)   The drilling of a wellbore in which, by the injection of drilling fluid into the wellbore while the drilling is underway, the degree of annular pressure in the wellbore is controlled so that it lies between overbalance and underbalance.

Management of change (MOC)   An operational methodology for the implementation of approved process changes in the performance of a petroleum project. *See also* **change order**.

Mandatory relinquishment   Under a concession, a compulsory obligation of the concession-holder to surrender certain parts of the contract area to the grantor. Contrast with **voluntary relinquishment**, whereby relinquishment is at the concession-holder's option. *See also* **relinquishment**.

Mandrel   A specialised tubular assembly around which other components are assembled in petroleum E&P operations.

Manifest   A document which recites the essential characteristics of a ship's cargo. Also called a **cargo manifest** and a **freight manifest**.

Manifold   An arrangement of onshore or sub-sea pipes and valves which is used variously to gather, control, commingle, monitor, segregate, distribute, and manage the flow of multiple petroleum streams.

Manning scale   A tool which is used to determine the minimum number of crew members which are necessary to meet the essential safe operational requirements of a ship or a marine-based petroleum facility. *See also* minimised manning and **safe manning and watchkeeping standards**.

Manometer   A mercury-filled u-shaped glass tube which measures differences in pressure between a gas stream and ambient conditions.

Manual of permitted operations (MOPO)   A defined set of procedures which describe the operational parameters of a petroleum project.

Manufactured gas   An alternative term for **synthetic natural gas**.

Margin play   The economic interest of a person under a simultaneous asset purchase and asset sale contract which is founded on the positive difference between the outgoing purchase costs and the incoming sale revenues.

Marginal resource/marginality   A petroleum resource which, at the time an economic assessment is made based on current and prospective applicable commercial and technical parameters, is identified as being unlikely to give a sufficient level of economic return if then developed as a production project. *See also* **paramarginal resource**.

Marigraph　A gauge which makes a continuous record of tidal heights in relation to time.

Marination/marinisation　The modification of terrestrially based petroleum infrastructure items or equipment for use in a marine environment. *See also* **proven offshore technology**.

Marine diesel oil (MDO)　An alternative term for **marine gas oil**.

Marine fuel oil (MFO)　A generic term for fuel oil which is used to power a ship.

Marine fuelling　An alternative term for **bunkering**.

Marine gas oil (MGO)　A fuel oil which is used as bunker fuel for ship engines. Also called **Bunker C** and **marine diesel oil**. *See also* **intermediate fuel oil**.

Marine growth prevention system (MGPS)　A system for the prevention of seawater-borne organic growths from accumulating on the exposed surfaces of a ship or a platform. Also called **anti-fouling programme**.

Marine loading arm (MLA)　An articulated mechanical arm which connects a marine loading or unloading terminal to a ship which is carrying petroleum.

Marine services　Services which are provided in respect of a ship's arrival to, residence at, and departure from a port (including the provision of tugs, bunkers, spares, victuals, pilots, line handling services, and fire-fighting services).

Marine spread　An alternative term for a **spread**.

Marine warranty survey　An independent survey, which is performed on behalf of an insurer, of the quality of the construction or of the operation of an offshore petroleum facility and guides the willingness of the insurer to insure the facility.

Mark-to-market (MTM)　An accounting practice whereby the full value of a contract (which is measured by price and duration) is reflected as income in the year when the contract is signed in order to reflect the contract's current real market value. Also called **fair value accounting**.

Marker　An alternative term for a **fuel dye**.

Marker crude　An alternative term for a **benchmark crude**.

Market area storage　A gas storage facility which is located near to a gas demand market.

Market of last resort　A market for the import and sale of petroleum which is the least attractive to sellers because of low prices compared to opportunities elsewhere. Also called a **balance market**.

Market-out clause　An alternative term for a **material adverse change** clause.

Marketing and sales (M&S)　The process by which petroleum is offered for sale (marketing) and by which that petroleum is sold under a contract (sales).

Marketing costs　All of the costs which are associated with making a quantity of petroleum fit for sale or transportation. Also called **handling costs**.

MARPOL　Derived from the words 'marine pollution', a description of the International Convention for the Prevention of Pollution from Ships (1973/1978), issued by the IMO in 1973 and modified in 1978.

Marsh gas   Sub-surface methane deposits which find their way to the surface naturally and may combust spontaneously. Also called **seepage**, **seeps**, **swamp gas**, and **will o' the wisp**.

Mass flow rate   A determination of the rate of flow of gas through a flow meter.

Mass-transport complex (MTC)   A seismic term for a particular formation in a deep-water setting which can only be completely imaged on a volumetrically large seismic survey.

Master choke line   A valve on a choke line which effects flow control through the choke line.

Master deed   An online portal, managed by OGUK, which is used for the management of UKCS asset transfers between registered persons (http://www.logic-oil.com).

Master sale and purchase agreement (MSPA)   An alternative term for a **master sales agreement**.

Master sales agreement (MSA)   A contract for the sale of petroleum which is made up of a detailed set of terms and a pro-forma confirmation notice. Also called an **enabling agreement** and a **master sale and purchase agreement**.

Master services agreement (MSA)   A contract for the provision of oil field services which is made up of a detailed set of terms and a pro-forma call-off sheet. Also called an **enabling agreement**. *See also* **federal contract**.

Mat/mattress   A mat, which is typically made of a steel mesh construction, which is placed on the sea-bed in order to give a stable foundation for a jacket to sit upon.

Matching right   An alternative term for a **right of first refusal**.

Material adverse change (MAC)   A provision in a petroleum project contract whereby the occurrence of a defined extraneous event leads to a defined protective relief for the affected party. Also called a **market-out clause** and a **material adverse event** clause.

Material adverse event (MAE)   An alternative term for a **material adverse change** clause.

Material balance equation (m-bal)   A tool for the analysis of the productivity of a formation which relates in-situ petroleum quantities to petroleum withdrawal quantities.

Material offloading facility (MOF)   A temporary facility which is used in the construction of a petroleum infrastructure item.

Material safety data sheet (MSDS)   A document which gives compositional information and handling suggestions in respect of a hazardous product. Also called a **product information bulletin**.

Materials take off (MTO)   A list of materials (including references to specifications and quantities) which are required to construct a petroleum infrastructure item.

Materials unaccounted for (MAF)   The difference between materials which are recorded on an inventory and materials which are actually accounted for.

Maturation   The ongoing upward re-categorisation of resources. Also called **resource maturation**.

Mature crude   A grade of crude oil which has been in production for a period of time and consequently is easier to market and to use as a price marker.

Maturity   A function of burial pressures, temperatures, and time which are used in determining whether a petroleum source will provide oil or gas. Also called **burial history**.

Maximising economic recovery (MER)   A state-sponsored initiative to maximise remaining petroleum production from the UKCS (http://www.gov.uk).

Maximum allowable annulus surface pressure (MAASP)   A measure of the upper limit of the absolute pressure which can exist within an annulus before a breakdown of the formation is caused, measured at the wellhead.

Maximum allowable operating pressure (MAOP)   An alternative term for **maximum pipeline operating pressure**.

Maximum allowable surface pressure (MASP)   The maximum degree of pressure which can be imposed on casing during a well kill, which is determined by the rating of surface equipment.

Maximum continuous rating (MCR)   The maximum output which a power-generation facility is capable of producing under normal conditions over a one-year period.

Maximum economic finding cost (MEFC)   An estimate of the maximum amount of costs which a petroleum company is willing to incur in exploring for a formation or in developing a discovered formation.

Maximum efficient rate/most efficient recovery (MER)   The maximum rate of petroleum production from a formation which it is possible to sustain without damaging the integrity of the formation.

Maximum pipeline operating pressure (MPOP)   The high end of the pressure range at which a gas pipeline can safely operate. Also called **maximum allowable operating pressure**.

Mazut   A low quality fuel oil with variable degrees of sulphur content which is typically used in Russia, where it is combusted in boilers as a feedstock for power generation or for district heating. Also called **M-100**.

Mean   The arithmetic average of a set of numbers (for example, the mean of 10, 15, 30, 45, and 50 is 30). *See also* **median**.

Mean high water neaps (MHWN)   The average over a year of the heights of two successive high waters during periods of twenty-four hours when the range of tidal movements is at its lowest.

Mean high water springs (MHWS)   The average over a year of the heights of two successive high waters during periods of twenty-four hours when the range of tidal movements is at its greatest.

Mean higher high water (MHHW)   The average height of the two high water heights at a particular place in respect of a particular tidal day.

Mean low water neaps (MLWN) The average over a year of the heights of two successive low waters during periods of twenty-four hours when the range of tidal movements is at its lowest.

Mean low water springs (MLWS) The average over a year of the heights of two successive low waters during periods of twenty-four hours when the range of tidal movements is at its greatest.

Mean lower low water (MLLW) The average height of the two low water heights at a particular place in respect of a particular tidal day.

Mean sea level (MSL) The average height of the sea's surface at a particular place in respect of a particular day, which is calculated as the mean of the mean higher high water and the mean lower low water.

Mean time to failure (MTTF) The recorded average period of time between unscheduled operational failures of a petroleum facility or an item of equipment, which is often used as an indication of reliability.

Means of rescue (MOR) A planned methodology for the rescue of personnel from a petroleum infrastructure item.

Measured depth (MD) The actual drilled vertical depth of a wellbore, which runs from a defined surface point to a defined end point. Also called **along-hole depth**. *See also* **bottomhole** and **true vertical depth**.

Measured distance (MD) The actual drilled horizontal distance of a wellbore, which runs from a defined surface point to a defined end point. Also called **along-hole distance**. *See also* **true horizontal distance**.

Measured variable Any potentially variable physical condition which can be measured (such as flow rate, pressure, temperature, or volume).

Measurement A combined term for the distinct activities of analysis and metering. Also called **measurement and testing**.

Measurement and testing (M&T) An alternative term for **measurement**.

Measurement range The range between the maximum and minimum reading, of a measurement device.

Measurement, reporting, verification (MRV) A methodology for tracking lifecycle emissions which are associated with a petroleum project (also known as **monitoring, recording, verifying**).

Measurement tank A petroleum storage tank which automatically measures petroleum inflows and outflows.

Measurement while drilling (MWD) The measurement of a number of different outputs from a wellbore while drilling is underway.

Mechanical integrity test (MIT) The act of setting a packer above the level of perforations in a wellbore and then applying pressure to the annulus in order to test the integrity of the casing down to the packed level.

Med-arb A dispute resolution mechanism whereby a dispute which the parties have not been able to resolve by mediation is referred back to the original mediator to act as a sole arbitrator.

Media bed   In the production of LNG, a sieve through which gas is flowed in order to remove water and other impurities. *See also* **media bed regeneration**.

Media bed regeneration   The cleaning of a saturated media bed by passing dehydrated gas through it to desorb accumulated impurities.

Median   The number at the midpoint of a set of numbers (for example, the median of 10, 15, 30, 45, and 50 is 30). *See also* **mean**.

Median line   The offshore boundary line between the jurisdiction of two or more states, which is drawn to be equidistant from the nearest points of land of each state.

Medium oil   Crude oil which has an API gravity in the range of 22.3° to 31.1°. *See* Appendix 6.

Medium range (MR)   A crude carrier which has a deadweight tonnage in the range of 25,000 to 45,000 metric tons.

Medium-range storage (MRS)   A gas storage facility in which the total quantity of working gas can be fully delivered to the storage facility users over a timeframe which, relatively, is somewhere between short-range storage and long-range storage.

Meet or release   Under a petroleum sales contract, a provision whereby: the buyer has a right to reduce the contract price, which the seller must accede to or the buyer may end the contract; or the seller has a right to increase the contract price, which the buyer must accede to or the seller may end the contract.

MEG regen   A process for the recovery and recycling for use of mono-ethylene glycol within a gas production process.

Megacalorie (Mcal)   One million calories.

Megapascal (Mpa)   One million Pascals.

Membrane tank   A particular form of containment vessel for LNG on an LNG carrier. *See also* **Moss spherical tank**.

Mercantile agent   An alternative term for a **commercial agent**.

Mercaptan   A sulphur-based compound which is added to natural gas in odourisation. Also called **methanethiol** and **thiol**.

Merchant–equity ratio (MER)   In a petroleum projects portfolio, the ratio of petroleum quantities which are contracted for purchase (merchant) to petroleum quantities which are produced from owned project interests (equity).

Merchant pipeline   A pipeline whereby the owner or operator trades in the petroleum which the pipeline transports, in contrast with a pipeline whereby the owner or operator only transports petroleum for other persons.

Merchant portfolio   A series of contracts and arrangements which are used for the buying and selling of petroleum.

Merchant risk   The risk of economic uncertainty which is associated with a petroleum project which sells its outputs into a market under spot sales or short-term contracts.

Merchantable gas   Gas which has a quality specification which makes it readily capable of being widely sold and/or transported by pipeline.

Mercury injection capillary pressure (MICP) A technique in which mercury is injected into a formation in order to measure porosity in the formation.

Meridians A series of imaginary longitudinal lines on the mapped surface of the Earth which run from the North Pole to the South Pole.

Metadata Data which is principally descriptive of other data.

Metagenesis The final stage of the sub-surface metamorphosis of organic matter to form petroleum, principally methane and certain non-methanous gases, occurring at the highest temperatures and pressures. *See also* **diagenesis** and **katagenesis**. *See* Appendix 2A.

Metals content A quality specification which describes the presence of trace elements of metals (principally vanadium and nickel) in a given quantity of crude oil.

Metamorphic rock One of the three main rock types (with igneous rock and sedimentary rock), formed by the pressure—and temperature-generated transformation of other rock types (which could be igneous, sedimentary, or other metamorphic rocks).

Meter A device which measures the quantity of petroleum which is passing through a point over a period of time.

Meter accumulator A meter which records and combines the readings of more than one individual meter. Also called a **meter combinator** and a **multi-meter**.

Meter combinator An alternative term for a **meter accumulator**.

Meter performance The difference between the volume of petroleum which is recorded as passing through a meter and the actual volume of petroleum which has passed through the meter, which could indicate an overdelivery or an underdelivery against a required delivery quantity.

Meter prover A device which is used to calibrate a meter.

Meter run point A point in a petroleum facility at which petroleum measurement equipment is installed.

Meter slippage A quantity of petroleum which passes through a meter without being measured.

Meter-line thermometer A thermometer which is located within a metering unit on a pipeline.

Metering A process for the determination of a quantity of petroleum. Contrast with **analysis**, as a determination of the quality of petroleum.

Metering by difference An allocation method whereby metered petroleum quantities are deducted from non-metered petroleum quantities in order to determine an allocated amount of petroleum.

Metering standard A defined technical protocol according to which petroleum is analysed or is metered. Also called a **fiscal standard**.

Metes and bounds A method of surveying onshore property which relies on physical geographical features and the distances between them in order to define a property's boundaries.

Methane clathrate   An alternative term for **hydrates**.

Methane drainage   A process of collecting methane as a natural by-product from active or disused coal mining operations, for which a methane drainage licence could be required. *See also* **abandoned mine methane** and **gob hole**.

Methane drainage licence (MDL)   A licence which is awarded to the operator of an active or a disused coal mine in order to capture and extract methane build-ups.

Methane slip   The unintended escape of methane into the atmosphere from a ship's engine which is fuelled by natural gas.

Methanethiol   An alternative term for **mercaptan**.

Methyl diethanolamine (MDEA)   A solvent which is used in acid gas removal.

Methyl tertiary butyl ether (MTBE)   An additive which is used in the production of gasoline to offer improved air quality benefits, which has largely been displaced by the preferred use of ethyl tertiary butyl ether.

Metocean   A description of the physical offshore environment. The term is derived from a combination of 'meteorology' and 'oceanography'. A metocean study will usually be undertaken as a precursor to the commencement of a marine engineering project. *See also* **morphodynamics**.

Metric ton (MT)   A metric weight measure of 1,000 kilogrammes. *See also* **long ton** and **short ton**.

Metrology   The science of measurement.

Mexican shoot-out   A colloquial alternative term for a **Texas shoot-out**.

Microseismic study   A technical analysis of the results of a fracking operation.

Micro-generation   The production of electric power and heat on a very small scale, as distributed generation and often using renewable energy sources.

Mid-body length (MBL)   An alternative term for **parallel body length**.

Mid office   The front office-supporting elements of an energy business (such as risk management and compliance). *See also* **back office** and **front office**.

Mid-reach lateral (MRL)   A horizontal wellbore of approximately 7,000 feet, which is a distance greater than a standard reach lateral but less than an extended reach lateral.

Middle distillate synthesis (MDS)   An alternative term for **gas-to-liquids conversion**.

Middle distillates   The range of products which are situated between light and heavy petroleum fractions (including gasoline, aviation fuel, and diesel).

Middle distillates lower olefins selective (MILOS)   A process which allows the maximised recovery of diesel and propylene in a refinery.

Middle sample   A spot sample which is taken from the middle layer of a petroleum storage tank. *See also* **upper sample**.

Midstream   The functions of petroleum storage, transportation, distribution, and wholesale marketing, which lie between the relative functions of upstream and downstream.

Migration   The natural movement of mobile petroleum in a formation from a source rock to a reservoir rock. *See also* **primary migration, secondary migration,** and **tertiary migration.**

Milk round   A scheduled delivery of petroleum products which is made to multiple destinations from a single ship.

Mill   A downhole cutting tool.

Millidarcy   One thousandth of a Darcy.

Milling   A process of cutting a window into the casing within a wellbore, which is typically done as a precursor to sidetracking the wellbore.

Millions of years before present (Mmybp)   A base reference point for the determination of geological time periods.

Mineral acre   A measure of the full value of petroleum which underlies an areal coordinate of one acre.

Mineral deed   A legal instrument which conveys the right to explore for and to produce sub-surface petroleum from one person to another.

Mineral estate   An alternative term for **ownership-in-place.**

Minifrac   A preliminary fracking operation which is performed on a formation prior to a main fracking operation to acquire operational data.

Mini-generation   The production of electric power and heat on a small scale but which is of a scale that is relatively greater than micro-generation.

Minimised manning (mini-man)   The process of reducing the need for human activity in a petroleum operation to the greatest extent possible. *See also* **manning scale.**

Minimum bill   Under a petroleum sales contract in which a take or pay provision applies, the minimum amount of petroleum which is required to be taken (and paid for) by the buyer in order to avoid making a take or pay payment.

Minimum breaking load (MBL)   The load that a crane can carry which is below the point at which the crane will fail.

Minimum facility platform (MFP)   An offshore petroleum production platform with minimum facilities topside, which is usually a normally uninhabited installation.

Minimum internal yield pressure (MIYP)   The lowest degree of pressure within a pipeline or a tubular which can exist before which yield pressure is achieved and deformation occurs.

Minimum pipeline velocity   A measure of the velocity of petroleum in a pipeline at the lowest possible operating flow rate of the pipeline.

Minimum royalty   Under a royalty agreement, a minimum payment which is required to be made to the royalty-holder regardless of the level of actual petroleum production from the concession.

Minimum stock to load (MSL)   In relation to petroleum storage, the most effective parcel of petroleum for the loading of an offtaking tanker.

Minimum work commitment (MWC)   Under a concession, a requirement that the concession-holder will during the exploration period perform a defined series of work obligations in exploration activities and will incur a minimum amount of expenditure in doing so. *See also* **commitment shoot, commitment well, minimum work expenditure, minimum work obligation,** and **work obligation payout.**

Minimum work expenditure (MWE)   Under a concession, a requirement that the concession-holder will during the exploration period incur a minimum amount of expenditure in performing exploration activities. *See also* **minimum work commitment** and **minimum work obligation.**

Minimum work obligation (MWO)   Under a concession, a requirement that the concession-holder will during the exploration period perform a defined series of work obligations in exploration activities. *See also* **minimum work commitment** and **minimum work expenditure.**

Miscibility   The commingling of separate phases of petroleum within a pipeline or a storage tank in order to produce a single, combined petroleum phase.

Miscibility pressure   The pressure at which gas which is injected into a formation can form a miscible transition zone with any resident crude oil in the formation.

Miscible drive   An alternative term for **miscible flood.**

Miscible flood   The injection of a diluent (such as carbon dioxide, methane, nitrogen, or propane) into a formation to lower the viscosity of any resident crude oil in the formation and so to improve the crude oil recovery rates. Also called **miscible drive.**

Missed pay   A payzone which has been identified by later or more sophisticated formation analysis (which could result from the application of new and improved technologies over time) and was previously overlooked during earlier E&A activities.

Missing barrels   The difference between a published inventory of estimated oil production and actual oil production as observed by state agencies and commercial entities.

Mist   Water particles which are entrained in a gas stream.

Mist drilling   Air drilling with the addition of water, soap, or chemicals in order to clean the wellbore.

Mist extractor   A filter which removes entrained water or liquid phase petroleum from a gas stream in a separator.

Mitigation response   A risk mitigation tool which is intended to minimise the consequences of a risk which has occurred. *See also* **protection response.**

Mitigation well   A wellbore which is drilled to allow the separate release of connate, rather than to have the connate brought to the surface within a petroleum stream.

Mixed naphtha   A combination of straight run and blended naphtha.

MM:HH:DD:MM:YYYY   A protocol which is used in contracts for the recording of a precise moment in time (which is broken down by minute, hour, day, month, and year—for example, 'notice shall be given no later than 59:23:31:12:2022').

Mob/mobilisation   Under a drilling contract, the movement of a drilling unit from its original location to a drill site prior to drilling a wellbore. Also called **moving in rig**. Contrast with **demobilisation**, as the return of the drilling unit to its original location (or to another agreed location) after the drilling of a wellbore.

Mobile Arctic caisson (MAC)   An alternative term for an **Arctic submersible rig**.

Mobile offshore drilling unit (MODU)   An alternative term for a **drilling unit**.

Mobile offshore production unit (MOPU)   A mobile petroleum facility which can be used for offshore petroleum production (such as an FPSO or a jack-up).

Mobile petroleum   Petroleum which is capable of migrating freely through the pore spaces within a formation.

Mobility
(1)   A measure of the ability of petroleum to flow through a formation.
(2)   A measure of the resistance to the flow of a fluid through a formation, where the intention is that a displaced oil phase will move more quickly through a formation than the water phase which is displacing it.

Modal share   The percentage of a particular form of petroleum which is moved by a particular form of transportation. *See also* **intermodal transport**.

Model clauses   Published standard terms which are supplied by law as the basis of a concession for petroleum E&P and apply equally to all concession-holders.

Modular construction   A form of construction whereby the bulk of assembly takes place off site, with only final assembly taking place on site. *See also* **stick built**.

Modular containment vessel (MCV)   A modified crude carrier which is stationed offshore to capture and to store crude oil leakages from sub-sea areas.

Modular dynamic testing (MDT)   A method of measuring the pressure in a formation through a wellbore.

Modular formation dynamics tester (MFDT)   An item of equipment which measures flowline resistivity in order to identify different fluid flows in a formation.

Module   A defined area in a petroleum infrastructure item in which a defined operational activity (for example, compression, processing, separation) is carried out.

Mogas   An alternative term for gasoline, which is derived from combining 'motor' and 'gasoline'.

Mol% (mole per cent)   The molar presence of a constituent within a sample of petroleum when expressed as a percentage of the whole.

Molecular sieve   A device for the separation of petroleum fractions by passing a petroleum stream through a succession of absorbent substances with differing degrees of resistance.

Monitoring well   A small-diameter well which is drilled principally in order to monitor groundwater levels.

Monobore completion   A simple completion design which uses the same ID from the bottom of the wellbore to the surface.

Monocline   A formation in which the stratigraphic layers which it contains are inclined in the same direction.

Mono-ethylene glycol (MEG)   A desiccant and anti-corrosion agent which is applied to protect the internal integrity of a pipeline and to prevent the formation of condensation in the pipeline. *See also* **tri-ethylene glycol.**

Monopod   An offshore platform which is used for the production of petroleum and rests on a single leg. Also called a **monotower.**

Monopoly   A commercial situation in which there is only a single seller of a commodity in a given market. *See also* **duopoly** and **oligopoly.**

Monopsony   A commercial situation in which there is only a single buyer of a commodity in a given market. *See also* **duopsony** and **oligopsony.**

Monotower   An alternative term for a **monopod.**

Monte Carlo   A mathematical modelling programme which randomly and repeatedly samples input distributions in order to generate an output model, which is often used in estimating petroleum reserves within a formation as part of an appraisal exercise.

Moon pool   An opening in the hull of a drill ship or other offshore drilling unit through which a drill bit passes for drilling into the sea-bed below.

Mooring analysis   A study which determines the optimum mooring model for an FSU.

Mooring dolphin   A dolphin which is used only for securing the mooring of a ship, typically by rope. *See also* **breasting dolphin.**

Mooring master   A pilot who has particular responsibility for the berthing or unberthing of a ship at a port.

Mooring tension damper (MTD)   Equipment which applies an automatic tension in order to reduce the movements of a ship or offshore petroleum infrastructure item which are caused by wave, wind, or other ship movements.

Moose pasture   A colloquial term for an exploration area which is expected to be of poor prospectivity for the discovery of petroleum.

Morphodynamics   A study of the interaction of metocean conditions, sea-bed topology, hydrodynamic effects, and sedimentary movements, which is usually undertaken as a precursor to the commencement of a marine engineering project (such as the installation of a port facility, a platform, or a pipeline) and often as part of a wider metocean study.

Moss spherical tank   A spherically shaped form of containment vessel for LNG on an LNG carrier. *See also* **membrane tank.**

Most-favoured nation (MFN)   Under a petroleum project contract, a provision whereby a contracting party will be entitled to enjoy the benefit of any better contract terms which have been offered by the contracting party's counterparty to any other person.

Mother Hubbard clause   A provision, typically found in a US petroleum lease, which covers minor defects in the property description by providing that the lessee is protected against deficiencies in the description of the demised interest by provision that the leased interest additionally covers all the land which is owned by the lessor within a specific area.

Motor octane number (MON)   A measurement of the anti-knock rating of a given quantity of gasoline, which is typically measured at higher temperatures. *See also* **research octane number**.

Mountweazel   A commodity which is owned by a person and is utilised by another person without permission.

Mousehole   An opening in the floor of a drilling rig in which a joint is placed temporarily for later use as part of a drill string.

Moveable modular liquefaction system (MMLS)   A small-scale modular LNG production plant which can be used to develop onshore stranded gas deposits.

Moveable oil originally in place (MOOIP)   An alternative term for **stock tank barrel**.

Moving in rig (MIR)   An alternative term for **mobilisation**.

Mud   An alternative term for **drilling fluid**.

Mud acid   A mixture of acids and surfactants which are used to remove mud cake from a wellbore.

Mud analysis   The examination of circulating drilling fluid from a wellbore in order to assess the existence of petroleum in the wellbore. Also called **mud logging**.

Mud cake   A solid residue which drops out from a slurry such as drilling fluid and builds up on the permeable parts of a wellbore. *See also* **filtrate**. Also called **filter cake**.

Mud density recorder   A device which automatically records the weight of circulating drilling fluid in a wellbore.

Mud flow indicator   A device which continuously records the flow rate of circulating drilling fluid in a wellbore in order to ensure a consistent flow.

Mud gas separator   A device which removes entrained gas from drilling fluid when a kick is being circulated out of a wellbore.

Mud house   An alternative term for a **shaker house**.

Mud logger   A person on a drilling unit who has responsibility for examining circulating drilling from a wellbore in order to assess the existence of petroleum in an underlying formation.

Mud logging   An alternative term for **mud analysis**.

Mud return line   On a drilling unit, a flexible line which takes circulated drilling fluid from the top of the wellbore to storage tanks, for processing or for storage.

Mudline
(1) The areal coordinates of the sea-bed in which drilling operations are being conducted.
(2) A pipe which carries drilling fluid.

Mudline suspension unit   A valve assembly which is located on the sea-bed at the mudline and from which a wellbore which has been drilled is suspended.

Multi-azimuth   Seismic imaging which uses a wider collection of input data than is customarily generated by a relatively narrow three D seismic shoot by using more than one source of input data. Also called **wide azimuth**.

Multi-client seismic   A seismic survey which is carried out by a survey company at its own cost, where the results are then licensed (for a fee) to any interested petroleum companies. Also called **speculative seismic**. *See also* **seismic contractor option**, **seismic contractor uplift**, and **seismic option agreement**.

Multi-meter   An alternative term for a **meter accumulator**.

Multi-pad production facility (MPPF)   A facility for handling petroleum which is produced from multiple pads.

Multi-pay zone   Two or more petroleum-producing formations which are penetrated by a single wellbore.

Multi-shipper pipeline   A petroleum-transporting pipeline which is operated by a person for the transportation of petroleum on behalf of other persons. Also called a **common-carrier pipeline**. Contrast with an **equity pipeline**, whereby the pipeline is operated principally for the transportation of a person's own petroleum entitlements.

Multi-stage fracking (MSF)   Fracking operations which are conducted at multiple intervals within a horizontal wellbore.

Multi-tiered dispute resolution   An alternative term for **tiered dispute resolution**.

Multi-zone completion   A completion of a wellbore which uses more than one packer in order to seal off different wellbore sections to create separated payzones. Contrast with a **single zone completion**, in which a single payzone is produced through a single wellbore using a single packer.

Multilateral   A horizontal wellbore which deviates into multiple directions through branches which radiate out from a main wellbore.

Multilateral investment treaty (MIT)   An accord which establishes the terms for the protection of an investment which is made between multiple states for private investor reliance. *See*, for example, the Energy Charter Treaty. Contrast with a **bilateral investment treaty**, which is made between a state and an investor. *See also* **inter-government agreement**.

Multiphase   A petroleum stream which contains more than one component (principally crude oil or natural gas but also possibly liquids or water) in a commingled whole. Contrast with **single phase**, as a petroleum stream which contains only one of those components.

Multiphase flow meter (MPFM)   A flow meter which measures the individual flow rates of the different components within a multiphase.

Multiphase pipeline   A single pipeline which is configured to transport refined petroleum products of different grades (which are typically sent as distinct batches and separated in the pipeline by cutterstock) in a single stream. *See also* **fluid holdup**.

Multiple completion   A single production wellbore which is configured to produce petroleum from more than one reservoir or zone.

Multiple on invested capital (MOIC)   A method of calculating the level of return which is made by an investor on its investment into a petroleum project and is based on the gross value of the return versus the investment.

Multiple product dispenser (MPD)   A single pump facility which dispenses various different grades of fuel.

Must-take   An alternative term for **take and pay**.

Mutual hold harmless (MHH)   A non-fault-based liability allocation regime in a petroleum project contract whereby a person assumes responsibility for any loss of or damage to its own property and personnel interests, regardless of how that loss or damage was caused. Also called **knock-for-knock**. *See also* **Industry Mutual Hold Harmless**. Contrast with **guilty party pays**, as a fault-based liability allocation regime whereby a person assumes responsibility for any loss or damage which that person caused.

My watch/your watch allocation   An alternative term for **point in time allocation**.

N

Naked lease   A concession which is held by multiple parties but in respect of which only one party is liable for the eventual decommissioning costs.

Nameplate capacity   The notional, declared operational capacity of a petroleum infrastructure item. Contrast with **effective capacity**, as the actual level of operational capacity. *See also* **headroom** and **sandbagging**.

Nameplate pricing   A notional, declared pricing level of petroleum which applies under a contract or in respect of a market. Contrast with **effective pricing**, as the real level at which petroleum is priced under a contract or in respect of a market.

Naphtha
(1)   A naturally occurring liquid petroleum form.
(2)   A product which results from the refining of crude oil, which is often further processed to produce gasoline. *See* Appendix 2E.

Naphthenic-base crude oil   One of the principal forms of crude oil, which is based on the presence of naturally occurring naphtha.

National Balancing Point (NBP)   A notional point on the NTS at which shippers trade gas quantities, using the short-term flat agreement.

National oil company (NOC)   A state-owned, or majority state-owned, oil company which is established principally to develop and manage domestic petroleum interests for the benefit of the state. *See also* **international national oil company** and **international oil company**.

National regulatory authority (NRA)   An alternative term for a **regulator**.

National Transmission System (NTS)   The United Kingdom onshore natural gas pipeline transmission network (http://www.nationalgrid.com/uk/gas).

Nationalisation   A process by which local employees displace expatriate employees in a petroleum project. Also called **fade-out** and **localisation**.

Nationality planning   The selection of the state of incorporation of a petroleum company which will allow that company to access investment treaty protection of its investment.

Native gas   Residual, unproduced gas which remains in a depleted natural gas deposit when it is later configured for use as a gas storage facility.

Natural drive   An alternative term for **capillary pressure**.

Natural gas   Naturally occurring gaseous phase petroleum, which consists principally of methane and the progressively heavier petroleum fractions of ethane, propane, butane, and pentanes to octanes (and other traces of nonpetroleum gases).

Natural gas liquids (NGLs)   The propane and butane fractions which are extracted from a natural gas stream and exist at reference conditions in a liquid form. *See also* **raw mix liquids**.

Natural gas vehicle (NGV)   A motor vehicle powered by gas (such as CNG).

Natural gasoline   A natural gas liquid which is similar to naphtha which can be extracted from a gas stream and blended into gasoline.

Naturally occurring radioactive material (NORM)   Radioactive elements which occur naturally in low concentrations in petroleum in a formation.

Nature-based carbon offset   A carbon offset which is effected by natural means such as an afforestation or a reforestation project.

Near miss incident (NMI)   A near miss incident which has to be logged as such.

Negative buoyancy   The tendency of an object to sink when it is denser than its surrounding medium. *See also* **API gravity** and Appendix 6.

Negative capture   A variation to the law of capture, whereby a landowner can: (i) inject substances (such as water or gas) into a formation which underlies its land which will migrate to neighbouring lands and could adversely affect the production of mobile petroleum from those neighbouring lands; and/or (ii) engage in competitive drilling to secure the mobile petroleum.

Negative degradation   The improvement over time of the operational efficiency of a petroleum infrastructure item because of technology advances and operational improvements which together offset degradation.

Negative pressure test   A leak test whereby fluid pressure in a wellbore is lowered to see if petroleum leaks into the wellbore through the casing. *See also* **positive pressure test**.

Negative royalty   A net profit interest whereby the costs which are due for payment exceed the income which is due under the royalty interest.

Negotiated third-party access (NTPA)   A regulatory position whereby third-party access rights are the subject of negotiation between a petroleum facility owner and a facility user. Contrast with **regulated third-party access**, whereby third-party access rights are applied between the facility owner and the facility user according to regulated terms.

Nelson complexity index (NCI)   A measurement tool which is used to assess refinery complexity. *See also* **equivalent distillation capacity**.

Net beach volume   Gas which is delivered onshore after water and other impurities have been removed from the gross gas stream.

Net calorific value   A determination of the calorific value of petroleum which is produced by combustion when water remains gaseous and is not combusted. Also called **lower heating value**. Contrast with **gross calorific value**, whereby the calorific value determination also accounts for the combustion of water.

Net debt   An accounting measure of a company's financial situation which is based on subtracting the total value of the company's debts and liabilities from the total value of the company's cash and other liquid assets.

Net export   In the context of a state's total petroleum production portfolio, the part of that portfolio which is free for export rather than which is committed to meeting domestic market petroleum demands.

Net interest billing   The amount which is payable by a JOA party as its net working interest share of a gross interest billing in respect of a joint operation.

Net pay   A measure of the feet of pay which contains ERR of petroleum, which is sometimes further distinguished as net gas pay or net oil pay (depending on the nature of the petroleum which occurs in the net pay).

Net petroleum   The total quantity of petroleum which is produced from a formation multiplied by a concession-holder's net working interest to give the net share of petroleum which accrues to the concession-holder.

Net present value (NPV)   In respect of a petroleum project, the difference between discounted cash inflows and discounted cash outflows, which is used to analyse the prospective economics of the project as an investment.

Net product worth (NPW)   A method of estimating the value of a barrel of crude oil based on the value of the products derived from it after it has been refined but also after deducting the associated costs of transportation and refining. Contrast with **gross product worth**, which takes no account of the associated costs. *See also* **yield**.

Net profit interest (NPI)   A royalty interest whereby the payee (the royalty-holder) is entitled to a defined share of the petroleum which is produced under the concession, but subject to paying a corresponding share of certain of the costs of production of the petroleum. Contrast with a **gross overriding royalty**, whereby the royalty-holder's entitlement is not reduced by a share of the production costs. *See also* **negative royalty**.

Net PSM   Under an LNG sale and purchase contract, a profit-sharing mechanism whereby the costs of a diversion are first deducted from the gross profits before the net profits are then divided between the parties. Contrast with a **raw PSM**, whereby the costs of a diversion are met by one party before division of the net profits between the parties.

Net registered tonnage (NRT)   A measure of a ship's cargo-carrying capacity which is expressed in register tons, excluding spaces in the ship which are not able to carry cargo. Contrast with **gross registered tonnage**, as a measure of the total internal volume of a ship. *See also* **register ton**.

Net sand   A measure of the vertical thickness of a payzone which is made up of sand.

Net spend above carry   A measure of the costs which are due for payment by a person who is party to a concession after the capped level of a carried interest in favour of that person has been reached.

Net standard volume (NSV)   In the measurement of oil in-tank, the total volume, excluding sediments and water (and free water), at prevailing temperature and pressure but with a correction for reference conditions. *See also* **gross observed volume**.

Net throughput   A metered quantity of petroleum whereby the meter reading is adjusted for meter errors and the application of extraneous factors such as temperature and humidity.

Net–gross ratio (NTGR)   A measure of the proportion of an entire rock volume which displays petroleum-bearing potential (expressed as the total depth of a payzone divided by the total thickness of the formation).

Net well test   A farm-in structure whereby the farmee's obligation is set as the requirement to complete a defined schedule of works, rather than to incur a defined amount of expenditure. *See also* **drilling dollar test**.

Net working interest   The percentage participating interest of an individual concession-holder in a concession which is held by multiple concession-holders. Also called a **participating interest**. Contrast with a **working interest**, which represents the aggregate percentage participating interests of all the concession-holders. *See also* **real working interest**.

Netback differential   The difference at any time between a spot market price and a rack price for a product.

Netback pricing   The construction of a price for petroleum which is generated by working backwards from an ultimate source of revenue, less any operational costs incurred, in order to determine whatever is left for the seller as profit. Also called **back value pricing**. Contrast with **cost plus pricing**, whereby the price is determined by working forwards from the seller's costs and profit expectations in order to determine the seller's revenue requirements.

Netback royalty A royalty payment which is based on a wellhead petroleum price with the deduction of processing and transportation costs downstream of the wellhead. *See also* **first marketable product** and **historical method**.

NetConnect Germany (NCG) A notional point on the German gas pipeline network at which shippers trade gas quantities.

Netting The set-off of amounts which are due for payment either way between two or more persons, which results in a single net balancing payment to be made either way between those persons.

Network-bound A sector (such as gas transmission) in which the applicable regulatory model is characterised by the existence of finite, monopolistic facilities and infrastructure items.

Network code A contractual framework which regulates the operation of a multi-shipper pipeline between the pipeline operator and the shippers. *See*, for example, the Uniform Network Code in the United Kingdom (http://www.ofgem.gov. uk/).

New field wildcat (NFW) An alternative term for an **exploration well**.

New Jason Under a charterparty trading through the United States, a provision for endorsement on a bill of lading that the shipowner can recover in general average from the charterer (as the cargo owner) even if loss or damage to the cargo owners' interests was caused by the shipowner's negligent navigation of the ship.

New oil Petroleum which is produced during a defined initial phase of a concession to which a defined set of fiscal terms will apply. *See also* **old oil**.

New pool wildcat (NPW) An exploration well which is drilled into an already-producing formation.

New York Convention The 1958 Convention on the Recognition and Enforcement of Foreign Arbitral Awards (http://www.newyorkconvention.org).

New York Mercantile Exchange (NYMEX) A New York-based trading exchange on which crude oil, natural gas, and products are traded (http://www.cmegroup. com/company/nymex.xhtml).

Newbuild A newly constructed ship or offshore petroleum infrastructure item. *See also* **hull accord**.

Next twelve months (NTM) The measurement of a company's financial condition which is measured on a leading indicator of the following twelve-month basis. Contrast with **last twelve months**, as a lagging indicator.

Ninety-day programme A specific lifting schedule, which is so-called because it covers a rolling next ninety-day period at any time.

Nipple up The assembly and pressure testing of a blowout preventer.

Nitrogen injection The injection of nitrogen into an LNG cargo or a gas stream in order to reduce calorific value. *See also* **inerts**.

Nitrogen oxides (NOX) A pollutant which results from the combustion of fossil fuels. *See also* **sulphur oxides**.

Nitrogen purging
(1) The injection of nitrogen into a gas pipeline in order to remove residual oxygen and water prior to pressuring up the pipeline with gas.
(2) The use of nitrogen in order to clean the interior of a gas or LNG storage tank.

Nitrogen rejection unit (NRU)   A facility which removes nitrogen from a gas stream (often as a precursor to LNG production).

No dues certificate   A certificate which is issued to a ship's master by a port owner, confirming the ship owes no fees prior to the ship's departure.

No further investment (NFI)   A decision which is made by a petroleum project developer to stop the further funding of a petroleum exploration or production project.

Nodding donkey   An alternative term for a **beam jack**.

Noise   Any unwanted disturbance in an electrical signal, which is more widely assumed to mean any unnecessary communication.

Nomination   Under a petroleum sales contract, a mechanism for determining the quantities of petroleum which are to be delivered from time to time by the seller to the buyer. *See also* **buyer's nomination regime, renomination**, and **seller's nomination regime**.

No-notice service   A gas pipeline delivery service which allows the shipper to receive extra gas on demand, without making a prior nomination for that gas to the transporter and without paying additional fees or penalties.

Non-associated gas (NAG)   Natural gas which is produced in its own right and without ancillary crude oil production. *See also* **associated gas** and **associated free gas**. Also called **free gas**.

Non-circumvention   A contractual provision whereby a person or business cannot be circumvented and denied an otherwise legitimate commercial return within a transaction.

Non-consent   Under a JOA, a petroleum project activity which is carried out by less than all of the parties to the JOA and is paid for by only the participating parties, where the operating committee has approved the operation as a joint operation but some parties elect not to participate. Also called an **exclusive operation** and a **stand-out right**. *See also* **sole risk**.

Non-destructive testing (NDT)   A method of testing the integrity of a petroleum facility or the composition of petroleum which takes place without destroying the test sample.

Non-disclosure agreement (NDA)   A contractual commitment to maintain the confidentiality of certain data which is given to a receiving person by a disclosing person. Also called a **confidentiality agreement**.

Non-discrimination provision   A stabilisation provision whereby equal treatment is afforded to all investors in a state's petroleum projects.

Non-governmental organisation (NGO)   A not-for-profit organisation, which is unaffiliated to a state, which operates principally to promote a social, political, or environmental agenda.

Non-joint operation   An alternative term for an **exclusive operation**.

Non-methane organic compounds (NMOC)   Hydrocarbon-based gases other than methane which exist in a natural gas stream.

Non-operated minority interest (NOMI)
(1)   Under a JOA, the interest which is held by a non-operating party.
(2)   Under a royalty agreement, the interests of a royalty-holder.

Non-operated venture (NOV)   A joint venture which is viewed from the perspective of a party which is not the operator.

Non-operating party   Under a JOA, any party other than the operator.

Non-ownership   A regulatory principle that sub-surface petroleum deposits cannot, before the point of their production, be licensed for private ownership. Contrast with **ownership-in-place**, whereby sub-surface petroleum deposits can be so licensed.

Non-participating royalty interest (NPRI)   A royalty interest which does not confer on the royalty-holder a right or an obligation to participate in the performance of the underlying concession.

Non-persistent oil   A grade of crude oil or oil product which dissipates quickly when spilled in a marine environment, not leading to the need for extensive pollution management techniques to be applied. *See also* **persistent oil**.

Non-petroleum gases (NPG)   Non-hydrocarbon-based gases which exist in a natural gas stream.

Non-petroleum use (NPU)   A covenant in an agreement for the sale of land that the land will not thereafter be used for defined petroleum sector purposes. *See also* **petroleum use**.

Non-producing facility (NPF)   A petroleum-producing infrastructure item which has ceased to be operational, as a precursor to eventual decommissioning.

Non-productive time (NPT)   No-progress time which is built into the schedule for the drilling of a wellbore as a contingency or occurs without being scheduled, and which in either case delays the intended drilling of the wellbore.

Non-prop trading   An alternative term for **book trading**.

Non-recourse finance   The use of third-party financing to meet the costs of developing a petroleum project (such as in project finance) in which the lenders have no direct recourse to the sponsor for repayment of the financing. *See also* **limited recourse finance**.

Non-renewable gas   Natural gas which is produced from non-renewable (that is, fossil fuel) sources. Contrast with **renewable gas**, which is produced from renewable sources.

Non-technical risks (NTRs)   Risks to the integrity of a petroleum project which are posed by external factors (such as social, economic, or political issues) rather than by technical factors which are inherent within the project.

Non-technical summary (NTS)   A summary of operational information which is written in a way such that it is accessible to a non-technical reader.

Non-unit operations   Petroleum E&P operations which are carried out in a unitised concession but not in the unit interval. *See also* **unit operations**.

NOPEC   An alternative term for **OPEC+**.

Norm price   A price at which petroleum is traded between independent parties in a free market.

Normal circulation   The circulation of drilling fluid through a wellbore whereby the drilling fluid flows down the drill string, out of the drill bit and up the annulus between the drill string and the wellbore. Contrast with **reverse circulation**, whereby drilling fluid flows down the annulus and up the drill string.

Normal litre per minute (NLPM)   A unit for measuring the volumetric flow rate of gas at standard temperature and pressure, which is typically used in Europe. *See also* **standard litre per minute**, which is used in the United States.

Normally uninhabited (unattended) installation (NUI)   An offshore petroleum production, processing, or storage facility which ordinarily has no personnel on board while in operation. Also called a **not permanently attended installation** and an **unmanned production facility**.

North American Energy Standards Board (NAESB)   A US-based trade organisation which monitors issues associated with wholesale and retail gas markets in the United States (http://www.naesb.org).

Northings   Geographic coordinates which measure distance across a northwards axis from a fixed point, often using the universal transverse mercator coordinate system. *See also* **eastings**.

Norwegian offshore operational standards (NORSOK)   A defined set of operational standards which apply to petroleum E&P operations on the Norwegian continental shelf (http://www.standard.no).

Norwegian Petroleum Directorate (NPD)   A Norwegian state-owned agency which is responsible for regulating participants on the Norwegian CS (http://www.npd.no/en).

Nostro account   An alternative term for **prop trading**.

Not permanently attended installation (NPAI)   An alternative term for a **normally uninhabited installation**.

Notice of readiness (NOR)   The notice which is given by a ship's master of the actual or impending arrival of ship at a defined point at a loading port or an unloading port.

Notice of readiness accepted (NORA)   The non-shipping party's acceptance of a notice of readiness which has been tendered by a shipping party.

Notice of readiness rejected (NORR)   The non-shipping party's rejection of a notice of readiness which has been tendered by a shipping party.

Notice of readiness tendered (NORT)   The issue of a notice of readiness by a shipping party.

Notional contract   A petroleum sales contract which assumes a traded quantity of petroleum without an actual physical delivery. Contrast with a **physical contract**, as a contract which is based upon a traded quantity of petroleum with a physical delivery.

Notional gas   An alternative term for **virtual gas**.

Notional path   The shortest route along which petroleum will travel between two defined points within a pipeline.

Novation   A transfer of contractual rights and obligations from a contracting party to another person. Contrast with an **assignment**, whereby rights only are transferred.

Nuclear log   An alternative term for a **radioactivity log**.

Nuclear magnetic resonance (NMR)   An analytical tool which is used in downhole testing which estimates petroleum presence, petroleum properties, and pore spaces within a formation.

Objective depth   An alternative term for a **target depth**.

Obligation well   An alternative term for a **commitment well**.

Observation well   An alternative term for a **parametric well**.

Obsolescing bargain model (OBM)   A description of the concession relationship between a state and an investor which suggests that the relationship inevitably changes in favour of the state over time. *See also* **political bargaining model**, as a suggested means of disapplying the risk of the model.

Ocean bottom current (OBC)   A sub-sea water current.

Ocean bottom node (OBN)   A marine seismic survey technique which relies on nodes which are placed on the sea-bed as seismic receivers.

Octane rating   An alternative term for **anti-knock rating**.

Octroi   A duty which is levied on imported goods and materials.

Odourisation   A process for giving odourless natural gas a smell for safety reasons by injecting small quantities of organic sulphur compounds, such as a mercaptan, into a gas stream.

Off-hire   Under a charterparty, a period of time when the charterer is not obliged to pay hire to the shipowner, despite the ship still being on hire to the charterer.

Off-peak   A level of demand for energy at any time other than peak demand.

Off-production   The temporary cessation of the production of petroleum through a wellbore. *See also* **on-production**.

Off-system supply   Under a petroleum sales contract, a quantity of petroleum which is purchased by the buyer and is transported to the buyer by a person other than the buyer's customary transportation service-provider.

Official sales price (OSP)   A sales price for petroleum which is mandated by a state.

Offramp termination   A right of suspension or termination in a contract which is applied in an open and predictable manner (such as a fixed expiry date or the achievement of a target quantity of petroleum).

Offset   A part of a transaction for the sale of petroleum or for the award of a concession whereby the seller or the grantor requires the transfer of additional non-petroleum sector trade benefits to the seller or the grantor as a condition of undertaking the transaction.

Offset pricing   Under a petroleum sales contract, a price for petroleum which is derived by the set-off of amounts which are due either way between the seller and the buyer, such that only a net price is paid by the buyer to the seller.

Offset vertical seismic profile (OVSP)   A vertical seismic profile in which the source of seismic energy is offset some way from the seismic receivers. *See also* **walkaway vertical seismic profile** and **zero offset vertical seismic profile**.

Offset well
(1) A new or existing wellbore which is drilled adjacent to an existing wellbore, which is used to assess an underlying formation—*see also* **parametric well**.
(2) A wellbore which is drilled as part of a competitive drilling programme.

Offshore installation manager (OIM)   A person who is designated to act as the senior manager of an offshore drilling unit or petroleum production facility.

Offshore pollution liability agreement (OPOL)   A voluntary arrangement between operators on the UKCS to take financial responsibility for offshore oil pollution and remediation (http://www.opol.org.uk).

Offshore service rig (OSR)   A moveable offshore platform or petroleum infrastructure item which is used to provide support services to offshore petroleum E&P facilities.

Offshore supply vessel (OSV)   A ship which is used to supply an offshore facility during petroleum exploration or production operations.

Offshore support vessel (OSV)   A ship which is used to provide support during offshore petroleum exploration or production operations.

Offshore title transfer point (OTTP)   Under a petroleum sales contract, an identified offshore international boundary point at which title to ship-transported petroleum transfers from the seller to the buyer, which is typically used to adjust the anticipated incidence of taxation between the seller and the buyer. *See also* **high-seas sale**.

Offtake dependent project   A petroleum production project which needs an offtaker in place prior to development. Contrast with an **open offtake project**.

Oil and gas services equipment (OGSE)   Equipment which is used in petroleum exploration or production operations.

Oil and Gas UK Limited (OGUK)   A trade organisation which represents the interests of the United Kingdom upstream petroleum industry (now known as Offshore Energies UK) (http://www.oeuk.org.uk).

Oil charge   The migration of crude oil into a reservoir rock.

Oil column   The measured vertical height of a crude oil deposit within a formation above the oil–water contact.

Oil Companies International Marine Forum (OCIMF)   A trade organisation of petroleum-handling companies which monitors issues associated with petroleum shipment and terminalling (http://www.ocimf.org).

Oil down-to (ODT)   A vertical measure of the vertical thickness of a crude oil deposit in a formation, as the deepest point at which oil saturation is measured. *See also* **gas down-to**.

Oil equivalent (OE)   A base reference point by which the calorific value of any petroleum other than crude oil is compared to the calorific value of a given quantity of crude oil.

Oil Exploration Licence (OEL)   A form of Nigerian petroleum concession.

Oil field services (OFS)   A composite term for the drilling, seismic, processing, and other services which are supplied by service-providers in support of petroleum E&P activities.

Oil gasification   A process for the conversion of crude oil into gas.

Oil initially in place (OIIP)   A measure of the amount of crude oil which is estimated to be contained in a formation before oil production begins. *See also* **gas initially in place**.

Oil Mining Licence (OML)   A form of Nigerian petroleum concession.

Oil offloading line (OOL)   A jumper which is used to transport oil from an FPSO or an FSU to a shuttle tanker.

Oil offloading system (OOS)   In a petroleum production project, the chosen method of evacuating produced crude oil from an offshore production facility.

Oil pledge   A promise by a potential concession-holder in a licensing round to supply the grantor with a certain percentage of the produced petroleum if the concession is awarded to the concession-holder.

Oil price escalation (OPE)   The indexation of a petroleum price to price movements of published crude oil or oil products prices.

Oil Prospecting Licence (OPL)   A form of Nigerian petroleum concession.

Oil rim   The crude oil part of a mixed crude oil and natural gas deposit, which is typically smaller than the gas part of the deposit and might or might not be economically recoverable in its own right.

Oil sales outlet   In an oil storage tank, an outlet valve at the base of the tank from which oil is released for sale.

Oil sands   Naturally occurring mixtures of sand, clay, water, and ultra-heavy oil, which are extracted typically by strip mining and are processed to give syncrude. Also called **tar sands**.

Oil sands off-gas   A mixture of hydrogen and light petroleum fractions which are produced when ultra-heavy oil is processed to give syncrude.

Oil shale   Shale rock which contains solid phase crude oil which has not yet been transformed into a liquid phase by geologic heat and time and is released by heating the rock.

Oil string   An alternative term for **production casing**.

Oil window   A measure of the sub-surface depth at which oil is formed from kerogen. *See also* **gas window**. *See* Appendix 2A.

Oil zone   A zone from which crude oil is principally produced.

Oil-backed loan (OBL)   An alternative term for a **reserves-based lending**.

Oil-based mud (OBM)   Drilling fluid which uses oil as the base component. *See also* **synthetic-based mud** and **water-based mud**.

Oil-cut mud (OCM)   Drilling fluid which is recirculated through a wellbore with entrained oil.

Oil-equivalent gas (OEG)   The quantity of natural gas which needs to be combusted to give the same energy equivalent as a barrel of combusted crude oil.

Oil-mature   Source rock which has been exposed to sufficient temperature and pressure over geological time to generate crude oil. *See also* **gas-mature**.

Oil–water contact (OWC)   The boundary in a formation above which predominantly crude oil occurs and below which predominantly water occurs (which is usually a blurred rather than defined boundary).

Oiler   A wellbore which principally or exclusively produces crude oil. Contrast with a **gasser**, as a wellbore producing natural gas.

Oilfield service company (OSC)   A company whose business is the provision of oil field services.

Old gas   Natural gas which has already been committed for sale prior to its production.

Old oil   Petroleum which is produced after a defined initial phase of a concession to which a defined set of fiscal terms will apply. *See also* **new oil**.

Olefins   A collective term for ethylene, propylene, and butadiene (which are used variously to manufacture industrial chemicals, plastics, synthetic rubber, and lubricants) which are produced by catalytic cracking.

Oleochemicals   Chemicals which are derived from processed plant and animal fats. Contrast with **petrochemicals**, which are derived from processed petroleum.

Oligopoly   A commercial situation in which there is only a small group of sellers of a commodity in a given market. *See also* **monopoly** and **duopoly**.

Oligopsony   A commercial situation in which there is only a small group of buyers of a commodity in a given market. *See also* **duopsony** and **monopsony**.

On ground maximum (OGM)   The maximum permissible weight of a petroleum infrastructure item which is to be crane-lifted.

On rate   The point at which a person who operates a petroleum infrastructure item for profit becomes entitled to be paid the intended amount for their efforts.

On the rig (OTR)   Monitoring or processing equipment which is mounted on the topsides of an exploration or production platform.

On title   The point at which a concession-holder becomes a party to, and fully entitled to the rights and exposed to the responsibilities which are associated with, a concession.

On trend   A positive indication that a sub-surface geological feature extends beyond the confines of a concession area and into adjacent acreage.

On-board quantity (OBQ)   An alternative term for **liquids and solids remaining on board**.

On-production   A formation which is currently producing petroleum. *See also* off-production.

On-the-pump   An alternative term for **artificial lift**.

One cancels other (OCO)   Two separate petroleum trades which are made between the same parties which, if executed, will cancel each other out.

One for one   An alternative term for **ground floor**.

One C (1C)   An estimate of contingent resources. *See* Appendix 7.

One P (1P)   An alternative term for **proved reserves**. *See* Appendix 7.

One ship company   A company which is incorporated specifically to own and finance a single newbuild.

One-way   A tool which is not intended to be retrievable from a wellbore once it is downhole.

One-way pigging   The pigging of a pipeline system whereby the pig travels in one direction from point to point. Contrast with **round-trip pigging**, where the pig is multi-directional.

Onshore licensing round (OLR)   A licensing round for onshore acreage. Contrast with a **seaward licensing round**, for offshore acreage.

OPEC fund for international development (OFID)   A multilateral development finance institution which is managed by OPEC.

Open access   A regulatory principle whereby ullage in a petroleum facility is made open to third-party access (whether based on NTPA or RTPA).

Open area   A marine or terrestrial area which is not at a point in time the subject of the award of a concession for petroleum exploration or production. Contrast with **acreage**, which is leased or licensed.

Open award   A process by which a state awards a concession to a person for the conduct of petroleum E&P activities as part of an ongoing, open-ended process. Contrast with a **licensing round**, whereby concessions are awarded within a defined time period.

Open book   A requirement that a person claiming reimbursement or compensation must demonstrate the underlying calculation of the amounts claimed.

Open book estimate (OBE)   A contracting methodology by which the contractor can recover identified incurred costs and also a specified margin which is added to those costs. Also called **cost plus contracting**.

Open book modelling   An analysis of the costs of a petroleum project or a contract which is made transparent to all interested parties.

Open credit   A petroleum sales commitment which is founded on the inherent credit risk of a payor, without the provision of collateral support in support of the payor.

Open cycle gas turbine (OCGT)   An alternative term for a **single cycle gas turbine**.

Open door award   A model for the award of a concession by a state whereby the concession terms are negotiated directly between the state and an investor on a bespoke basis. Also called **direct negotiation**. Contrast with **competitive bidding**, whereby concession terms are made transparent and are contestable items in a licensing round.

Open flow   The production of petroleum through a wellbore from a formation which occurs naturally through inherent capillary pressure and without the intervention of artificial lift. Also called a **flowing well**.

Open flow potential (OPF)   The theoretical maximum flow rate of a wellbore if the bottomhole pressure could be lowered to match atmospheric pressure.

Open formation   A formation which has good porosity and permeability.

Open hole (OH)   A wellbore which is drilled without casing. Contrast with a **closed hole**, whereby a wellbore is cased.

Open hole completion   The completion of an open hole without the use of casing to protect the integrity of the wellbore. Also called **barefoot completion**.

Open hole gravel pack (OHGP)   An open hole which is completed with a gravel pack in place.

Open loop
(1) A process system which regulates a process to a defined outcome with manual intervention. *See also* **closed loop**.
(2) An LNG regasification system which uses sea water as a heat source to regasify LNG. Contrast with **closed loop**, which uses steam as a heat source.

Open offtake project   A petroleum production project which is developed without having first secured an offtaker. Contrast with an **offtake dependent project**.

Open procedure   A model for the award of a concession by a state whereby all interested persons are free to submit a bid to the state. *See also* **limited procedure** and **selective procedure**.

Open rack vaporiser (ORV)   A marine-located heat exchanger which is used to convert LNG to regas, using the ambient temperature of adjacent seawater to raise the temperature of the LNG to the point at which vaporisation occurs. *See also* **ambient air vaporiser** and **submerged combustion vaporiser**.

Open season   A defined time period in which, at the request of a regulator, ullage in a petroleum facility or a petroleum sale/purchase opportunity is offered to third parties. *See also* **third-party access**.

Operated-by-other (OBO)   A non-operated interest in a petroleum project which is held by a person under a JOA (where the other person is the operator under the JOA).

Operating (operation) and maintenance (O&M)   The provision of defined operating and maintenance services.

Operating (operation) and maintenance agreement (O&MA)   A contract which provides for the operation, maintenance, and management of a petroleum facility on behalf of the facility owner by another person. Also called an **asset management agreement**.

Operating basis earthquake (OBE)   An earthquake which could be expected to affect the site of a petroleum infrastructure item but through which the item would remain safe and functional. *See also* **safe shut-down earthquake**.

Operating committee (OpCom)   A committee which is representative of all the parties to a JOA and is intended to supervise the performance of the joint operations.

Operating committee meeting (OCM)   A meeting of the operating committee under a JOA.

Operating expenditure (OPEX)   The costs which are associated with providing the intangible operating (consumable) elements of a petroleum project. Contrast with **capital expenditure**, which relates to tangible item costs.

Operating lease   A leasing arrangement for a petroleum infrastructure item whereby the lessor takes the risks and returns which are associated with ownership of the item, and the lease is treated as an expense on the lessee's balance sheet. *See also* **finance lease**.

Operating rate   Under a drilling contract, the basic dayrate which is payable by the employer company to the drilling contractor for the hire of a drilling unit.

Operating ratio   The ratio of petroleum project operating expenses to petroleum project gross revenues (where a lower operating ratio indicates greater project profitability).

Operational balancing agreement (OBA)   An agreement between the owners of different but interconnected gas transportation pipelines whereby service level discrepancies are balanced between the pipelines and are thereby mitigated or eliminated. *See also* **wheeling**.

Operational flow order (OFO)   Under a pipeline transportation agreement, an order which is issued to the shipper by the transporter to balance petroleum inputs and/or offtakes in order to protect the operational integrity of the pipeline.

Operational risk assessment (ORA)   An alternative term for a **risk register**.

Operations management system (OMS)   A systematic requirement for petroleum operations which standardises practices and procedures and implements risk identification and mitigation processes.

Operations readiness   A transitional process for preparing a person to assume ownership and operational responsibility for a petroleum infrastructure item which is being constructed.

Operator   Under a JOA, the person who is designated and appointed to act as the operator. *See also* **hybrid operator**, **incorporated operator**, **party operator**, and **split operator**.

Operator's extra expense (OEE)   A policy of insurance which is available to an operator which covers the additional costs of regaining control of a runaway well.

Opex-covered transportation   An alternative term for **zero-tariff transportation**. **Optimum flow rate** The flow rate of petroleum from a wellbore which will over time result in the maximum ultimate recovery.

Ordinary oil   An alternative term for **black oil**.

Organic-rich rock   An alternative term for a **kitchen**.

Organisation of Petroleum Exporting Countries (OPEC)   An association of oil-exporting countries which coordinates global crude oil production levels and pricing (http://www.opec.org).

Organisation of Petroleum Exporting Countries + (OPEC+)   An association of oil-exporting counties which, although not members of OPEC, will sometimes collectively restrict oil production levels in support of OPEC's ambitions.

Organisation pour l'Harmonisation en Afrique du Droit des Affaires (OHADA)   An African multi-state treaty organisation which promotes a harmonised investment climate for projects within its membership (http://www.ohada.com).

Orientation   The positioning of a deflection tool in a wellbore so that a required degree of drift is achieved during the drilling of a wellbore.

Oriented core   A core which is obtained from a precisely defined location within a wellbore.

Oriented drill pipe   A drill string in a wellbore which is positioned to undertake directional drilling.

Orifice   An aperture through which a petroleum stream flows.

Orifice meter   A device which measures the flow rate of petroleum through a pipeline, whereby the flow rate is determined from the pressure differential which is created by the petroleum passing through a choke. *See also* **differential drop**. Also called an **orifice well tester**.

Orifice well tester   An alternative term for an **orifice meter**.

Original equipment manufacturer (OEM)   The original manufacturer of an item of equipment. Contrast with a **pattern manufacturer**, as a non-OEM.

Orimulsion   A fuel that is used for combustion in an electric power-generating facility which is manufactured by the combination of ultra-heavy oil, water, and a surfactant.

Orogeny   The geological deformation of a landmass which can bring a formation to prominence.

Orphan well   A wellbore which does not have an identifiable owner.

Orsat analyser   Portable equipment which is used for the analysis of gas mixtures.

Oslo and Paris Conventions (OSPAR)   A multi-state legislative instrument which regulates marine environmental protection in the north east Atlantic (http://www.ospar.org).

Other condition materials (Con-O)   Under a JOA, materials which are transferred by the operator to the parties for use in joint operations which cannot be classified as any of Condition A, Condition B, or Condition C materials.

Other peoples' money (OPM)   A colloquial term for the monetary amounts which are spent by the operator in the performance of joint operations under a JOA.

Out   A circumstance in which a party to a contract becomes contractually or legally entitled to avoid a liability for a breach of the contract to which that party would otherwise be exposed.

Out-of-the-money (OTM)   Under a petroleum sales contract, an assessment that current price performance and future price expectations will combine to make the contract likely to be loss-generative at the point of assessment. Contrast with **in-the-money**, whereby the assessment indicates that the contract is likely to be profit-generative.

Out-turn   The measured quantity of petroleum unloaded from a ship (which could lead to a buyer paying for actual quantities of petroleum rather than contracted quantities of petroleum).

Outage
(1)   Space which is left in a tank in order to allow for the expansion of a quantity of petroleum where the temperature of the petroleum increases over time.
(2)   A measure of the space which is unoccupied in a product container. Contrast with **innage**, as a measure of occupied space.
(3)   A period of time when a petroleum facility is out of use (which is typically unscheduled).

Outage measurement   A measurement of the ullage at any time in a petroleum storage tank.

Outboard block   A block which lies in the seaward direction from another block. *See also* **inboard block**.

Outcrop   The surface appearance of a formation. *See also* **field mapping** and **topography**.

Outer continental shelf (OCS)   An area of the continental shelf of the United States which does not fall under the jurisdiction of an adjacent US state.

Outpost well   An alternative term for a **step-out well**.

Outside battery limit (OSBL)   Any operational activities which take place outside a battery limit. Contrast with **inside battery limit**, whereby operational activities take place within the battery limit.

Over, short, and damage (OS&D)   A report which is issued by the recipient of a cargo as to whether the cargo has been over-delivered, under-delivered, or delivered damaged by the carrier.

Over the rack   An alternative term for **rack delivery/rack market**.

Over the tide (OTT)   The loading of a cargo into or the unloading of a cargo from a ship which continues throughout ebb and flow tidal cycles.

Overbalance   The drilling of a wellbore whereby drilling fluid in the wellbore will counterbalance the pressure of petroleum or water in the formation. Also called a **stabilised well**. Contrast with **underbalance**, whereby the pressure in the wellbore is less than the formation pressure.

Overbought   A petroleum-trading situation whereby prices are trading beyond the fair value of the underlying commodity and a sell-off occurs as trading positions are liquidated. *See also* **oversold**.

Overburden   Rock and soil deposits which lie above a formation. *See also* **interburden**.

Overfunded operation   A planned joint operation for which the operator has more funding from all the JOA parties than is needed. Contrast with an **underfunded operation**.

Overgauge hole   A borehole which has a diameter which is greater than the diameter of the drill bit which drilled the wellbore.

Overhead product/overhead stream   In a crude distillation unit, gaseous petroleum products which condense above the unit and leave it as a vapour. *See* Appendix 2E.

Overlift   The lifting of a quantity of petroleum by a producer at a defined point which is greater than the quantity of petroleum which the producer is entitled to lift. Contrast with **underlift**, as the lifting of a quantity of petroleum which is less than the producer's lifting entitlement.

Overmature   A formation which, because of its geological and compositional maturity, has gone beyond the point of being capable of producing petroleum.

Overproduction   A quantity of petroleum which is produced through a wellbore which exceeds a defined allowable quantity.

Overproving   The incurrence of costs which are associated with excessive amounts of appraisal and planning work in respect of a prospective petroleum project which does not proceed.

Overranging   A flow of petroleum through a meter which exceeds the maximum permissible flow rate of the meter.

Over-relinquishment   The voluntary relinquishment by a concession-holder of a greater part of a contract area than would be required by mandatory relinquishment.

Overriding royalty interest (ORRI)   An alternative term for a **gross overriding royalty**.

Overrun   Under a pipeline transportation agreement, a charge which is payable by the shipper to the transporter if the shipper exceeds its booked transportation capacity rights within the pipeline.

Overselling   The commitment of a petroleum producer to sell more petroleum to a buyer or a number of buyers than it has in store or has access to.

Overshot   A tool which is used for fishing, to recover items lost or stuck in a wellbore.

Oversizing   A deliberate increase in the planned capacity of a petroleum infrastructure item which is beyond the item owner's own capacity needs, so that incremental capacity in the item can be offered to third parties.

Oversold   A petroleum-trading situation in which prices are trading below the fair value of the underlying commodity and a buy-up occurs as trading positions are generated. *See also* **overbought**.

Overspend   Under a JOA, the making of expenditures by the operator which exceed the amounts of expenditure which the operator was properly authorised to make in respect of a work programme and budget or an authorisation for expenditure. *See also* **permitted overspend**.

Overtake   Under a gas sales contract, the offtake of a quantity of gas by the buyer at a defined point which is greater than the quantity of gas which the buyer is entitled to offtake. Contrast with **undertake**, as an offtake of gas which is less than the quantity which the buyer is entitled to offtake.

Overtopping   The tendency for wave movements to swamp an offshore petroleum infrastructure item.

Overtrawlable   A sea-bed mounted installation which is engineered to be protected from the risk of being snagged in a trawler's fishing nets.

Outpost well   An alternative term for a **step-out well**.

Outsourcing   The sourcing of resources for a petroleum project by a petroleum company from outside the company. *See also* **insourcing**.

Own-use gas   Gas which is used in a petroleum production or processing facility as fuel or for gas lift. Also called **fuel gas**.

Owner-controlled insurance programme (OCIP)   An insurance package in respect of a petroleum project which is arranged by the project owner. Contrast with a **contractor-controlled insurance programme**, whereby a project contractor arranges the insurance package.

Owner's costs   The aggregation of all the costs which are associated with the development of a petroleum project (including design and engineering, permitting, and financing costs).

Owner's lien   Under a charterparty, a provision which entitles the shipowner to exercise a lien on the cargo until any freight or demurrage which is owing by the charterer to the shipowner is paid.

Ownership-in-place   A regulatory principle that sub-surface petroleum deposits can, before the point of their production, be licensed for private ownership. Also called **absolute ownership** and a **mineral estate** and *see also* the **ad coelum doctrine**. Contrast with **non-ownership**, whereby sub-surface petroleum deposits cannot be so licensed.

Ownership unbundling (OU)   The disaggregation of energy generation, supply, and transportation interests at the corporate ownership level. Contrast with **functional unbundling**, whereby disaggregation takes place at the operational level.

Ozokerite   A naturally occurring semi-solid paraffin-based mineral. Also called **earthwax**.

P

P zero   An alternative term for **base price**.

Pacific Basin   A trading area for LNG production and consumption which is bounded by countries having a maritime coastline on the Pacific Ocean. *See also* **Atlantic Basin**.

Packer   An expanding plug which is used downhole in order to seal off a section of a wellbore.

Packer test   A test which is used to measure formation permeability from a section of a wellbore by using packers in order to isolate the target wellbore section.

Packing   The loading of linefill into a pipeline.

Pacta sunt servanda   A legal doctrine which suggests that the terms of a contract are inviolable and cannot be ignored except by agreement between the parties. *See also* **rebus sic stantibus**.

Pad
(1)   An onshore drill site.
(2)   An area from which multiple wells are drilled and operated.

Pad gas
(1)   An alternative term for **cushion gas**.
(2)   The gas which is injected during **padding**.

Padding   The injection of an inert gas into the vapour space of a tank in order to prevent the forming of an ignitable vapour-air mixture. Also called **inerting**.

Paid-up contract   A contract for the provision of services in which all the fees which are due to be paid to the service-provider are fully paid in advance of the provision of the services by the service-provider.

Paid-up lease   A concession under which the rentals which are due for the entire term are paid to the grantor in advance by the concession-holder.

Pairing agreement   An agreement which regulates the relationship between a group of persons to jointly develop petroleum exploration and/or production opportunities in a defined area of mutual interest for a defined period of time, typically with more detail than an AMIA and with the appearance of a non-concession specific JSBA. Also called an **alliance agreement**.

Paleogeography   The geography of a geological period.

Palynology   The study of fossilised plant remains in order to determine the origin and nature of sedimentary rock.

Pan tilt zoom   (PTZ) An internet-controlled camera where the user can control the movement and position of the lens from a remote location.

Panamax
(1) The largest ship size which can pass through the Panama canal (and any larger ship is a post Panamax).
(2) A crude carrier which has a deadweight tonnage of 65,000 metric tons.

Pancaking The accumulation of tariffs which are paid by a petroleum facility user that is using more than one facility.

Paper barrel A notional barrel of traded oil, in respect of which title is not backed up with a physical commodity. *See also* **wet barrel**.

Paper cargo In a daisy chain, a cargo of petroleum which is bought and sold at any time prior to final delivery to an end-user but without actual delivery of the cargo. Contrast with a **wet cargo**, as a cargo actually which is delivered to an end-user.

Paper well A time and cost estimate for the drilling of a wellbore.

Paradox of plenty An alternative term for **Dutch disease**.

Paraffin A petroleum product which can assume a liquid form or which can be found as a wax-like substance. *See also* **kerosene**.

Paraffinic-base crude oil One of the principal forms of crude oil, based on the presence of naturally occurring paraffin, which produces relatively high concentrations of lubricating oils when refined.

Parallel body length (PBL) A measure of the middle side area of a ship's exterior hull which is in contact with a pier when the ship is docked. Also called **mid-body length**.

Paramarginal resource A petroleum resource which could be removed from classification as a marginal resource if market conditions improve sufficiently.

Parameter A composite term for any of the geological or compositional characteristic of a formation.

Parametric well A wellbore which is drilled principally to gain information about a formation, rather than to find petroleum or to produce petroleum from a formation. Also called a **control well**, a **strat well/stratigraphic well**, and an **observation well**. *See also* **offset well**.

Parcel A part of a ship-transported cargo of petroleum which results from **parcelling**.

Parcelling The loading or unloading of a ship-transported cargo of petroleum which is undertaken in separate parts (typically for safety or inventory conflict management reasons), rather than as a single continuous activity.

Parent company guarantee (PCG) A written undertaking, which is issued by the parent company of a contracting party, whereby the parent company offers a formal guarantee of the contracting party's payment or performance obligations under a contract.

Pari passu The principle that the incidence of a cost, liability, or opportunity will be borne equally between all of the members of a defined group of persons.

Park and loan service (PALS)   A gas storage service whereby the storage facility user stores a given quantity of gas in a facility for later redelivery ('park') or borrows gas from the storage facility owner for later return ('loan').

Parking   A service for the provision of the temporary storage of gas by a gas storage facility owner.

Part cargo (PC)   A cargo in a ship which is anything less than a full cargo lot. *See also* **stub quantity**.

Partial purchase contract   A hybrid gas sales contract which displays certain elements of a depletion contract and a supply contract.

Partial risk guarantee (PRG)   A guarantee which is offered by a guarantor in respect only of a defined portion of the overall project risk of a petroleum project.

Participating interest   An alternative term for a **net working interest**.

Participation agreement
(1)   A contract which provides for the future provision of a carried interest to a person.
(2)   A contract which provides for the future right of a person to an earn-in or to a farm-in.
(3)   A rudimentary JOA which contains the essential terms of a proposed joint venture.

Participation doctrine   A regulatory principle which focuses on establishing and increasing the degree of state involvement in the state's own domestic petroleum sector.

Participation petroleum   A measure of the petroleum entitlement of a state which accrues to the state as a consequence of the state's direct or indirect equity participation in a petroleum project.

Participation right   The right of a state to participate at the concession-holder level in a concession (possibly also with the benefit of a carried interest) and also to become party to the underlying JOA. Also called a **back-in**.

Partly knocked down (PKD)   An item of equipment which is supplied in a partially assembled state and requires further assembly to be fully functional. *See also* **knocked down**.

Partner drag   Under a JOA, the situation whereby the operator needs the non-operating parties to follow its recommendations in order to effect a joint operation. Partner drag could be positive (if the non-operating parties follow the operator's recommendations) or could be negative (if the non-operating parties do not).

Party operator   Under a JOA, a term which is used to designate one of the parties which has been appointed to act as the operator.

Pascal (Pa)   A unit of pressure. *See* Appendix 1.

Pass   A single line seismic shoot of a survey vessel.

Passive survey   A survey of a petroleum facility which is conducted without active intervention (such as the visual inspection of the exterior of a pipeline). Contrast with an **active survey**, as a survey which is conducted with active intervention.

Passmark   Under a JOA, a provision whereby votes are taken by the parties according to their percentage participating interests, to count towards a defined percentage level to secure agreement between the parties.

Pass-through factor   In a petroleum pricing formula, a factor which passes through a defined part but not all of an indexation package to determine the petroleum price.

Passive acoustic monitoring (PAM)   The use of hydrophones to detect and monitor marine mammals around offshore petroleum infrastructure items.

Passive fire protection (PFP)   Materials and systems which are intended to prevent or to slow the effects of a fire.

Pattern manufacturer   A manufacturer of an item of equipment which is anyone other than the OEM.

Pay now argue later   Under a JOA, the principle that each party pays its share of the operator's cash calls without argument, subject to the later right of a party to audit the joint account or to dispute the cash call.

Paying quantity   A producing formation which generates a positive cash flow for the producer (where operating revenues exceed operating expenses) at a particular time, even if the overall economics of developing the formation are negative. *See also* **producible well**.

Payment in-kind interest (PIK)   Interest on a loan which is not repaid in cash by the borrower but which is added to the principal sum and is repaid to the lender when the principal sum is repaid. Also called **capital interest**.

Payoff point   A point in time at which the production of petroleum from a formation begins to generate revenues for the producer.

Payout   A point in time at which the aggregate costs of drilling a wellbore are recovered from revenues from the sale of petroleum which has been produced from the wellbore. Also called **upside**.

Payout point   In relation to a back-in after payout provision, the point at which the before payout regime ceases to apply and the after payout regime begins to apply.

Paysand   An alternative term for a **payzone**.

Payzone/pay   A part of a formation which contains ERP quantities. Also called a **paysand**.

Peak day   A day in a defined period of time upon which market demand for petroleum is at its highest.

Peak demand   A maximum level of demand for energy at a particular time. Contrast with **off-peak**, as the demand for energy at any time other than peak demand.

Peak gas   A supply of gas which is reserved to meet anticipated peak demand profiles.

Peak lopping   Gas production which can be ramped down from rapid response facilities. Contrast with **peak shaving**, as gas production which can be ramped up.

Peak oil
(1) A point in time at which the maximum rate of petroleum extraction from a formation is reached, after which extraction rates from the formation will decline.
(2) A theory, principally attributed to M. King Hubbert in 1956, that a point of maximum global oil production would eventually be reached, after which global oil extraction rates would decline irreversibly.

Peak shaving   Gas production which can be ramped up from rapid response facilities. Contrast with **peak lopping**, as gas production which can be ramped down.

Peak supply contract   A gas sales contract which reflects peak and off-peak demand profiles. Also called a **seasonal supply contract**.

Peaking plant   A gas-fired electric power-generating facility which is operated specifically in order to respond to peak demands for electric power.

Penalty clause   Provision in a contract whereby a party which is in default is exposed to a monetary or other liability which significantly exceeds the loss which is suffered by a contracting party as a consequence of the default.

Penalty gas   An alternative term for **excess gas**.

Pendulum arbitration   An alternative term for **baseball arbitration**.

Pendulum assembly   A bottomhole assembly which is used to eliminate drift in a wellbore during drilling.

Penetration   The entry of a drill bit into a formation.

Penetration point   The initial point of intersection of a wellbore with a formation.

Pentanes plus   An alternative term for **condensate**.

Per diem   A daily allowance which is paid to a person to cover costs (for transportation, accommodation, and incidentals) which are incurred while on a business trip.

Percentage of proceeds contract (POP)   A petroleum sales contract whereby a marketing entity aggregates and sells petroleum on behalf of a group of producers and remits an agreed percentage of the sales proceeds to the producers.

Percentage participating interest (PPI)   Under a JOA, an ascribed individual percentage value of a party's several interests in the underlying concession. *See also* **net working interest** and **working interest**.

Perched water   A sub-surface water accumulation which exists in an otherwise unsaturated zone.

Percolation   The tendency of entrained gas to rise in drilling fluid in a wellbore.

Percussion drilling   A method of drilling a wellbore which involves repeatedly lifting and dropping a cable tool into the bottom of a shaft in order to fracture a geological structure.

Perforating gun   An alternative term for a **bullet perforator**.

Perforation   The activity of punching a hole through casing (using a perforating gun) in order to test a wellbore or to facilitate improved formation connectivity. *See also* **jet perforation**.

Performance bond   An undertaking which is issued by a bank or other financial institution in support of a person's obligations under a petroleum contract or project. *See also* **demand guarantee**.

Performance concession award   The award of a concession whereby continued maintenance of the concession is made conditional upon the concession-holder achieving specified levels of performance.

Period   A sub-division of geological time. *See* Appendix 4. Also called a **system**.

Perimeter well   A wellbore which is drilled close to the line of delineation of the contract area of a concession, often in order to establish the dimensions of a unit.

Permanent cessation of production (PCOP)   An alternative term for **cessation of production**.

Permanent completion   The completion of a wellbore whereby the tubing and the wellhead are installed and after that completion and remedial work is performed through the wellhead and the tubing.

Permanent establishment (PE)   A suggestion by a national taxation authority that an entity has, by certain activities which it has undertaken, established a taxable presence in the authority's country.

Permanent gas   A gas which remains as such and cannot easily be liquefied.

Permanent International Association of Navigation Congresses (PIANC)   An international technical association which offers advice in relation to the use of waterborne transport infrastructure, now known as the **World Association for Waterborne Transport Infrastructure** (http://www.pianc.org).

Permanent sovereignty over natural resource (PSONR)   The principle that a state has an inherent right to own, control, and licence the natural resources within its territory for the benefit of its people.

Permeable/permeability   The extent to which a formation allows petroleum deposits to migrate freely through the pore spaces of its internal structure.

Permeability height (Kh)   The product of the permeability and the thickness of a producing formation.

Permissible exposure limit (PEL)   A measure of the maximum concentration of a toxic gas which a person can be exposed to without adverse effect.

Permit to operate (PTO)   A formal written system which facilitates and manages the operation of a petroleum facility.

Permit to work (PTW)   A formal written system which facilitates and manages the conduct of works on a petroleum facility.

Permitted overspend   Under a JOA, an overspend which the operator is permitted to make if the terms of the JOA give some latitude to the operator to do so. Also called an **authorised overrun**.

Persistent oil   A grade of crude oil or oil product which does not dissipate quickly when spilled in a marine environment, which leads to the need for the application of extensive pollution management techniques. *See also* **non-persistent oil**.

Personal protective equipment (PPE)   Safety clothing which is worn by personnel when working in petroleum project operations.

Personnel on board (POB)   A schedule of the total number of persons who are on board a ship or an offshore petroleum facility at any time.

Petcoke   A solid which is derived from crude oil and is used in metallurgy applications, such as the manufacture of aluminium and steel products.

Petrochemicals   Chemical products which are produced primarily from crude oil but also from natural gas in a refinery or in a bespoke petrochemical production facility. *See also* **oleochemicals** and **specialties.**

Petrodollar   A US dollar which is earned by a state from the sale of produced petroleum.

Petrol   An alternative term for **gasoline.**

Petrol station   A facility from which gasoline (and other comestibles) is retailed. Also called a **gas station** and a **retail outlet.**

Petrolatum   A semi-solid product which is made up of residual fuel oils and paraffin wax, which is used principally as a feedstock for pharmaceutical applications.

Petroleum   A composite term for natural gas, crude oil, and liquids. Also called **hydrocarbon.**

Petroleum exploration development licence (PEDL)   A licence for onshore petroleum E&P which is particular to the United Kingdom.

Petroleum investment fund (PIF)   A state-owned fund which promotes national growth in the petroleum sector, into which a petroleum company makes payments under the terms of a concession.

Petroleum, oil, lubricants (POL)   A composite term for refined petroleum products.

Petroleum system   A composite term which is used to describe the essential components of a conventional petroleum-bearing deposit: source rock, a reservoir (which is porous and permeable), and a seal. *See* Appendix 2A.

Petroleum use (PU)   A covenant in an agreement for the sale of land that can or will thereafter be used for defined petroleum sector purposes. *See also* **non-petroleum use.**

Petroleum window   The combined circumstances of chemistry, geology, and time which lead to the sub-surface formation of petroleum. *See* Appendix 2A.

Petroliferous   Containing petroleum (said of rock formations).

Petrology   The branch of geology which is concerned with the origin, structure, composition, and distribution of rocks.

Petrophysics   The study of physical and chemical rock properties and their interactions with the movement of fluids such as petroleum.

Phantom costs   A colloquial term for cost recovery claims under a PSC which are made by the concession-holder but where the relevant costs have not been incurred by the concession-holder.

Phase Any component of crude oil, natural gas, liquids, or water within a formation or within a petroleum stream.

Phase behaviour An indication of how different petroleum phases behave in sub-surface conditions with variables of pressure and temperature.

Phase envelope A pressure and temperature projection which plots the phases within a formation, which is made up of a bubble curve and a dew curve.

Photogeology The examination of aerial photography in order to assess surface features which could indicate a petroleum-bearing sub-surface formation.

Physical contract A petroleum sales contract which assumes a traded quantity of petroleum for physical delivery. *See also* **actuals**. Contrast with a **notional contract**, as a contract which is based upon a traded quantity of petroleum without a physical delivery.

Pig A specialist scouring and inspection tool which is used for pigging a pipeline. Also called a **go-devil**, a **rabbit**, and a **scraper**.

Pig launcher A device which launches a pig into a pipeline for pigging.

Pig receiver A device which recovers a pig from a pipeline after pigging.

Pigging The forcing of a pig through a pipeline to clean or inspect the pipeline. Also called **sphering**. *See also* **one-way pigging** and **round-trip pigging**.

Piggyback pipeline A secondary, small-diameter pipeline which is attached to a main pipeline and transports injection water or chemicals to a point of petroleum production.

Pilot A small-scale test or trial operation which is used to assess the potential for a greater commercial application.

Pilot boarding station (PBS) A point in the ocean at which a pilot boards an incoming ship, and is dropped off from an outbound ship, for piloting the ship into and out of a port. Also called a **boarding station**.

Pilot boarding station inbound (PBSI) The point at which a ship reaches the PBS on its voyage into a port. *See also* **pilot boarding station outbound**.

Pilot boarding station outbound (PBSO) The point at which a ship reaches the PBS on its voyage from a port. *See also* **pilot boarding station inbound**.

Pilot string A small-diameter drill string which is used to drill a pilot hole in a formation.

Pin The solid (male) end of a section of drill pipe. *See also* **box**.

Pinch-out The tapering out of a reservoir's thickness against a seal.

Pinnacle reef A sub-surface conical structure, which is usually higher than it is wide and is usually composed of limestone, in which petroleum could be present.

Pipe jack A hand tool which is used to lift and position a length of drill string when a wellbore is being drilled. Also called a **slip**.

Pipe rack A horizontal or vertical storage unit which is used to hold lengths of drill pipe on a drilling unit. *See also* **carousel**.

Pipe ram A blowout preventer which, when closed, forms a seal in a wellbore which closes around a drill string in the wellbore.

Pipe within a pipe   A pipeline operational model whereby a defined portion of the capacity of a pipeline is given over to a shipper, which might then act as a transporter in its own right in respect of that capacity.

Pipelay support vessel (PLSV)   A ship which is used to support a pipelay vessel in laying pipeline lengths offshore.

Pipelay vessel (PLV)   A ship which is used to lay pipeline lengths offshore.

Pipeline end manifold (PLEM)   A manifold at the end of a pipeline.

Pipeline end termination (PLET)   A flange at the end of a pipeline.

Pipeline natural gas (PNG)   Raw gas which is transported by pipeline.

Pipeline oil   Crude oil which has a chemical composition (particularly with reference to basic sediment and water) which makes it acceptable for pipeline transportation without prior processing. *See also* **bad oil**.

Pipeline operating principles (POP)   The published rules which govern the operation of a pipeline. Also called **pipeline system rules**.

Pipeline pre-commissioning   The preparation of a newbuild pipeline for petroleum transportation by confirming the pipeline's integrity. *See also* **dry pre-commissioning** and **wet pre-commissioning**.

Pipeline quality gas (PQG)   Natural gas which has a quality specification that makes it suitable for transportation in a pipeline.

Pipeline safety limit (PSL)   The upper pressure limit at which a gas pipeline is designed to operate.

Pipeline string   A pipeline which is made up of two parallel pipeline lengths (where each length is a string). *See also* **looping**.

Pipeline system rules (PSR)   An alternative term for **pipeline operating principles**.

Piper Alpha incident   The explosion of the Piper Alpha production and processing platform in the North Sea on 6 July 1988, resulting in 167 deaths. *See also* **Cullen Report**.

Pipes, flanges, and fittings (PFF)   A term for the common connective elements of pipework in a petroleum infrastructure item.

Piping and instrumentation diagram (P&ID)   A detailed schematic which is used to illustrate the interconnection of flow, instrumentation, and control devices in an operational process flow.

Piracy risks clause   Under a charterparty, a provision which sets out the course of action which is open to the master of a ship in the event that the ship, its cargo, or its crew would be put at risk because of piracy affecting the intended voyage.

Pitch   The bow to stern pitching movement of a ship or offshore petroleum infrastructure item.

Pivot point   An alternative term for a **kink point**.

Plan of development (POD)   An alternative term for a **field development plan**.

Planned for development (PFD)   A discovery which is intended for development but in respect of which the POD has yet to be completed.

Planned preventative maintenance (PPM)   A programme of scheduled maintenance which is agreed in respect of a petroleum facility.

Planned shut-down (PSD)   An alternative term for a **routine shut-down**.

Planning gain agreement (PGA)   An alternative term for a **community development agreement**.

Plant condensate   An alternative term for **stripped condensate**.

Plant gate sale   An alternative term for a **tailgate sale**.

Plat   An official map which indicates the areas of the petroleum exploration and/or production concessions which have been awarded by a state.

Plateau period   Under a petroleum sales contract, a defined period of time in respect of which the deliverable quantity of petroleum from the seller to the buyer is set at a defined level for a defined period of time.

Platform jacket   An alternative term for a **jacket**.

Platform LNG   LNG production facilities which are located on an offshore platform (rather than as an FLNGV or as an onshore LNG production facility).

Platform rig   A permanent offshore production platform from which development wells are drilled and tied back for production. *See also* **slot**.

Platform supply vessel (PSV)   A ship which is used for the ongoing supply of consumables to an offshore petroleum exploration or production platform.

Platformate   A petroleum fraction which results from the platforming process.

Platforming   The transformation of initially produced naphtha into lighter petroleum fractions in a refinery.

Play   A geographically and geologically referable petroleum system which has a distinct physical identity. *See also* **type curve**.

Play-based exploration (PBE)   A petroleum exploration philosophy which focuses on looking for plays rather than prospects to exploit.

Play opener   The first successful exploration well in a play.

Playaway provision   Under a petroleum sales contract, a provision whereby an exclusive sale or purchase commitment which binds the seller or the buyer is disapplied for defined circumstances (such as force majeure or suspension).

Plenum   A chamber within which air or another gas is at a pressure greater than outside the chamber, to prevent infiltration from outside.

Plimsoll line   A series of markings which are painted on the side of a ship or an offshore drilling unit which indicate the maximum depth to which the ship or the unit can be loaded or ballasted when afloat.

Plug and abandon (P&A)   The permanent sealing up of a wellbore. Also called **plugging**. Contrast with **plug and make safe**, whereby a wellbore is temporarily sealed up.

Plug and make safe (P&MS)   The temporary sealing up of a wellbore. Also called **plugging**, **suspension**, and **temporary abandonment**. Contrast with **plug and abandon**, whereby a wellbore is permanently sealed up.

Plug back   The activity of blocking off a lower section of a wellbore in order to exclude bottom water or to seal off a completion interval, so that certain operations can be performed in the upper section of the wellbore.

Plug n'perf   A method of multi-stage well completion in a cased wellbore which involves plugging and perforating different stages.

Plug trap   A geological configuration by which an intrusive body of rock has penetrated and deformed the surrounding rock layers, which could indicate the presence of a petroleum-bearing formation.

Plugging   An alternative term for **plug and abandon** and **plug and make safe**.

Plugging back   When a wellbore has been drilled to a certain depth, the abandonment of the bottomhole in favour of a completion in a shallower payzone.

Pogo plan   A financing plan for the conduct of a petroleum exploration programme through the incorporation of a special purpose vehicle company, which is often undertaken if high cost/low return and low cost/high return cycles are predicted.

Point   One percentage point of the notional value of a concession, which consists overall of one hundred points (and from which percentage participating interests and working interests are determined).

Point d'Échange de Gaz (PEG)   A notional point on the French gas pipeline network at which shippers trade gas quantities.

Point in time allocation   A liability allocation mechanism in a contract which allocates liability to persons by reference to either side of a defined point in time. Also called **my watch/your watch allocation**.

Point pricing   A method of fixing the price of petroleum by reference to a traded regional market price plus a factor to cover transportation costs.

Point-to-point   An alternative term for **tramline**.

Policy bank   A bank that lends money to petroleum projects which promote the national interest of that bank's home state.

Political bargaining model (PBM)   A description of the concession relationship between a state and an investor which recognises how the investor can use political and commercial pressure to maintain stability in the relationship. *See also* **obsolescing bargaining model**, which suggests that the relationship changes in favour of the state over time.

Political risk insurance (PRI)   A policy of insurance which pays out to an insured person when a defined political risk event (which could include an expropriation event) affects the insured person's interests in a petroleum project.

Polluter pays   A principle of liability allocation whereby a person who is responsible for a polluting event is solely responsible for remedying it.

Poly refrigerant integral cycle operation (PRICO)   A proprietary single mixed refrigerant process for gas liquefaction to produce LNG.

Polyformation   The combination of propane and butane with naphtha to produce gasoline.

Polymer flooding   A miscible flood which relies on the injection of a polymer into a formation.

Pontoon   A hollow buoyancy tank which gives support to an offshore petroleum infrastructure item.

Pool
(1)   A cooperative arrangement between different shipowners whereby their ships are administered collectively in order to maximise trading opportunities.
(2)   An alternative term for a **reservoir**.

Pooling/pooled   The aggregation of multiple petroleum deposits (which are not necessarily connected) for a joint development. Contrast with **unitisation,** as the development of contiguous cross-block petroleum deposits as a single unit. *See also* **forced pooling.**

Poorboy   A colloquial term for the development of a petroleum project whereby the developer has an incentive to use the lowest-priced available materials. Also called **shoestring**. Contrast with **goldplating**, whereby the project developer uses the highest-priced available materials.

Pop valve   An alternative term for a **pressure relief valve.**

Pop-up buoy   A submerged marker buoy which floats to the sea surface on receipt of a coded acoustic signal, which is often used to mark the presence of a sub-sea wellhead. Also called a **yoo-hoo buoy.**

Pore gas   Natural gas which is stored in the pore spaces of a formation. Also called **interstitial gas**.

Pore volume (PV)   The total volume of pore space in a formation.

Porosimeter   A device which determines the porosity of a formation.

Porosity and permeability (P&P)   The characteristics of a formation which is porous and permeable. *See also* **reservoir quality.**

Porous/porosity   A measure of the percentage of PV within a formation which is capable of retaining petroleum.

Port congestion charge   In petroleum shipping, the measure of compensation which is due for payment from the shipping party to the non-shipping party from the point at which used laytime exceeds allowed laytime. Also called **excess berth occupancy** and **damages for detention**. Contrast with **demurrage**, as payment which is due from the non-shipping party to the shipping party.

Port liability agreement (PLA)   An agreement between a port owner and the master of a visiting ship which regulates the allocation of liability for loss or damage between them.

Portfolio sale   A sale of generic (not source-specific) LNG which is procured by the seller from various different sources. *See also* **branded LNG.**

Poseidon Principles   An agreed set of principles which are used to track the carbon intensity of a ship or a fleet in order to assess compliance with decarbonisation targets (www.poseidonpriciples.org).

Position mooring (POSMOOR)   The maintenance of a ship's position on station using chain and anchor connections. *See also* **dynamic positioning**.

Positioning fee   Under a charterparty, a fee which is payable by the charterer to the shipowner to cover the costs of positioning the ship at a place ready for the start of the hire period.

Positive pressure test   A leak test whereby fluid pressure in a wellbore is increased to see if fluid leaks out of the wellbore through the casing. *See also* **negative pressure test**.

Possible reserves   Quantities of petroleum which are less likely to be recoverable than probable reserves. Essentially, there should be at least 10 per cent probability that the quantities of petroleum actually recovered will equal or exceed the possible reserves estimate. Also called **three P (3P)** and **P10** reserves. *See* Appendix 7.

Post-closing price adjustment accounting   Under a contract for the sale of a business, a provision whereby post-completion accounts are prepared in respect of the business and post-completion payment adjustments are made between the parties. Contrast with **locked box accounting**, whereby the price for the business is fixed at completion without post-completion adjustments.

Post hole   An ineffective wellbore (by distance, depth, or expense) which is drilled in order to comply with the requirement of a concession that a wellbore has to be drilled.

Post Panamax   *See* **Panamax**.

Post-production carry   A carry right which applies even after petroleum production revenues have started to accrue to the carried person. *See also* **pre-production carry**.

Post-salt   Petroleum deposits which overlie (and were deposited after) a naturally occurring layer of salt. *See also* **pre-salt**.

Post-stack   The creation of a stack whereby seismic migration is effected after the stack has been created. Contrast with **pre-stack**.

Postage stamp tariff   Under a pipeline transportation agreement, a tariff which is payable by the shipper to the transporter, is set at a flat rate, and is payable irrespective of the distance over which petroleum is transported. *See also* **distance tariff** and **zonal tariff**.

Posted price   An officially declared price for the wholesale or retail sale of a grade of petroleum. *See also* **tax reference price**.

Potable water (PW)   Water which is fit for drinking.

Potential impact radius (PIR)   A defined surface area which would likely be affected by an explosion at a petroleum facility.

Pound per square inch (psi)   A unit of pressure. *See* Appendix 1.

Pour-point   The lowest point of temperature at which crude oil will flow naturally without some manual intervention such as dilution or heating.

Pour-point depressant (PPD)   A chemical compound which is added to high-viscosity crude oil in order to prevent low temperature thickening.

Power gas   Gas which is released from the combustion of coal with a low calorific value (which consists principally of carbon monoxide and hydrogen), which is typically further combusted as a feedstock for electric power generation. Also called **producer gas**.

Power of attorney (POA)   A legal instrument by which one person appoints another person to act as its lawful attorney to perform certain acts, with either general or specific powers to so act being granted in favour of the attorney.

Power purchase agreement (PPA)   A contract which is made between a generator and an end-user for the sale and purchase of electric power.

Powerback   A gas to electric power project which is driven by downstream electric power demand and economics rather than by upstream production costs.

Powered emergency release coupling (PERC)   An alternative term for **emergency release coupling**.

Prairie dog plant   A topping refinery which is constructed in a remote area.

Pre-assembled rack (PAR)   A pipe rack which is assembled in a modular fashion in order to reduce on-site construction time.

Pre-assembled unit (PAU)   A process unit which is assembled in a modular fashion in order to reduce on-site construction time.

Pre-commissioning   In the development of a new petroleum facility, the initial proving of the ability of the facility to produce, process, transport, or store petroleum which takes place prior to commissioning.

Pre-emption   The right of a party to a joint venture or of a shareholder in a company to assume the rights and interests of a co-venturer or of another shareholder when that co-venturer or shareholder intends to sell or otherwise dispose of those interests. *See also* **right of first offer** and **right of first refusal**.

Pre-FEED   A process step which provides for the conceptual design of a petroleum project, as a precursor to the front-end engineering design stage.

Pre-heater   A heater which warms a sample of crude oil prior to conducting a centrifuge test.

Pre-heating   In pipeline construction, the heating of the ends of the pipeline sections prior to welding them together.

Pre-inversion   A description of a formation before inversion has taken place.

Pre-production carry   A carry right which applies before petroleum production revenues have started to accrue to the carried person. *See also* **post-production carry**.

Pre-salt   Petroleum deposits which underlie a naturally occurring layer of salt. *See also* **post-salt**. Also called **sub-salt**.

Pre-stack   The creation of a stack whereby seismic migration is effected before the assembly of the stack. Contrast with **post-stack**.

Pre-stack depth migration (PSDM)   A seismic processing technique which determines data on a depth-related basis.

Pre-stack time migration (PSTM)   A seismic processing technique which determines data on a time-related basis.

Pre-start-up audit (PSUA)   An alternative term for **start-up assurance review**.

Pre-start-up safety review (PSSR)   An audit check which is performed prior to an equipment item coming into operation to ensure adequate process safety management has been performed.

Pre-unit agreement (PUA)   An agreement which is entered into prior to a UUOA, to record the basic transaction terms of the intended unitisation between the participating parties.

Precautionary principle   A risk management tool whereby early steps are taken to mitigate the consequences of a potentially harmful act at a time when those consequences are not fully understood. Also called **caution in advance**.

Predictive emissions monitoring system (PEMS)   A process for the monitoring of emissions from a petroleum infrastructure item which is carried out on a selective basis from the item and is used to extrapolate assumed emissions levels. *See also* **continuous emissions monitoring system**.

Predictive preventative maintenance (PPM)   The scheduled maintenance of a petroleum facility which is intended to reduce or to eliminate the risk of future operational failures.

Pref right   A preferential right of a person to buy an asset or an interest. *See also* **right of first offer** and **right of first refusal**.

Premise demand factor   An alternative term for **dedicated design day factor**.

Premium   Under a JOA, a monetary amount which is payable by a non-participating party to the participating parties to back-in into an exclusive operation and is expressed as a percentage of the base costs which were incurred by the participating parties.

Premium motor spirit (PMS)   An alternative term for **gasoline**.

Prenflo   A term which is derived from a combination of 'pressure', 'entrained', and 'flow' to describe a proprietary industrial process for converting coal or coke into synthetic natural gas.

Prepay   A basic credit risk management whereby a buyer of petroleum or a user of services makes payment in advance to the seller or the service-provider for that petroleum or for those services.

Present value 10 (PV 10)   An SEC-mandated methodology for determining the present value of a petroleum company's estimated future petroleum revenues, reduced by direct expenses, and discounted at an annual rate of 10 per cent.

Presentation   The status of the tanks of an LNG carrier which arrives for the loading of a cargo. *See also* **cooled with heel**, **under vapour**, **warm and inerted**, and **warm under vapour**.

Pressed up   A tank on a ship which contains liquid which is full or almost full. *See also* **free surface effect** and **slack tank**.

Pressure activated circulating valve (PACV)   A valve arrangement which, when triggered, permits the flow of produced petroleum.

Pressure control   The ability to effectively control the movement of gas into or out of a pipeline through managing pressure differentials between the pipeline and any non-pipeline facilities.

Pressure depletion   The production of gas from a formation until the capillary pressure in a formation has dropped to the point at which all naturally recoverable gas has been recovered.

Pressure drop   A loss of pressure within a petroleum stream which is caused by the stream passing through a choke or a reduced diameter pipeline section. *See also* **Venturi effect**.

Pressure habitat   An alternative term for a **hyperbaric chamber**.

Pressure integrity test (PIT)   An alternative term for a **leak test**.

Pressure maintenance   An alternative term for **artificial lift**.

Pressure reducing station   A choke point which reduces the pressure of a flow of gas through a pipeline.

Pressure relief valve   A safety valve in a pipeline which closes when a pre-set level of pressure is achieved in the pipeline in order to protect the pipeline from the consequences of excessive pressure. Also called a **pop valve**.

Pressure swing adsorption (PSA)   A technology which is used to separate certain gas elements from a mixture of gases under pressure according to the molecular characteristics of the gas elements to be separated.

Pressure transient analysis (PTA)   The analysis of pressure changes over time in a formation, which are especially associated with small variations in the volume of petroleum, which can indicate the deliverability of petroleum from the formation.

Pressure vacuum valve (PVV)   A valve arrangement which maintains in-tank pressure within defined limits.

Pressure, volume, temperature (PVT)   The essential operational characteristics of a gas stream. *See also* **equation of state** and the **Gas Laws**.

Pressure while drilling (PWD)   A form of **measurement while drilling**.

Prewash   The cleaning of the tanks in a crude carrier before the loading of a cargo.

Price maker   A person, product, or project which can dictate the petroleum prices that it charges or pays because of a lack of effective competition. Contrast with a **price taker**, as a person, product, or project which can only accept prevailing petroleum prices in the market.

Price premium
(1)   In petroleum project development economics, the expectation of the developer that the realised petroleum price will be higher than the corresponding costs of exploration, extraction, and delivery, as a condition of undertaking a project.
(2)   Under a petroleum sales contract, an additional element to the base price which is payable to the seller in certain circumstances.

Price reporting agency (PRA)   An independent agency which compiles trade data (such as S&P Global Platts or Argus Media).

Price review clause   Under a petroleum sales contract, a provision for a periodic review of the pricing of petroleum. *See also* **adaptation clause.**

Price taker   A person, product, or project which can only accept the prevailing petroleum prices in the market because it lacks market influence. Contrast with a **price maker**, as a person, product, or project which can dictate the petroleum prices which it charges or pays.

Pricing corridor   An analysis of market price differentials for petroleum which apply between different countries.

Primary capacity   Capacity in a pipeline which is let by the transporter to the shipper. Contrast with **secondary capacity**, which is let by the shipper to a sub-shipper.

Primary cementing   The first run of cementing which takes place after casing has been inserted into a wellbore.

Primary containment   A principal means of containing a petroleum leak or spillage, such as a containment boom (offshore) or a firewall (onshore).

Primary fuel   A fuel which directly produces energy upon initial combustion (such as natural gas). Contrast with a **secondary fuel**, which is processed from a primary fuel.

Primary migration   The movement of mobile petroleum out of a source rock. *See also* **secondary migration** and **tertiary migration**. *See* Appendix 2A.

Primary recovery   The first stage of petroleum production, in which capillary pressure displaces petroleum from a formation into a wellbore without intervention. Also called the **flush phase**. *See also* **secondary recovery** and **tertiary recovery**.

Primary target   A formation which is principally targeted when an exploration wellbore is drilled. *See also* **secondary target.**

Prime mover   The principal source of power for a drilling unit.

Prior appropriation rule   A principle whereby the first person to undertake drilling in a formation has the right to extract the entirety of mobile petroleum from the formation. Also called **common property**. *See also* the **law of capture.**

Probabilistic modelling   A method of petroleum project modelling which, from all of the variable inputs, yields a range of possible solutions with weighted likelihoods of outcome. Contrast with **deterministic modelling**, which yields a single solution.

Probability of success (POS)   A percentage ranking of the likelihood of success of an activity (such as drilling a wellbore). Also called **chance of success.**

Probable reserves   Quantities of petroleum which are less likely to be recovered than proved reserves but more certain to be recovered than possible reserves. Essentially, there should be at least a 50 per cent probability that the quantities of petroleum actually recovered will equal or exceed the probable reserves estimate. Also called **two P (2P)** and **P50** reserves. *See* Appendix 7.

Process energy requirement (PER)   A measure of the energy which is required to enable a petroleum production, processing, or transportation process.

Process gain   The volumetric amount by which the total output of a refinery is greater than the corresponding feedstock input for a given period of time. *See also* **process loss**.

Process heat   Heat which is used in an industrial process in order to achieve a particular operational outcome.

Process loss   The volumetric amount by which the total output of a refinery is less than the corresponding feedstock input for a given period of time. *See also* **process gain**.

Processed gas   Raw gas which is processed for the removal of impurities.

Processing and operating services agreement (POSA)   A contract whereby the owner of a petroleum facility offers a petroleum-processing service to a person as a facility user. Also called a **hosting agreement**, a **production handling agreement**, and **up and over**. Related to a **transportation, processing, and operating services agreement**.

Processing rights   An alternative term for **seller's reservations**.

Produce the limit (PTL)   A review methodology which examines how best to maximise the future production of petroleum from a formation.

Produced and saved (P&S)   A measure of the gross quantity of petroleum which is produced from a defined interest or area, prior to the processing, consumption, storage, or transportation of that petroleum.

Produced into pipeline (PIP)   Petroleum which is sold at the point when it has been produced and delivered into a pipeline for transportation.

Produced water (PW)
(1)   Water which is extracted from the sub-surface with petroleum (which could include connate and water which has been injected into the formation during the petroleum production process). Also called **brine** and **waste water**.
(2)   Water which results from an industrial process for the processing or the production of petroleum.

Produced water overboard (PWO)   A measure of the produced water which is discharged into the sea from an offshore petroleum infrastructure item.

Producer gas   An alternative term for **power gas**.

Producers 88   The most widely used form of mineral rights lease in the United States.

Producible well   A wellbore which is capable of delivering a paying quantity.

Producing zone   The part of a formation from which petroleum is produced.

Product cracks   The gross margin which is realised from the sale of individual products which result from processed petroleum from a refinery.

Product information bulletin (PIB)   An alternative term for a **material safety data sheet**.

Product sales agreement (PSA)　A contract for the sale and purchase of products.

Product slate　An alternative term for a **crude slate**.

Production　The production of petroleum from a discovery which has been developed.

Production analogue　An alternative term for an **analogue**.

Production baseline　An estimate of the quantity of petroleum over time which is expected to be produced from a formation (against which the actual performance of the formation can be assessed).

Production bonus　A monetary sum which is payable by a concession-holder to the grantor under the terms of a concession as a bonus payment at the point at which a specified level of petroleum production has been attained by the concession-holder.

Production casing　The last string of casing which is set in a wellbore, which usually contains a liner, as a precursor to the production of petroleum through the wellbore. Also called **long string** and **oil string**. *See* Appendix 2C.

Production ceiling　The maximum producible amount of petroleum from a formation, which is set by operational or regulatory constraints.

Production enhancement contract (PEC)　A form of concession which is granted particularly to facilitate enhanced oil recovery from mature fields.

Production gap　The extent of unproduced petroleum which results from a throughput cutback.

Production handling agreement (PHA)　An alternative term for a **processing and operating services agreement**.

Production logging　A generic term for wellbore services including cement monitoring, formation fluid measurement, and plug n'perf.

Production logging tool (PLT)　A tool which is used for production logging.

Production maintenance　An activity which is undertaken in order to arrest the decline in production from a wellbore (including acid wash and enhanced oil recovery).

Production-only farm-out　A form of farm-out agreement whereby the farmee will only secure its interests if the drilling of an exploration well results in a commercial discovery.

Production payment　A royalty interest which subsists only until a defined amount of petroleum or money has been recovered by the royalty-holder and does not subsist for the lifetime of the underlying concession.

Production possibility curve　A predictive tool which indicates the maximum petroleum output possibilities from a discovery.

Production quotas　Agreed levels of crude oil production which each OPEC member state commits to produce up to in order to regulate the international market price of crude oil.

Production services　An alternative term for **lease and operate**.

Production sharing contract (PSC)   A concession for petroleum E&P whereby the concession-holder first recovers its cost oil (or cost gas) expenditures from petroleum production revenues and then receives a share of profit oil (or profit gas). Contrast with a **revenue sharing contract**, whereby the concession-holder pays those expenditures from its agreed share of the gross revenues. *See also* **EPSA**, **ExplrPSA**, and **ExpltPSA**.

Production tail   An alternative term for a **tail**.

Production test   A test which is used to determine the likely daily petroleum production rates from a formation.

Production tubing   A liner which sits within production casing through which petroleum flows to a wellhead. *See* Appendix 2C.

Productivity index (PI)   A mathematical means of expressing the ability of a reservoir to deliver petroleum. *See also* **productivity index curve**.

Productivity index curve   A graph which is used to map a productivity index.

Products   Petroleum products which result from the refining and processing of crude oil or natural gas. *See also* **cuts**, **finished products**, and **semi-finished products**.

Profile   The forecast production levels of petroleum from a formation.

Profile testing   A technique for the simultaneous sampling of a petroleum stream in a pipeline which takes samples from across the pipeline's diameter in order to identify stratification within the pipeline.

Profit à prendre   A non-possessory interest in land which gives the holder a right to extract mineral and other interests from the land.

Profit oil (profit gas)   Under a PSC, the proportion of produced oil (or gas) which is available for distribution between the grantor and the concession-holder. Contrast with **cost oil (cost gas)**, which is used to pay back the concession-holder for its capital investment. Also called **equity oil (equity gas)**.

Profit-sharing mechanism (PSM)   Under an LNG sales contract, a provision whereby the incremental costs and revenues accruing in respect of a diversion are shared between the seller and the buyer according to an agreed ratio. *See also* **net PSM** and **raw PSM**.

Progressive cavity pump (PCP)   A type of pump for the transfer of a fluid which gives a relatively fixed flow rate.

Progressive regime   A fiscal regime which is applicable to a concession whereby state take increases as project profitability improves. Contrast with a **regressive regime**, whereby state take remains static or decreases relatively as project profitability improves.

Project bond   A long-term, specific-purpose bond which is issued by the developer of an identified petroleum project, where the amount of money which is raised by the bond is committed to meet the development costs of the project.

Project finance   The use of third-party financing to meet the costs of developing a petroleum project whereby the principal means of repaying the lenders is through the revenues which are generated by the project. *See also* **limited recourse finance non-recourse finance, sponsor,** and **sponsor support.**

Project life coverage ratio (PLCR)   A ratio test which is used in project financing and reserves-based lending which assesses the ability of the borrower to repay the debt over the anticipated life of the project. *See also* **field life coverage ratio.**

Project management team (PMT)   A team of personnel which is assembled in order to undertake a task on a petroleum project.

Promissory note   A written promise by a person to make payment of a defined sum of money on demand or at a defined time from one person to another person.

Promote   Under an earn-in agreement or a farm-in agreement, the relationship between the percentage of the seller's total required work costs which are assumed by the buyer and the net working interest which is sold down by the seller to the buyer, which is expressed as a percentage of the working interest. *See* Appendix 3.

Promote licence   A concession for petroleum E&P activities which is typically intended for small and start-up companies, with reduced expectations for up-front financial competence from the concession-holder.

Promoter   A petroleum company with a business model which is focused on acquiring assets and interests and disposing of them for profit as soon as possible, with minimal development costs being incurred between the acquisition and the disposal.

Prompt delivery   Traded petroleum which is intended for delivery later in the same month as the month in which the trade is made.

Proof of concept (POC)   A realisation of a conceptual design idea to prove its suitability for application in an operational context.

Prop trading   Energy trading which is carried out on a speculative basis. Also called **nostro account** trading. Contrast with **book trading**, where energy entitlements within a portfolio are traded for optimisation.

Propane/butane ratio (P/B)   The ratio of propane to butane in any quantity of LPG.

Propane dehydrogenation (PDH)   A process for the production of polymer grade propylene from propane, without cracking.

Propane grades   Varying grades of propane, which are reflective of their commercial uses. *See also* **HD5, HD10, commercial grade.**

Proppant   Microscopic sand or ceramic particles which are intended to preserve the porosity which is created by fracking to allow petroleum migration through pore spaces. Also called **propping agent.**

Propped length   The length of a horizontal wellbore in which proppants are intended to travel.

Propping agent   An alternative term for **proppant.**

Prorationing   An alternative term for **curtailment.**

Prospect   A formation in which economically recoverable reserves of petroleum are predicted to exist.

Prospective resources   Estimated quantities of petroleum which are less certain of recovery than contingent resources. *See* Appendix 7.

Protection acreage   A concession area which is secured by a petroleum company for E&A if an adjacent petroleum deposit might extend into it.

Protection and indemnity club (P&I)   A not-for-profit cooperative insurance association which offers protection and indemnity marine insurance cover to shipowners.

Protection casing   A string of casing which is set in a wellbore within the outermost casing in order to permit the wellbore to be deepened with a reduced risk of collapse from external pressure.

Protection response   A risk mitigation tool which is intended to prevent a risk from occurring. *See also* **mitigation response**, which minimises the consequences of an occurrence.

Proved developed not producing (PDNP)   Reserves which have been proved and developed ready for production but which are not presently in production.

Proved developed producing (PDP)   Reserves which have been proved and developed and are presently in production.

Proved not developed (PND)   Reserves which have been proved but which are not presently developed.

Proved reserves   Quantities of petroleum which can be estimated with reasonable certainty to be commercially recoverable, from a given date forward, from known reservoirs and under defined economic conditions, operating methods and regulations. Essentially, there should be at least a 90 per cent probability that the quantities of petroleum actually recovered will equal or exceed the proven reserves estimate. Also called **one P (1P)** and *P90* reserves. *See* Appendix 7.

Proved reservoir (PR)   A reservoir which contains proved reserves.

Proved undeveloped (PUD)   Petroleum resources which have been proved and developed ready for production but where relatively major expenditure is required for recompletion.

Proven offshore technology (POT)   Terrestrially developed technology which has been proved to operate successfully in a marine environment. *See also* **marination**.

Proximity   The situation where a new pipeline or cable runs close to (but does not cross) another existing pipeline or cable (and the proximation is regulated by a proximity agreement).

Proximity agreement   A contract which is made between the owner of an existing pipeline or cable and the owner of a new pipeline or cable which regulates rights, obligations, and liabilities between the parties in respect of the intended proximation.

Proxy invoice   An invoice which is issued by a seller for a quantity of petroleum which the seller has failed to deliver so that a buyer's remedy can be established.

Pseudo-regulation   An alternative term for **crypto-regulation**.

Pseudo-steady state (PSS)   A model for petroleum flows from a formation which has boundaries to natural flow-out such that the formation effectively acts as a tank.

Public service obligation (PSO)   A requirement which is imposed on a person to provide services which might not otherwise be profitable but which are judged to be in the public interest (such as the supply of diesel and gasoline to remote areas).

Puddling   The agitation of cement which takes place before cementing starts in order to remove entrained air bubbles.

Pugh clause   Under a petroleum lease, a provision whereby production which is attributed to a unitised or pooled interest will not include production from any part of the interest within the lease area but is not included within the defined unitised or pooled interest, which is sometimes further defined as a vertical Pugh clause or a horizontal Pugh clause (depending on the stratas to which it relates). Also called a **freestone rider**.

Pull/pulling   The activity of withdrawing casing, drill pipe, or tools from a wellbore.

Pull back   An alternative term for a **trip**.

Pull dry   The removal of a drill string from a wellbore without first filling the wellbore with drilling fluid.

Pump jack   An alternative term for a **beam jack**.

Pump pressure   The pressure of drilling fluid in a wellbore which results from the use of a pump.

Pump rate   The speed at which a drilling fluid pump is run during the drilling of a wellbore.

Pump station   An installation on an oil pipeline which contains pumps, heaters, filters, and other devices which are necessary to maintain the flow of oil through the pipeline.

Pumpkin provision   A colloquial alternative term for a **Cinderella clause**.

Punch through   The unintended penetration of the legs of a jack-up through the sea-bed.

Punchlist   An alternative term for a **snaglist**.

Punto di Scambio Virtuale (PSV)   A notional point on the Italian gas pipeline network at which shippers trade gas quantities.

Pup joint   A length of drill string which is of a shorter length than a conventional single.

Purchase order (PO)   A written record of the terms of an intended sale and purchase transaction which is issued by a buyer to a seller.

Pure service contract   A service contract whereby the concession-holder performs defined services and receives a service fee even if there is no corresponding petroleum production. Contrast with a **risk service contract**, whereby the concession-holder does not receive a service fee if there is no petroleum production.

Pure transporter   A petroleum pipeline transportation project formulation whereby the transporter receives certainty of revenue for the transportation service which it provides to the shipper, without regard to the risks associated with any of the underlying petroleum sales arrangements.

Purging   The cleaning of a petroleum infrastructure item through the use of inert gases. *See also* **nitrogen purging**.

Push-up area   A defined area on a ship's hull at which a tug can make direct contact for pushing purposes.

Pushed cost   A petroleum project cost which can be deferred for payment to a future date.

Put option   The right, but not the obligation, of a person to sell a particular asset or interest when certain exercise conditions have arisen. *See also* **call option**.

Pycnometer   A device which conducts a comparative determination of the relative densities of different fluids.

Pygas   A term which is derived from the combination of 'pyrolysis' and 'gasoline' to describe naphtha which is produced in an ethylene production facility.

Pyrolysis   The heating of a core sample which is taken from source rock in an inert gas atmosphere in order to assess the total organic compounds of the source rock.

Pyrometer   A device which measures temperatures which are ordinarily not measurable by a conventional thermometer.

Q

Q unit   A unit of energy which is equivalent to 10^{18} British thermal units.

Qatar Flex (Q-Flex)   An LNG carrier which has an approximate cargo-carrying capacity of 216,000m^3 of LNG.

Qatar Max (Q-Max)   An LNG carrier which has an approximate cargo-carrying capacity of 266,000m^3 of LNG.

Quadrant   A fixed geographical description of a marine or terrestrial area in respect of which a number of concessions might be awarded. On the UKCS a quadrant contains approximately thirty blocks.

Quadratic transfer function (QTF)   A methodology which is used for assessing the movement of waves.

Qualified person's report   An alternative term for a **competent person's report**.

Quality activity plan (QAP)   A petroleum project development plan which is assembled by reference to the achievement of defined quality assurance targets.

Quality assurance/quality control (QA/QC)   A process in which quality assurance is used to assess the quality of a product and quality control is used to ensure that a product meets defined expectations.

Quantitative risk assessment (QRA)   A form of assessment of the risks which are associated with a particular activity.

Quantities-based contract   A contract for the pipeline transportation of petroleum which is founded on the principle of the transportation by the transporter, on a shipper's behalf, of a defined quantity of petroleum between defined points. Contrast with a **capacities-based contract**, which is founded on the shipper's reservation of a defined amount of capacity in the pipeline (whether or not the shipper uses that capacity).

Quanto   A commodity which is denominated in a currency other than the currency in which that commodity is usually traded.

Quarantine costs   Storage and processing costs which apply to goods and materials that are held in storage at the point of being imported into a state.

Quarter days   A set of dates which mark the passing of one quarter of a calendar year: 31 March, 30 June, 30 September, and 31 December. *See also* **lady days**.

Quarters and utilities (QU)   The part of an offshore petroleum infrastructure item which houses crew living quarters and utility services such as power and water.

Quench oil   Oil which is introduced as a coolant into a high temperature petroleum stream during a refining process.

Questionnaire 88 (Q88)   A form which is used for the assessment of a ship's condition and suitability when chartering a ship.

Quick release mooring hooks (QRMH)   A method for anchoring a ship to a petroleum infrastructure item but with the ability to release the ship quickly in emergency situations.

Quicklook   Data which is obtained from the drilling of a wellbore and is processed at the drill site for immediate inspection.

Quiet enjoyment   Under a loan agreement, an undertaking from a lender to a borrower that the lender will not interfere with the ordinary operation of a petroleum facility which has been pledged by the borrower as security for the loan. *See also* **letter of quiet enjoyment**.

Quitclaim   The documentary evidence of the surrender of a concession-holder's interest in a concession.

Quittance   A release or discharge from a debt or obligation.

R

R-factor   Under a concession, a sliding scale which is used to determine the level of royalty which is payable by a concession-holder to the grantor, where the R (ratio) factor is determined by cumulative production revenues divided by cumulative production costs.

Rabbit   An alternative term for a pig.

Rack delivery/rack market   Products which are sold from a refinery export loading rack. Also called **over the rack**.

Rack price   The posted sale price of products at a refinery export loading rack.

Radial drilling   The drilling of several wellbores and sidetracks which all radiate from a common starting point.

Radioactive densitometer (RAD)   A densitometer which measures specific gravity by reference to the presence of naturally occurring radioactive materials in a sample stream.

Radioactivity log   A record of the radioactive characteristics of a formation, which can indicate the presence of petroleum. Also called a **gamma ray log** and a **nuclear log**.

Raffinate   The part of a petroleum stream which remains after a refining process has taken place and the desired products have been extracted from the petroleum stream.

Ragged edge rule   A principle that in a relinquishment the relinquished part of a contract area should not be of a dimension nor have boundaries which are so irregularly defined that the relinquished part is not readily capable of being re-licensed as a contract area within another concession.

Ram blowout preventer   A blowout preventer which uses a ram as a closing and sealing component to seal off pressure in a wellbore. *See also* a **blind ram**, a **pipe ram**, and a **shear ram**. Contrast with an **annular blowout preventer**, which seals a borehole or tubing diameter by the use of a rubber plug.

Ramp rate   The rate by which an existing level of petroleum production can be increased from a formation.

Range
(1)   The time which is taken by a gas storage facility to deliver the total quantity of working gas to the storage facility users.
(2)   The distance between two kink points on an S-curve.

Rank wildcat   An exploration well which is drilled in a previously undrilled area.

Rapid phase transition (RPT)   The sudden transition of LNG from a liquid to a gas.

Raster graphic   A spatial data model which is made up of miniature dot matrix structures, which is used for seismic analysis. Also called a **grid and raster**.

Rate of penetration (ROP)   The rate at which a drillbit drills a wellbore (which is recorded in feet or metres per hour). Also called **footage**.

Rate of return (ROR)   A measure of the gain (or loss) which is made by a petroleum project investor on an investment over a specified time period, with investment income expressed as a percentage of the total cost of the investment.

Rate sensitivity   The influence which the rate of petroleum production from a formation will have upon the recovery factor and the rate of production decline from the formation.

Rateable delivery   Under a petroleum sales contract, a provision whereby the amount of petroleum which the seller delivers to the buyer will be a rateable proportion of the petroleum which the seller is delivering to a wider series of buyers.

Rateable take   Under a petroleum sales contract, a provision whereby the amount of petroleum which the buyer takes delivery of from the seller will be a rateable proportion of the petroleum which the buyer is taking delivery of from a wider series of sellers.

Rathole
(1)   An enlarged extra hole which is drilled at the bottom of a wellbore in order to allow tools to be abandoned there.
(2)   A storage hole on the drilling floor of a drilling unit in which a kelly and kelly bushing are stored when not in use.

Ratio cutback   A cutback in the production of petroleum from a wellbore which is required when defined gas–oil or water–oil ratios are exceeded.

Raw gas   Natural gas as it is produced directly from a formation. Also called **green gas** and **wellstream gas**.

Raw make   An alternative term for **raw mix liquids**.

Raw mix liquids   NGLs which exist in a mixed form (principally propane and butane) prior to extraction from a natural gas stream. Also called **raw make**.

Raw PSM   Under an LNG sale and purchase contract, a profit-sharing mechanism whereby the costs of a diversion are met by one party before the net profits are then divided between the parties. Contrast with a **net PSM**, whereby the costs of a diversion are first deducted from the gross profits before division of the net profits between the parties.

Ready for operation (RFO)   A declaration that a petroleum facility or an item of equipment is ready to commence operational use.

Ready to produce (RTP)   A declaration that a petroleum-processing or production facility is ready to commence processing or production operations.

Real gas   An actual gas which less obviously demonstrates the application of the Gas Laws. Contrast with an **ideal gas**, as a hypothetical gas which better demonstrates that application.

Real working interest   An adjustment to a net working interest which recognises the impact of an adjustment factor such as a carry.

Reamer   A tool which is used to smooth the sides of a wellbore or to enlarge the originally drilled diameter of a wellbore.

Reaming   The drilling of a wellbore beyond its originally drilled diameter. Also called **underreaming**.

Reasonable and prudent operator (RPO)   An objective standard of perform-ance which could be required of a party in the performance of a contract, whereby actual performance is measured against the analogue of what a similar person in similar circumstances would do.

Rebus sic stantibus   A legal doctrine which suggests that the agreed terms of a contract could be disregarded if there has been a change in the fundamental circum-stances surrounding the contract. *See also* **fait du prince** and **pacta sunt servanda**.

Recap   A short message which confirms the terms of an agreed petroleum trade. Also called a **trade recap**.

Recapture   The process of recovering produced petroleum and reinjecting it into a formation for processing or storage purposes.

Recip   Reciprocating engine: a simple gas-fuelled or oil-fuelled piston engine which is used to generate electric power.

Reclamation   An alternative term for **decommissioning**.

Recompletion   The entry into an existing wellbore through an alternative method (including deepening or sidetracking) in order to enhance petroleum production.

Recompletion petroleum   Petroleum that is produced from a formation through a wellbore which has previously penetrated a different formation.

Recon crude   An alternative term for **reconstituted crude**.

Reconciliation   An after-the-event adjustment of petroleum quantities or mon-etary amounts, which is intended to reconcile short-term performances with longer term expectations. Also called a **true-up**.

Reconnaissance permit   A short-term concession which is awarded by a state which allows for the gathering of seismic and similar data from a defined area. Also called a **hunting licence**.

Reconstituted crude   Crude oil which is used as feedstock in a refinery which has been specifically blended from a number of crude oil sources to meet the needs of the refinery. Also called **recon crude**.

Recoupment   The recovery of make-up.

Recoverable   The total quantity of petroleum which is expected to be recovered from a formation.

Recoverable gas lift   Gas which is used in gas lift operations and returns to the surface but which is not reinjected. Also called **spent gas lift**.

Recoverable oil   Oil which is injected into an oil-bearing formation in order to stimulate production and is expected to be recovered as part of the eventual production of oil. *See also* **load oil**.

Recovery   The total quantity of petroleum which has been recovered from a formation.

Recovery efficiency   An alternative term for **recovery factor**.

Recovery factor (RF)   The ratio of recoverable petroleum from a formation to the total petroleum in place in the formation. Also called **recovery efficiency**.

Recurrence interval   An alternative term for a **return period**.

Recycle rates   A measure of a petroleum company's efficiency which is based on its levels of petroleum production compared to petroleum exploration costs.

Recycling   The injection of make-up gas into a formation.

Red diesel   Diesel to which a red dye has been added, to indicate that it is intended for use as a fuel for industrial applications rather than for use as a motor vehicle fuel. *See also* **dyed fuel**.

Redelivery
(1) The function of withdrawing gas from a gas storage facility. Contrast with **injection**, as the function of depositing gas into a gas storage facility.
(2) The return of a ship by the charterer to the owner at the end of the charter period. *See also* **delivery**.

Redelivery point   Under a pipeline transportation contract, the point at which the transporter delivers petroleum from the pipeline back to the shipper. *See also* **delivery point**.

Redeployment   The relocation of an FLNGV, FPSO, FSRU, or FSU into a different location.

Redetermination   A periodic revision to the agreed participating interests in a UUOA. Also called a **unit adjustment**. *See also* **fixed interest**.

Redrill
(1) If damage is caused to a wellbore by a drilling contractor's negligence, the drilling by the drilling contractor of same or a replacement wellbore. *See also* **redrill rate**.
(2) A second wellbore which has been drilled in close proximity to a first failed wellbore.

Redrill rate   Under a drilling contract, the revised dayrate which is payable by the employer company to the drilling contractor in consideration of a redrill.

Re-drive   An e-drive process which is powered by electrical energy which is derived from a renewable energy source.

Reduced circulating pressure (RCP)   A degree of pressure on a drill string which is generated by the circulation of drilling fluid at a pressure lower than the degree of pressure that is ordinarily applied during drilling, as a technique applied where a kick is being circulated out of a wellbore.

Reduced crude oil   An alternative term for **topped oil**.

Reduction   An alternative term for **relinquishment**.

Reel barge   A specialist ship which is used in a reel pipelay.

Reel pipelay   A method of offshore pipeline construction whereby pre-completed pipeline sections are loaded onto a reel barge and are paid out onto the sea-bed. *See also* J-lay and S-lay.

Re-entry   An alternative term for a **workover**.

Reference conditions   Standardised conditions of temperature and pressure within which petroleum measurement operations are carried out. Also called **standard temperature and pressure**.

Reference month   In a petroleum price review provision, the calendar month to which reported data applies and used as a basis for the review process.

Refi   A shorthand term for **refinancing**.

Refinancing   The provision of third-party financing for the development costs of a petroleum infrastructure item after the item has already been built. Also called **refi**.

Refinery   A petroleum-processing facility which takes in, stores, and processes crude oil in order to give various finished petroleum products. *See also* **complex refinery**, **conversion refinery**, **hydroskimming refinery**, and **topping refinery**. *See* Appendix 2E.

Refinery acquisition cost of crude (RACC)   The price that is paid by a refinery for crude oil feedstock, which is relevant to the determination of the gross product worth.

Refinery complexity   A measure of the complexity of a refinery by reference to the range of petroleum refining processes which are performed at the refinery site. Also called **complexity**. *See also* **equivalent distillation capacity** and **Nelson complexity index**.

Refinery gas   An alternative term for **still gas**.

Refinery margin   In a refinery, the difference between the costs of feedstock and processing and the revenues which are earned from the sale of the resultant products.

Refinery own consumption   Petroleum which is consumed within a refinery as fuel for the various petroleum-processing equipment.

Refinery swing   The ability of a refinery to process a wide range of crude oil qualities.

Refinery yield   An alternative term for **yield**.

Refining and marketing (R&M)   The activity of processing crude oil and selling the resultant products. *See also* **downstream**.

Refining slate   An alternative term for a **crude slate**.

Refluxing/reflux   In atmospheric distillation, a return of part of the condensed vapour (the reflux) to the CDU in order to improve the separation of components from the feedstock.

Reformate   A petroleum fraction which results from the reforming process.

Reforming   The transformation of initially produced naphtha into lighter petroleum fractions in a refinery. Also called **catalytic reforming**.

Reformulated blendstock for oxygenate blending (RBOB)   A specially produced form of gasoline blendstock which is intended for blending with oxygenates downstream of a refinery to produce reformulated gasoline. *See also* **conventional blendstock for oxygenate blending**.

Reformulated fuel   A fuel which has been chemically modified in order to reflect any applicable environmental concerns, performance standards, or regulatory requirements.

Reformulated gasoline (RFG)   Gasoline which has become a reformulated fuel.

Refuge   A designated safe area in a petroleum infrastructure item in which personnel can gather during an emergency event. Also called a **safety zone**, a **temporary refuge**, and a **temporary refuge area**.

Refund guarantee   A financial instrument by which a bank undertakes to repay the amounts which have been paid by the owner of a newbuild if the shipyard becomes insolvent during construction.

Regas   LNG which has undergone regasification in order to transform it from a liquid to a gas.

Regasification   The transformation to the gaseous state of a quantity of LNG. Contrast with **liquefaction**, as the transformation of gas into a liquid form (as LNG) through refrigeration. Also called **vaporisation**.

Register ton   A unit of volume which equals 100 cubic feet, which is used to measure the cargo-carrying capacity of a ship. *See also* **gross registered tonnage** and **net registered tonnage**.

Regressive regime   A fiscal regime which is applicable to a concession whereby state take is static or decreases as project profitability improves. Contrast with a **progressive regime**, whereby state take increases as project profitability improves.

Regret costs   A measure of the costs which are incurred in order to replace or to enhance a tactical solution which is no longer fit for purpose.

Re-guarantee   A term in a petroleum project contract which allows for a person to call for the provision of collateral support in respect of a counterparty at a future date.

Regulated third-party access (RTPA)   A regulatory position whereby third-party access rights are applied between a petroleum facility owner and a facility user according to regulated terms. Contrast with **negotiated third-party access**, whereby third-party access rights are the subject of negotiation between the facility owner and the facility user.

Regulation below cost (RBC)   A state-set petroleum price which is below the average cost of petroleum production and transportation and is intended to give a subsidy to the consumer population.

Regulation cost of service (RCS)   A state-set petroleum price which covers a petroleum producer's costs.

Regulator   A person who operates independently of governmental direction in ensuring that a sector and its participants operate in accordance with defined regulatory principles. Also called a **national regulatory authority**. *See also* **crypto-regulation**.

Regulatory capture   The situation in which a regulator promotes the interests of participants in the industry which it is intended to regulate, rather than applies the required regulation to those persons.

Regulatory convergence   The situation in which gas sector regulation and electric power sector regulation converge in the office of a single regulator.

Regulatory out   A contractual provision whereby a person is excused from performance because of the intervention of a regulatory agency.

Reid vapour pressure (RVP)   A measure of the level of pressure which is required to prevent a quantity of petroleum from freely evaporating, which is calculated against a fixed temperature and is used to determine the volatility of gasoline and other products. *See also* **true vapour pressure**.

Reimbursable contract   An alternative term for a **time and materials contract**.

Reinstatement
(1)   The application of an insurance payment towards the reinstatement of the insured facilities.
(2)   An alternative term for a **back-in**.
(3)   The remediation of a land surface area after the cessation of petroleum project operations.
(4)   A decision of a non-consenting party in an exclusive operation to later participate in that operation.

Re-insurance   Insurance whereby a primary insurer lays off part of the insured risk to a secondary insurer. *See also* **co-insurance**.

Reinterpretation   The reprocessing and re-examination of existing seismic data, often using interpretive techniques which are more advanced than the techniques which existed at the time when the seismic data was originally generated.

Reinvestment   A term in a concession which requires the concession-holder to spend a certain amount of money on local projects that benefit the state of operations.

Relative density   An alternative term for **specific gravity**.

Release gas   Gas which is sold to third parties by an incumbent market supplier, rather than gas which is kept or controlled by that supplier, as a regulatory condition that is intended to open up a market to third-party access. *See also* **remedy sale**.

Reliability   The ability of a petroleum facility to operate to specification for a defined period of time without unscheduled interruption. *See also* **mean time to failure**.

Reliability, availability, maintenance (RAM)   An analysis of the anticipated and the actual performance efficiency of a petroleum infrastructure item.

Relief well   A wellbore which is drilled into the bottomhole of a runaway well and is intended to relieve blowout pressure. Also called a **killer well**.

Relinquishment   Under a concession, a principle that certain parts of a contract area are removed from the scope of the concession and are surrendered back to the grantor by the concession-holder. *See also* **mandatory relinquishment** and **voluntary relinquishment**. Also called **area reduction, reduction**, and **surrender**.

Re-liquefaction   The process by which boil-off and other petroleum vapours are liquefied to give LNG within an LNG processing plant or on-board an LNG carrier.

Reload   The loading of an empty LNG carrier which has just discharged a cargo with a fresh cargo of LNG at the unloading port.

Remaining established   At any point in time in respect of a formation, the initially established reserves of petroleum in the formation minus the cumulative production of petroleum to date from the formation.

Remedial cementing   Cementing in a wellbore which takes place after primary cementing has occurred, in order to fill any voids which are apparent.

Remedy sale   A sale of a commodity or of capacity in a facility which is required by a regulator in order to improve market competition. *See also* **capacity release** and **release gas**.

Remote gas   An alternative term for **stranded gas**.

Remote-operated vehicle (ROV)   A tethered, remotely operated undersea vehicle which is used for petroleum facility inspection or maintenance works.

Remote power generation (RPG)   Localised power generation for a location which does not benefit from connection to a power distribution network. Also called **stranded area power generation**.

Rendering   A process by which tallow is produced.

Renewable energy source (RES)   Energy which is generated from natural resources such as wind, wave, and solar.

Renewable gas   Natural gas which is produced from renewable sources, such as biogas. Contrast with **non-renewable gas**, which is produced from non-renewable sources. *See also* **renewable natural gas**.

Renewable natural gas (RNG)   Biogas which has been modified to give a methane content similar to that of natural gas. Also called **sustainable natural gas**.

Renomination   Under a petroleum sales contract, a variation to an existing nomination, or a nomination which is given to confirm an existing nomination.

Rent   The income which is earned by a state from a concession-holder (such as through taxation and royalties). Also called **government take** and **state take**. *See also* **effective royalty rate**.

Rentier state   A state which derives all or a significant part of its income from the sale of its natural resources to non-national persons.

Re-opening   An alternative term for a **workover**.

Repair   Under a drilling contract, the remediation of the failure of drilling contractor-provided equipment.

Repair rate   Under a drilling contract, the revised dayrate which is payable by the employer company to the drilling contractor during a repair.

Repeat formation testing (RFT)   A particular method which is used for measuring the pressure in a formation through a wellbore.

Repeatability and reproducibility (R&R)   The extent to which the characteristics of a batch of petroleum match the characteristics of an earlier sample of that petroleum.

Reperforation   The secondary perforation of existing perforation holes in casing in order to further improve reservoir connectivity.

Reporting standard   A specified standard for classifying and reporting petroleum deposits according to geological and economic certainty (*see*, for example, SPE PRMS).

Repowering   The replacement of existing electric power-generation facilities with newer, more efficient facilities.

Re-pressure   An alternative term for **artificial lift**.

Reprocessing   The further interpretation and reorganisation of existing geophysical data which has been acquired by a seismic survey.

Reputation equity   Publicly quoted shares in a company which are exposed to the risk of having their values diminished by an event which damages the company's reputation.

Requirements contract   Under a petroleum sales contract, a commitment whereby the seller undertakes to supply all of the buyer's petroleum requirements and the buyer undertakes to buy its petroleum exclusively from the seller.

Requisition   The temporary expropriation by a state of a ship which is on hire under a charterparty.

Res nullius   A Latin term for something which belongs to nobody because it has not been appropriated (which is applicable, for example, to mobile petroleum which is not subject to the law of capture).

Resale price maintenance (RPM)   An arrangement between a seller and a buyer of petroleum whereby the buyer undertakes to only resell that petroleum for a specified price.

Research octane number (RON)   A measurement of the anti-knock rating of a given quantity of gasoline. *See also* **motor octane number**.

Reserve tail date (RTD)   In reserves-based lending, usually the point at which 25 per cent of the borrowing base remains unproduced, which will be relevant to the assessment of the borrower's compliance with a field life coverage ratio or a project life coverage ratio.

Reserves   Estimated quantities of petroleum which are believed to be recoverable from a formation, which are classified according to the degree of certainty associated with the estimate. *See also* **possible reserves, probable reserves**, and **proved reserves**. *See* Appendix 7.

Reserves-based lending (RBL)   A type of financing whereby a loan is secured by the unproduced reserves of petroleum of the borrower and is repaid to the lender from the proceeds which are derived from petroleum sales from those reserves. Also called an **oil-backed loan.**

Reserves reinforcement   Under a petroleum sales contract, the obligation of the seller to carry out works for the recovery of further reserves.

Reserves replacement ratio (RRR)   A measure of the amount of proved reserves which are added to a petroleum company's reserve base during a year relative to the amount of petroleum which is produced by the company in the year (where the ratio must be at least 100 per cent for the company to stay in business over the long term).

Reservoir   A sub-surface accumulation of petroleum which is formed within a stratigraphic layer within a formation. Also called an **accumulation**, a **horizon**, and a **pool.**

Reservoir compaction   An alternative term for **rock compaction.**

Reservoir drive   An alternative term for **capillary pressure.**

Reservoir height   An alternative term for **column height.**

Reservoir management plan (RMP)   A defined plan for the most effective development, ongoing operation, and general stewardship of a reservoir.

Reservoir pressure   The pressure at the face of a producing formation when a wellbore is shut in, which is made up of the shut in pressure and the weight of the column of accumulated petroleum in the wellbore.

Reservoir quality (RQ)   A measure of the quality of a formation by reference to its petroleum storage and deliverability parameters, which is determined respectively by the formation's porosity and permeability.

Reservoir rock   A rock formation into which petroleum has migrated from a source rock. *See* Appendix 2A.

Reservoir simulation model   A computer-generated model which predicts the physical characteristics of a formation.

Reservoir unit (RU)   An area which results from the sub-division of a formation into a number of defined units, each of which will be the subject of evaluation and assessment.

Residence petroleum   An alternative term for **linepack.**

Residual fuel oil (RFO)   A heavy fuel oil which forms part of the bottoms in a refinery.

Residual oil   An alternative term for **trapped oil.**

Residual oil zone (ROZ)   Part of a formation containing crude oil which is ordinarily of too low a level of saturation to be recoverable but which could become a recoverable resource through secondary recovery or with improved technology and/or lower lifting costs.

Residuals   An alternative term for **bottoms**.

Residue gas   The part of a gas stream which remains after any liquids have been stripped and any impurities have been removed.

Residuum   An alternative term for **bottoms**.

Resilience   A measure of the ability of a petroleum project facility, project, or contract to withstand a major disruption and still to be able to function as originally intended.

Resistivity   Well logging which measures the degree of resistance of a formation to the flow of an electric current, where high resistivity readings indicate the presence of petroleum.

Resistivity log   The recorded evaluation of a formation's resistivity.

Resource curse   An alternative term for **Dutch disease**.

Resource maturation   An alternative term for **maturation**.

Resource nationalism   The tendency of a state to assert control over the natural resources which are located with its territory.

Resources   A composite term for contingent resources, prospective resources, and reserves. *See* Appendix 7.

Resources other than reserves (ROTR)   A composite term for contingent resources and prospective resources. *See* Appendix 7.

Response amplitude operator (RAO)   An engineering methodology which refers to the movement of a ship or offshore petroleum infrastructure item by reference to the six degrees of motion in reaction to wave movements.

Responsibly sourced gas (RSG)   Raw gas which is independently certified as having been produced in accordance with defined environmental and/or carbon abatement standards.

Restoration quantity   An alternative term for **make good**.

Restricted bidding   A procurement model which limits participation in the procurement process to invited or pre-qualified bidders.

Restricted cash   A monetary payment which is held back from the recipient and is instead applied to specific purposes.

Restricted catenary mooring (RCM)   An alternative term for a **catenary mooring**.

Retail outlet (RO)   An alternative term for a **petrol station**.

Retainage   In a petroleum-processing facility, any quantity of petroleum which belongs to the facility user but which is not redelivered by the facility owner to the facility user.

Retainer head   An alternative term for a **cementing head**.

Retention licence   A permission whereby a concession which has not moved into development and production can be retained by the concession-holder without relinquishment, pending a later determination of commerciality.

Retention of title (ROT)   Under a petroleum sales contract, a provision whereby, despite the delivery of petroleum by the seller to the buyer, title to the petroleum does not pass from the seller to the buyer until the buyer has paid the agreed purchase price.

Retention on board (ROB)
(1)   A measurement of the quantity of bunkers which are found on-board a ship at the start and/or at the end of the term of a charterparty.
(2)   Any part of a cargo which is left in a ship's tanks upon completion of the discharge of the cargo. *See also* **liquids and solids remaining on board**.

Reticulation   The transmission of gas through a small-diameter, low pressure gas pipeline system serving residential and commercial end-users.

Retired carbon offset   A carbon offset which has been fully applied to a carbon emission.

Re-transfer   A provision in an asset transfer agreement whereby the seller can unwind the transaction after completion of the transfer.

Retrofitting   The addition of new technology or equipment to an existing petroleum infrastructure item.

Retrograde condensate/retrograde gas   Gaseous-phase condensate which, in a formation, is close to the dewpoint and where minute adjustments to formation pressure caused by petroleum production could cause condensation to occur.

Retrograde reservoir   A reservoir which is principally gas-filled and in which retrograde condensate forms as ongoing production depletes the gas.

Return period (RP)   An estimated average time between the estimated probable occurrence of different events of the same nature (such as storms or earthquakes). Also called a **recurrence interval**.

Returns   Drilling fluid and cuttings which are circulated to the surface during the drilling of a wellbore.

Revenue rent   The revenues which accrue from a petroleum project after the associated costs have been accounted for.

Revenue sharing contract   A concession for petroleum E&P whereby the concession-holder pays the operational expenditures from its agreed share of the gross revenues from the sale of produced petroleum. Also called a **gross split contract**. Contrast with a **production sharing contract**, whereby the concession-holder first recovers its expenditures from petroleum production revenues.

Reverse break   A sudden decrease in the rate of penetration. Contrast with a **drilling break**, which indicates an increase in the rate of penetration.

Reverse circulation   The circulation of drilling fluid through a wellbore whereby the drilling fluid flows down the annulus between the drill string and the wellbore, out of the drill bit and up the drill string. Contrast with **normal circulation**, whereby drilling fluid flows down the drill string and up the annulus.

Reverse flow   A planned change in the direction in which petroleum ordinarily flows through a pipeline.

Reverse valorisation   The conversion of a monetary entitlement to an in-kind quantity of petroleum. *See also* **valorisation**.

Reversionary interest   A current right to future possession of an interest in a property (such as the right of a lessor to assume control of leased property at the end of a lease term or the right of a state to assume possession of petroleum facilities at the end of a PSC).

Revision   Any upwards or downwards change to an earlier estimate of reserves.

Rework   Additional work which has to be performed on a petroleum infrastructure item. *Also called* **carry-over work**.

Rheology   A study of the flow of gases and liquids, which is related to viscosity.

Rho   The rate at which the value of a commercial position changes relative to a change in interest rates.

Rhumb line   An alternative term for a **loxodrome**.

Rich gas   Gas which is composed principally of the heavier petroleum fractions (ethane, propane, butane) and consequently has a relatively higher calorific value. Contrast with **lean gas**, as gas which is composed principally of methane.

Rig award   The award of a contract for the drilling of a wellbore to a drilling contractor.

Rig count   The number of drilling units which are in active use at any time, which is often used as an indicator of the health of the petroleum exploration market.

Rig-down   The disassembly of a drilling unit. Contrast with **rig-up**, as the assembly of a drilling unit.

Rig of opportunity   A drilling unit which unexpectedly becomes available for hire.

Rig-share agreement   A contract which provides for the sharing of a drilling unit between different persons according to an agreed sequence.

Rig-time   The length of time which is taken by a drilling unit in drilling a wellbore.

Rig-to-reef   An option for decommissioning an offshore petroleum facility which sees a jacket toppled in-situ in order to provide a permanent artificial reef for marine wildlife.

Rig-up   The assembly of a drilling unit. Contrast with **rig-down**, as the disassembly of a drilling unit.

Rig wash   A degreasing and detergent agent which is used to clean metalwork and surface areas.

Right of first offer (ROFO)   A pre-emption right by which the holder has the first option to acquire the interests of a disposing person. *See also* **right of first refusal**.

Right of first refusal (ROFR)   A pre-emption right by which the holder has the option to match the terms of a purchase offer which has been made to a disposing person. Also called a **matching right**. *See also* **right of first offer**.

Ring fence   A fiscal separation of a petroleum project and its associated costs and revenues.

Ring wall foundation   A concrete base for a petroleum storage tank.

Riser   Pipework which connects a sub-sea wellhead or pipeline to an offshore platform.

Riser balcony   A structure on an FSO or an offshore platform at which a series of incoming risers congregate.

Riser platform   An offshore platform which accommodates multiple incoming and outgoing risers.

Risk   A calculated assessment of the probability of success or failure of a defined activity.

Risk-based inspection (RBI)   A protocol for determining required levels of maintenance in respect of a petroleum infrastructure item which ranks individual failure likelihoods and impacts in order to determine an appropriate level of maintenance.

Risk reduction factor (RRF)   A target for the degree of mitigation of a risk which has been recognised within a risk assessment in order to make the risk acceptable.

Risk register   A matrix of identified project risks, assessed likelihood and impact factors and suggested risk mitigants. Also called an **operational risk assessment**.

Risk service contract   A service contract whereby the concession-holder performs defined services but does not receive a service fee if there is no petroleum production. Contrast with a **pure service contract**, whereby the concession-holder receives a service fee even if there is no corresponding petroleum production.

Risk-weighted portfolio   A portfolio of prospective petroleum production projects which is ranked according to individual project risks in order to reflect the likelihood of each project proceeding.

Risked   An assessment of resources which takes into account certain defined geological, commercial, financial, or development risks which would qualify the estimation of recoverable petroleum. Contrast with **unrisked**, whereby the assessment does not account for the risks.

Roads   A safe anchorage area offshore at which a ship can lie awaiting entry into port. Also called a **roadstead**.

Roadstead   An alternative term for **roads**.

Robber well   A wellbore which is drilled deliberately in order to drain petroleum from neighbouring lands. *See also* the **law of capture**.

Roberts Torpedo   An explosive device which is dropped to the bottom of a wellbore and is detonated in order to fracture a formation.

Rock armour   Concrete structures which are used to protect the surface of an offshore pipeline. Also called **rock damping** and **rock dump**.

Rock compaction The subsidence of a formation which results from petroleum production (which can increase capillary pressure and petroleum production rates). Also called **reservoir compaction**.

Rock damping An alternative term for **rock armour**.

Rock dump An alternative term for **rock armour**.

Rocky Mountain Mineral Law Foundation (RMMLF) A US-based educational body which is focused on the study of oil and gas law, which in 2022 became known as The Foundation for Natural Resources and Energy Law (http://www.fnrel.org).

Rod pumping An alternative term for **sucker rod pumping**.

Roll The side-to-side rolling movement of a ship or offshore petroleum infrastructure item.

Rollback The relaxation of a legislative limit or condition.

Rolling mortality A series of accrued commercial entitlements which expire over time according to a defined order.

Rollover
(1) Under a petroleum sales contract for a defined term, a provision which allows the duration of the contract to be extended beyond its original basic term.
(2) The rapid release of vaporised LNG which occurs when LNG streams of different densities are mixed in-tank, leading to over-pressurisation and tank failure.

Roof rock A layer of impervious rock which overlies a formation.

Root cause analysis (RCA) A problem-solving methodology which addresses the fundamental cause of a particular problem.

Rotameter A form of flow meter.

Rotary drilling A method of drilling whereby the drill string is powered by a drive motor at the rotary table level. Also called **kelly drilling**. Contrast with **topdrive drilling**, whereby the drive motor is suspended from a derrick above the drill string.

Rotary hose In drilling, a high-pressure flexible line which connects to a swivel.

Rotary sidewall core (RSC) A sidewall core which is extracted as a plug from the sidewall in a wellbore.

Rotary table A revolving section of the drill floor on a drilling unit which provides power to turn a drill string in rotary drilling.

Rotational gas lift A closed loop system for gas lift in which gas is injected, recovered, and reinjected in a formation without the need for the introduction of gas from an extraneous source.

Roughneck A crew member on a drilling unit who has responsibility for assembling drill strings during a drilling operation.

Round-down Under an LNG sales contract, a downward adjustment to the ACQ for a contract year in order to reflect a closing quantity of LNG which is less than a full cargo lot. *See also* **round-up**.

Round trip The removal of a drill string from a wellbore and the later return of the drill string to the wellbore. *See also* **short trip** and **trip**.

Round-trip pigging The pigging of a pipeline system whereby the pig travels round in a loop. Contrast with **one-way pigging**, where the pig is unidirectional.

Round-up Under an LNG sales contract, an upward adjustment to the ACQ for a contract year which is made in order to reflect a closing quantity of LNG that is less than a full cargo lot. *See also* **round-down**.

Roustabout A crew member on a drilling unit who has responsibility for general operational activities.

Routine core analysis (RCA) A set of measurements which are carried out on a core in order to determine porosity, permeability, and lithology. *See also* **special core analysis**.

Routine shut-down (RSD) A predicted shut-down of a petroleum facility. Also called a **planned shut-down**. Contrast with **emergency shut-down**, as an unpredicted shut-down of a facility.

Royalty A payment that is due (in money or in-kind) from a concession-holder to the state, which is calculated by reference to quantities of petroleum produced under or sold from the concession. *See also* **blind royalty, dollar-denominated petroleum production interest, net profit interest, overriding royalty interest**, and **volumetric production payment interest**.

Royalty holiday Under a royalty, a defined period (which is set by reference to time or levels of petroleum production) during which the concession-holder is not obliged to pay all or part of the royalty to the royalty-holder.

Royalty/tax (R/T) An alternative term for a **licence**.

Royalty trust A company which is created to own multiple resource royalty interests (as the royalty-holder) that passes earned royalty income back to shareholders as dividends, which is more usually found in the United States and Canada. *See also* **fixed R/T** and **growth R/T**.

RPI-x A regulatory formula by which a regulated monopoly is allowed to increase its prices each year by the retail price index (RPI) minus a defined amount or percentage (x).

R:P ratio Reserves to production ratio: a representation of the number of years for which current sub-surface stocks of petroleum would last at current petroleum production levels.

Rugosity A measure of the roughness or the irregularity of a solid surface.

Run-in-hole An operation whereby a drill bit or a wireline logging tool is lowered into a wellbore.

Run out A transfer of petroleum from a storage tank to a pipeline for transportation.

Run ticket A document which evidences the composition of a delivered quantity of crude oil.

Runaway well An uncontrolled expulsion of petroleum from a reservoir through a wellbore to the surface, despite sub-surface and surface control. Also called a **gusher** and a **wild well**. *See also* **kick**.

Rundown tank   A storage tank which receives a cut in a refinery for temporary storage prior to further processing.

Running hours   The period of time which elapses between the start and the end of a period of laytime.

Running-in pipe   The process of casing a wellbore.

Running room   A measure of the upside potential in a prospective petroleum development project.

Russian export blend crude oil (REBCO)   A crude oil blend from Russia.

Russian roulette   A colloquial term for a joint venture deadlock resolution provision whereby one party offers to sell its interests to the other party and offers to buy the other party's interests at a specified price, and the other party can accept the sale offer or the purchase offer. *See also* **Dutch auction** and **Texas shoot-out**.

S

S-curve   Under a petroleum sales contract, an agreed economic boundary for petroleum price movements where different prices apply at defined kink points.

S-lay   A method of offshore pipeline installation, so-called because of the pipeline's S-shaped profile as it comes off the pipelay vessel. *See also* **J-lay** and **reel pipelay**.

Sack   A unit of measure of cement, chemicals, and other solids used in drilling operations which are delivered to a drilling unit.

Sacrificial anode   An alternative term for a **tethered goat**.

Saddle   A sub-surface geological feature which indicates a low point between two peaks.

Saddle volume   Unproduced petroleum which is trapped in a saddle.

Safe harbour   A contractual promise of a port area at which a ship can safely lie afloat while loading or unloading its cargo.

Safe harbour provision   A contractual term by which an identified future activity or arrangement will be expressly permitted.

Safe manning and watchkeeping standards   Prescribed minimum standards which relate variously to manning (crew levels), training, certification, and watchkeeping on ships. *See also* **manning scale**.

Safe shut-down earthquake (SSE)   The maximum earthquake potential up to which a petroleum infrastructure item is designed to remain safe and functional. *See also* **operating basis earthquake**.

Safe trading limits   An alternative term for **trading limits**.

Safe working load (SWL)   The maximum load which a crane can safely carry. Also called **working load limit**.

Safety boat   A ship which is stationed to provide support for offshore petroleum E&P operations.

Safety case   A document which is produced by the operator of a petroleum infrastructure item and identifies the item's associated operational risks and a methodology for managing those risks.

Safety critical element (SCE)   An aspect of a petroleum infrastructure item or a process whose principal purpose is to mitigate or to remove the risk of the occurrence of an identified hazard.

Safety, health, and environment (SHE)   An alternative term for **health, safety, and environment**.

Safety instrumented function (SIF)   An alternative term for a **safety instrumented system**.

Safety instrumented system (SIS)   A series of engineered hardware and software controls which are intended to provide safety limits in a particular petroleum operation. Also called a **safety instrumented function**. *See also* **layer of protection analysis**.

Safety integrity level (SIL)   A measure of the level of risk reduction which is achieved by the implementation of a safety instrumented system. *See also* **layer of protection analysis**.

Safety Of Life At Sea (SOLAS)   An international maritime treaty which provides for minimum safety standards for ship operation, which is issued by the IMO.

Safety shut-down (SSD)   An alternative term for an **emergency shut-down**.

Safety stock (SS)   In relation to petroleum storage, a certain petroleum stock level which is held in-tank at all times as a buffer in order to allow continued send-out while unplanned events have interrupted tank filling.

Safety ullage (SU)   In relation to petroleum storage, a certain petroleum stock level which is held in-tank when an offtake is interrupted for any reason.

Safety zone   An alternative term for a **refuge**.

Saffir   Simpson Scale A 1 to 5 rating which is based on a hurricane's sustained wind speed.

Sag bend   In offshore pipeline laying, an unsupported span of pipeline which lies between the stinger and the sea-bed.

Sail-away   The point of readiness of a ship or offshore petroleum infrastructure item to be taken away from its place of construction by its owner.

Sale and purchase agreement (SPA)
(1) A contract for the long-term sale and purchase of LNG quantities.
(2) A contract for the sale and purchase of assets or shares.

Sales gas   Raw gas which has been processed and meets the quality specification requirements of a petroleum sales contract or of a pipeline.

Sales line   A pipeline through which petroleum flows to a delivery point.

Sales point   An alternative term for a **delivery point**.

Salt cavern   An excavated underground cavern in a salt-based geological structure which is capable of use as a gas storage facility.

Salt dome   A particular geological feature which indicates the potential presence of petroleum in a formation.

Salt water disposal well (SWDW)   A wellbore which is used for the disposal of brackish water which has been produced during exploration or production.

Salvable equipment   Materials or equipment which have been used in petroleum exploration or production operations and can be reconditioned and reused.

Sample log   A recorded analysis record of the cuttings which are produced during the drilling of a wellbore.

Sampler   A device which is attached to a pipeline and permits the continuous sampling of petroleum that is passing through the pipeline.

Sanction package   A suite of agreed contracts or principles which underpin an investment decision.

Sand control   A process for minimising sand production during petroleum production in order to prevent sand fill.

Sand cyclone   A filtration vessel which removes sand particles from petroleum or water streams.

Sand fill   A column of sand which has accumulated in a wellbore.

Sand out   The inadvertent plugging of a wellbore with proppant during fracking.

Sandbagging   A colloquial term for a deliberate degree of underperformance which is applied in order to conceal the true measure of greater possible performance.

Sanded up   A wellbore through which petroleum production is restricted because of sand fill.

Satellite   A formation which is remote from, but is capable of being connected to, petroleum production, processing, or transportation infrastructure.

Satellite well   A single well which is drilled to tie in to a mobile offshore production unit, in order to produce petroleum from a remoter part of a formation or from a satellite.

Saturation
(1)   A measure of the percentage pore space of a formation which is occupied by crude oil, natural gas, liquids, or water. *See also* **water saturation**.
(2)   A state of thermal equilibrium at the interface between a gas and a liquid.

Savings index   A measure of how much profit is made by a company in relation to a corresponding costs reduction.

Saybolt Second Universal (SSU)   A measurement methodology for the viscosity of lighter petroleum fractions.

Scalability/scalable   In the context of a petroleum project, a project which is capable of being increased in size and/or scope from its original dimensions.

Scale inhibitor   A chemical agent which prevents the accumulation of contaminants and deposits in a petroleum infrastructure item.

Scavenged heat   Waste heat that is produced from an industrial process which is saved and applied to other purposes (such as power generation). *See also* **heat recovery**.

Scf   An alternative term for **standard cubic foot**.

Scope 1   According to the GHG Protocol Corporate Standard (www.ghgprotocol. org), GHGs which are direct emissions from sources that are owned or controlled by a reporting entity or which are under its control.

Scope 2   According to the GHG Protocol Corporate Standard (www.ghgprotocol. org), GHG emissions from the consumption of purchased electricity, heat, or steam that are a consequence of the activities of a reporting entity.

Scope 3   According to the GHG Protocol Corporate Standard (www.ghgprotocol. org), all other indirect GHG emissions related to the activities of a reporting entity from sources not owned or controlled by the reporting entity (including from the combustion of products).

Scour blanket   A rock or metal construction which overlies an area of the sea-bed in order to prevent the occurrence of scouring.

Scouring   Changes which occur naturally over time to the level of the sea-bed.

Scout group   A group of petroleum companies which collate and share market information and opportunities in a scout report.

Scout report   A periodic report about market information and opportunities which is issued by and between the members of a scout group.

Scouting   The general assessment and analysis of the viability of a petroleum exploration project.

Scouting report   An initial field survey which is conducted by a contractor in an area in which drilling or seismic works are to be performed.

Scraper   An alternative term for a **pig**.

Screen   A sieve which is inserted into a wellbore in order to hold back debris and cuttings while petroleum enters the wellbore.

Screening   A preliminary assessment of the suitability of a formation for further development.

Scrubber   A filter which is used to remove oil, water, and other impurities from a gas stream, leading to a knock-out.

Scrubber liquids   Liquid phase petroleum which is recovered from a scrubber.

Scrubber oil   Oil which is recovered from a scrubber.

Scrubbing   An alternative term for **absorption**.

Scuf   An alternative term for a **standard cubic foot**. *See also* **cubic foot**.

Sea-bed gas diverter (SBGD)   A diverter line which handles a shallow gas kick which occurs offshore, allowing gas to be released from a wellhead on the sea-bed.

Sea island   A man-made offshore structure without a direct connection to the shore at which tankers can berth (on one or several sides).

Seal   An impermeable rock layer which prevents the further migration of petroleum from a reservoir. *See* Appendix 2A.

Seal off
(1)   Any closing off of a wellbore.
(2)   The inhibition of a formation's ability to produce petroleum (including where this results from an invasion).

Seal pot   A small catchment chamber which is attached to a gas meter and traps water to prevent inaccurate measurements.

Seal risk   The risk that a seal will fail, leading to the migration of petroleum from a reservoir.

Seasonal normal weather   The normal pattern of weather which is to be expected in any given season.

Seasonal storage/seasonality   Gas storage which is used to meet seasonal demand variations, whereby gas is sourced and injected during periods of low price and low demand (the summer) and is redelivered during periods of higher price and higher demand (the winter). Contrast with **diurnal storage**, whereby injection and redelivery takes place on a daily basis.

Seasonal supply contract   An alternative term for a **peak supply contract**.

Seasonalised contract   A contract for the sale of petroleum whereby the quantities and the pricing are variable across the seasons of the year.

Seaward licensing round (SLR)   A licensing round for offshore acreage. Contrast with an **onshore licensing round**, which applies to onshore acreage.

Second-cycle exploration   Petroleum exploration which takes place in an area that has previously been explored and dismissed for development.

Secondary capacity   Capacity in a pipeline which is let by the shipper to a sub-shipper. Contrast with **primary capacity**, which is let by the transporter to the shipper.

Secondary fuel   A fuel which is processed from a primary fuel (such as LNG which is derived from natural gas). Also called a **derived fuel**. Contrast with a **primary fuel**, which directly produces energy upon initial combustion.

Secondary migration   The movement of mobile petroleum to a reservoir rock. *See also* **primary migration** and **tertiary migration**. *See* Appendix 2A.

Secondary porosity   Further porosity which is created in a formation as a result of natural dissolution or distortion of the formation.

Secondary recovery   The second stage of petroleum production, in which water or gas is injected into a formation in order to displace petroleum into the wellbore. Also called **advanced recovery**. *See also* **primary recovery** and **tertiary recovery**.

Secondary target   A formation which is targeted after the primary target when an exploration wellbore is drilled, in order to maximise the cost effectiveness of hiring a drilling unit.

Secondee   A person who is the subject of a secondment.

Secondment   The temporary location of an employee of one company within another company in order to assist with certain activities of that other company (such as under a JOA, under which non-operating party employees could be seconded to the operator). *See also* **secondee** and **secondment agreement**.

Secondment agreement   A contract which sets out the terms of a proposed secondment between the relevant persons.

Section   A stratigraphic section within a formation. *See* Appendix 2A.

Sediment inhibitor   A chemical agent which prevents the accumulation of contaminants and deposits in a petroleum infrastructure item.

Sedimentary rock   One of the three main rock types (with igneous rock and metamorphic rock), which is formed by the accumulation and compression of mineral and organic particles.

Sedimentary zone   A zone which is made up of sedimentary rock.

Seep study   A study of seeps in order to assess the prospectivity of sub-surface petroleum.

Seepage/seeps   Crude oil which finds its way naturally to the surface from a sub-surface formation.

Segregated condensate   Condensate which can be stripped from crude oil and sold separately.

Segregated storage   Petroleum which is stored in-tank and is distinct and identifiable (in contrast to commingled petroleum stocks).

Segregation   Under a JOA, the operator's distinction of its own monies from the monies in the joint account.

Seismic contractor option   A pre-emption right in respect of a farm-out which is given by the farmor to the seismic contractor under a multi-client seismic agreement.

Seismic contractor uplift   A right to an additional payment in respect of a farm-out which is given by the farmor to the seismic contractor under a multi-client seismic agreement.

Seismic migration   A process by which raw seismic data is rearranged (in space or in time) in order to give a data set which more accurately portrays an actual sequence of sub-surface events.

Seismic option agreement   A contract between a state and a seismic contractor which permits the seismic contractor to generate seismic data for use as multi-client seismic.

Seismic option farm-out   A form of farm-out agreement whereby the farmee acquires and processes seismic data, leading to the drilling of an exploration well.

Seismic profile (sprof)   The profile of a formation which is generated from the conduct of a seismic survey.

Seismic survey   A technique which is used to acquire geophysical data by projecting sound waves in order to create an image of a formation. *See also* **acoustic logging, amplitude, reprocessing, two D, three D**, and **four D**. Also called **imaging**.

Seismic survey vessel (SSV)   A ship which is used to conduct an offshore seismic survey.

Selective procedure   A model for the award of a concession by a state whereby the state identifies a list of potential bidders and invites them to bid. *See also* **limited procedure** and **open procedure**.

Self-billing   Under a petroleum sales contract, a provision whereby the buyer invoices itself for petroleum delivered by the seller, rather than the conventional seller-invoicing route.

Self-elevating drilling unit   An alternative term for a **jack-up**.

Self-insurance   The election of a person not to arrange third-party-provided insurance cover in respect of petroleum project liabilities, losses, and risks, in favour of assuming those liabilities, losses, and risks directly itself.

Self-potential   An alternative term for **spontaneous potential**.

Self-propelled multi-wheel trailer (SPMT)   A multi-wheeled, independently powered platform vehicle which is used for transporting a petroleum infrastructure item onshore.

Seller's nomination regime   Under a petroleum sales contract, a provision whereby the seller nominates the quantities of petroleum for delivery to the buyer. Contrast with a **buyer's nomination regime**, whereby the buyer's nominations prevail.

Seller's reservations   Under a petroleum sales contract, certain operational rights and entitlements in respect of produced petroleum which are reserved in the seller's favour. Also called **processing rights**.

Semi-finished products   Products which need further processing in order to be ready for sale or use. *See also* **finished products**.

Semi-sub/semi-submersible   An offshore drilling unit or production platform which floats with a significant part of the structure being submerged, which is capable of drilling or producing in water depths up to 10,000 feet. *See also* **column-stabilised drilling unit**. *See* Appendix 2B.

Semi-sub pipelay vessel   A pipelay vessel with a submersible pontoon hull.

Semi-submersible crane vessel (SSCV)   A heavy-lift crane which is used for the installation and decommissioning of an offshore petroleum infrastructure item.

Send-out   Under a petroleum sales contract, the quantity of petroleum which is delivered by the seller to the buyer during a specified period of time.

Senior supervisory personnel (SSP)   Under a JOA, a defined group of persons, which is used typically in relation to arrangements for the allocation of liabilities.

Separation   A process of separating oil, gas, liquids, and water from within a single petroleum stream.

Separator   A vessel which is used to separate oil, gas, and water. *See also* **two-phase separator** and **three-phase separator**.

Separator gas   Gas which remains after separation from condensate in a separator.

Series   An alternative term for an **epoch**.

Service contract   A concession whereby the concession-holder performs defined services relating to petroleum E&P in return for a defined fee which is paid by the grantor. *See also* **pure service contract** and **risk service contract**.

Service factor   An alternative term for **uptime**.

Service speed   Under a charterparty, the sailing speeds which a ship commits to (for ballast voyages and for laden voyages).

Service user risk management (SURM)   A risk and uncertainty analysis tool.

Service well   A wellbore which is drilled in order to support ongoing petroleum production, including a disposal well, a parametric well, and a wellbore, which is drilled for enhanced oil recovery purposes.

Servitude   An alternative term for an **easement**.

Set aside   A contractual or regulatory arrangement whereby revenues which are generated from a petroleum project are designated and directed for a specific purpose.

Set-back   A regulatory requirement for an onshore well site to be set at a minimum distance away from a defined point.

Set-back/pickup   The placing of tubulars in a rack after use (set-back) and the removal of tubulars from a rack prior to use (pickup).

Set-off   The deduction from amounts which are due to be paid by one person to another person of amounts which are due for payment to the first person by the other person, such that a single net payment results between those persons.

Setting depth   The depth to which the bottom edge of casing is set in a wellbore when it is ready to be cemented.

Setting point   A measure of the vertical depth to which casing has been set and cemented in a wellbore.

Settled production   Petroleum production from long-established fields which continues at relatively static production rates on a daily basis.

Settling tank   A tank into which used drilling fluid is deposited in order to allow solid materials in the fluid to settle out by gravity.

Settlor   Under a trust, a person who places property in trust.

Seven Sisters   A colloquial term which describes the multinational oil companies which dominated the global oil industry between the 1940s and the 1970s, which is made up of the following: Anglo-Iranian Oil Company (now BP), Gulf Oil (now part of Chevron), Royal Dutch Shell, Standard Oil Company of California (now

Chevron), Standard Oil Company of New Jersey (later Exxon, now ExxonMobil), Standard Oil Company of New York (later Mobil, now ExxonMobil), and Texas Oil Company (Texaco, now part of Chevron).

Seven-day rule   An extension of time to allow certain slippage in an LNG shipping programme, whereby LNG deliveries which are made in the first seven days of a contract year will count as deliveries having been made in the preceding contract year.

Several liability   The liability of a group of persons which, as between them, is individual to another person (such that each person within that group will assume only a defined share of the entire liability). *See also* **joint liability**, which creates joint liability between a group of persons.

Severance   The legal separation of an entitlement to exploit sub-surface petroleum rights from a wider land title interest.

Severance tax   A state-imposed tax upon the profits of a petroleum company from the exploitation of natural resources, which is more commonly found in jurisdictions which permit the ownership-in-place of petroleum resources.

Shadow JOA   A secondary JOA which is made between the non-state parties when a participation right applies and the state has become party to the concession and the principal JOA.

Shake-out   An alternative term for a **centrifuge**.

Shaker house   The area within a drilling unit which houses shale shakers. Also called a **mud house**.

Shaker well   A wellbore from which the recycled drilling fluid is filtered through a shale shaker.

Shale   A finely grained, fissile metamorphic or sedimentary rock which is formed by the consolidation of clays and silts within thin, relatively impermeable layers. *See also* **oil shale, shale gas, shale oil**.

Shale gas   Gaseous phase natural gas that is produced directly from shale rock, which is released by hydraulic fracturing.

Shale oil   Liquid phase crude oil that is produced directly from shale rock, which is released by hydraulic fracturing.

Shale shaker   A sieve which is used to remove cuttings from recycled drilling fluid in a shaker well. *See also* **shaker house**.

Shallow cut plant   A gas processing facility which is located close to a gas transportation pipeline, removes liquids from a raw gas stream, and has reduced processing capacity compared to a deep cut plant. Also called an **extraction plant**.

Shallow gas   A sub-surface gas deposit which is located sufficiently close to the surface that conductor casing will penetrate the deposit.

Shallow gas kick   Kick from a shallow gas deposit which cannot ordinarily be controlled by shutting in the wellbore. *See also* **sea-bed gas diverter**.

Shallow pool well (SPW)   A wellbore which is drilled into an already-producing formation but down to a shallower payzone. Contrast with a **deep pool well**, which is drilled into an already-producing formation but down to a deeper payzone.

Shallow water (SW)   Water depths of up to 400 metres. Contrast with **deep water**, as water depths in the range of 400 to 1,500 metres.

Shape   The definition of a gas storage facility's working gas and cushion gas capacity, injection rates, redelivery rates, and cycle rates, all of which combine to give an overall profile of the facility's size and flexibility.

Shared data repository (SDR)   A common-access repository for seismic data.

Shared interest   An alternative term for a **carry**.

Shear ram   A blowout preventer which, when closed, cuts through a drill string in the wellbore and seals the wellbore.

Shell and tube   An assessment and analysis of the ability of a petroleum infrastructure item to withstand an ingress of smoke in the event of an emergency.

Shell storage capacity   The design capacity of a petroleum storage tank (which is usually greater than the actual working capacity of the storage tank).

Shell UK Oil (SUKO)   Known as SUKO90, a set of standard contractual terms which are used for the trading of Brent crude oil.

Shifting   The activity of moving a ship between berths at a port.

Ship inspection report programme (SIRE)   A ship condition inspection and assessment programme which is managed by the OCIMF. (http://www.ocimf.org/sire).

Ship of opportunity   A ship which unexpectedly becomes available for hire.

Ship or pay   Under a capacities-based contract, the payment which is made by the shipper to the transporter in consideration of a capacity reservation.

Ship-shore compatibility study (SSCS)   A review of the compatibility of a ship and its intended loading port or unloading port.

Ship-to-ship (STS)   The transfer of petroleum between ships offshore. Also called **transhipment** and **vessel-to-vessel**. *See also* **cryo-transfer** and **lightering**.

Ship tracking   A satellite-based system for tracking the movements of a ship during its voyage.

Shipowner's lien   A lien which is exercised by a shipowner over a charterer's cargo if the charterer has failed to pay the hire due under a charterparty.

Shipowner's option (SHOPT)   An election in respect of the operation of a ship which can be made solely by the shipowner.

Shipper   A person who contracts to receive the provision of petroleum transportation rights in a pipeline. Contrast with a **transporter**, as a person who contracts to provide those rights.

Shoe   A strengthened section at the lower end of a section of casing, which is intended to protect tubulars downhole.

Shoestring   An alternative term for **poorboy**.

Shooting   The firing of an explosive charge into a petroleum-producing formation as part of a fracking process.

Shooting right   The right of person (which could be established by law or by contract) to enter into a marine or terrestrial area in order to conduct a seismic survey.

Shoreline deposit   A sedimentary environment which is found in shallow-water depths at the near shore. Also called a **littoral deposit**.

Short delivery   Under a petroleum sales contract, a measure of petroleum which is delivered by the seller less than the quantity of petroleum which the buyer had contracted to buy from the seller.

Short position   Under a petroleum sales contract, the position of the seller which does not own or have access to the petroleum which it has contracted to sell. Contrast with a **long position**, whereby the seller already owns or has access to the petroleum which it has contracted to sell.

Short-range storage (SRS)   A gas storage facility whereby the total quantity of working gas can be fully delivered to the storage facility users over a relatively short timeframe. *See also* **long-range storage** and **medium-range storage**.

Short term flat agreement (NBP 1997/2015)   A trading agreement, known fully as 'Short Term Flat NBP Trading Terms & Conditions', which is used for the sale and purchase of gas at the NBP (http://www.gasgovernance.co.uk).

Short ton (ST)   An imperial weight measure of 2,000 pounds. *See also* **long ton** and **metric ton**.

Short trip
(1)   A trip of only a section of, and not the entirety of, a drill string.
(2)   A trip during which the drill string does not reach the surface.

Shortfall   Under a petroleum sales contract, the extent of the failure of the seller to deliver petroleum in satisfaction of the buyer's entitlements to receive petroleum.

Shortfall price discount   Under a petroleum sales contract, the remedy of the buyer for a shortfall, which is expressed as a discount to the price that is otherwise payable by the buyer to the seller for subsequent deliveries of petroleum.

Shot hole   A wellbore in which an explosive charge is set off downhole in order to fracture a formation and to enable petroleum to flow into the wellbore.

Shot on goal   The drilling of a wellbore in an attempt to discover petroleum in a formation.

Shot point   A precise point from which a seismic survey is conducted. Also called a **drop point**.

Shotgun arbitration   A colloquial term for a method of dispute resolution whereby the appointed arbiter decides its preferred outcome, after receiving the disputing parties' submissions. *See also* **baseball arbitration** and **golf arbitration**.

Shoulder   A slope up to, or down from, a particular plateau position; or a slope down to, or up from, a particular floor position.

Shoulder months   Under a petroleum sales contract, the months which lie on either side of a peak demand period.

Show   Evidence of the existence of petroleum in a formation which is obtained from cores, circulated drilling fluid, or cuttings. Also called a **drilled show**.

Shrink   A measure of the reduction to a quantity of stored petroleum which results from theft, leakage, or compositional changes.

Shrinkage
(1)   An alternative term for **extraction loss**.
(2)   An alternative term for **lost gas**.

Shrinkage purchase   Gas which is bought by a transporter in order to make up for lost gas quantities in a pipeline transportation system.

Shut-down   A period when a petroleum facility ceases to operate as intended while essential maintenance work is undertaken (whether as an emergency shut-down or as a routine shut-down).

Shut-down valve (SDV)   A valve configuration in a pipeline or a petroleum facility which can stop the flow of petroleum.

Shut in
(1)   The temporary cessation of the production of petroleum through a wellbore for operational or commercial reasons which is effected by a shut-down valve or by another mechanism at the surface.
(2)   The temporary cessation of the production or the processing of petroleum through a petroleum infrastructure item.

Shut in bottomhole pressure (SIBHP)   Bottomhole pressure when a wellbore is shut in.

Shut in casing pressure (SICP)   The pressure on the casing in a wellbore when the wellbore is shut in.

Shut in drill stem pressure (SIDSP)   The pressure on a drill string in a wellbore when the wellbore is shut in.

Shut in pressure   The pressure which is measured at the wellhead when a wellbore is shut in.

Shut in rate   A reduced rate of payment (whether as a royalty or as a contract rate) which applies when a petroleum production operation is closed down because of defined circumstances.

Shut in royalty   A cash payment which is received by a royalty-holder in lieu of produced petroleum (if a royalty is paid in petroleum) when the production of petroleum is suspended.

Shut in temporarily abandoned (SITA)   An alternative term for **suspension**.

Shuttle mode FSRU   An alternative term for a **shuttle regasification vessel**.

Shuttle regasification vessel (SRV)   An FSRU which trades LNG between markets, rather than being a fixed FSRU for LNG imports. Also called a **shuttle mode FSRU**.

Shuttle tanker   A crude carrier which lifts crude oil from an FPSO, an FSU, or a spar.

Sidecut   A generic term for the intermediate petroleum products which emerge from the side (rather than the top or the bottom) of a crude distillation unit. Also called **sidestream**. *See* Appendix 2E.

Sidecut stripper A further column which sits alongside a crude distillation unit for improved refinery yield.

Sidestream
(1) An alternative term for **sidecut**.
(2) A revenue stream for a petroleum company which is generated by an activity that is ancillary to the company's core business.
(3) A device which monitors the presence of microbial organisms which can cause corrosion in a petroleum infrastructure item.

Sidetrack A secondary wellbore which is drilled to emanate away from an original wellbore.

Sidewall The vertical wall of a wellbore.

Sidewall core (SC) A core which is extracted perpendicular to a wellbore.

Signal/noise ratio (S/NR) In a seismic shoot, a measure of the strength of a seismic signal relative to the amount of background noise (where a higher ratio indicates more accurate results).

Signature bonus Under a concession, a monetary sum which is payable by the concession-holder to the grantor at the point of awarding the concession.

Siltation The contamination of a port area or berth which is caused by the accumulation of sedimentary deposits, leading to the need for dredging.

Siltation rate The rate at which siltation takes place within a port area or berth.

Simple refinery An alternative term for a **topping refinery**.

Simultaneous operations (SIMOPS) Two or more activities which are conducted at the same time and place within a petroleum project process or a petroleum infrastructure item and could by virtue of their interaction present risks that were not presented by the individual conduct of the activities.

Single A single length of drill pipe. Also called a **joint**. *See also* **double, fourble,** and **thribble**.

Single anchor loading/leg mooring (SALM) An alternative term for a **single buoy mooring**.

Single buoy mooring (SBM) A floating chamber which is moored near to an offshore petroleum production platform, providing a connection to a ship for the offtake of petroleum but without capacity for petroleum storage. Also called a **catenary anchor leg mooring** and a **single anchor loading leg/mooring**. Contrast with a **spar**, which provides for the storage of petroleum prior to offtake.

Single buyer model An alternative term for **aggregation**.

Single containment tank (SCT) An LNG containment tank which is surrounded by an external safety bund. *See also* **double containment tank** and **full containment tank**.

Single cycle In a gas storage facility, cycling which takes place once a year.

Single cycle gas turbine (SCGT) An electric power-generating facility with a single turbine to generate electric power, without provision for waste heat recovery.

Also called an **open cycle gas turbine**. Contrast with a **combined cycle gas turbine**, which incorporates gas and steam turbines to generate electric power and with provision for waste heat recovery.

Single fold   A fold which records a single seismic data point. *See also* **full fold**.

Single hull   A crude carrier which has a single-skinned hull. *See also* **double hull**.

Single mixed refrigerant (SMR)   A process for the liquefaction of gas to give LNG which uses a single phase of a mixed refrigerant agent. *See also* **dual mixed refrigerant**.

Single phase   A petroleum stream with only one component (principally crude oil or natural gas but also possibly liquids or water). Contrast with a **multiphase**, as a petroleum stream which contains more than one of those components in a commingled whole.

Single point mooring   A form of mooring whereby a ship is anchored to the sea-bed using multiple mooring lines which are connected to a single point on the ship. Also called a **catenary anchor leg mooring** and a **single anchor loading/leg mooring**.

Single point responsibility (SPR)   The assumption by a person of all of the responsibility to design, construct, and commission a petroleum project under an EPCI.

Single sourcing   The procurement of goods and services from a selected single supplier when competitive supply options exist. Contrast with **sole sourcing**, as the procurement of goods and services from the only supplier capable of supplying them.

Single string dual completion   A method of well completion in which two payzones are produced through a single wellbore by channelling production from one payzone through a tubing string and from the other payzone through the annulus.

Single zone completion   A method of completion in which a single payzone is produced through a single wellbore using a single packer. Contrast with a **multizone completion**, as the completion of a wellbore which uses more than one packer to seal off different wellbore sections to create separated payzones.

Sink   A physical repository for waste products or a notional repository for costs.

Sinking fund   A fund which is constituted by the set-aside of money over time in order to service a future debt obligation.

Sister ship inspection   An inspection of a single ship within a batch of identical ships which is accepted as an effective inspection of all the ships in that batch.

Six degrees of motion   The movement of a ship or offshore petroleum infrastructure item by reference to any of heave, pitch, roll, surge, sway, and yaw.

Skid   A steel frame which supports an item of petroleum-processing equipment, for mounting on a petroleum production platform.

Skidding   The movement of a petroleum facility from one place to another but with minimal dismantling beforehand.

Skidding the rig   The skidding of a drilling unit.

Skimmer   A machine which removes floating oil deposits from the surface of water.

Skin damage   The surface area of a formation relative to a wellbore which has been damaged by an invasion.

Skin value   A factor which determines the production efficiency of a formation through a wellbore by comparing actual conditions with theoretical conditions (where a positive skin value indicates adverse influences which impair productivity).

Slack loop   An alternative term for an **expansion joint**.

Slack tank   A liquid-containing tank on a ship which is not full. *See also* **pressed up** and **free surface effect**.

Slant drilling   An alternative term for **directional drilling**.

Sleeve/ sleeving
(1)   The provision of contract coverage or credit support for a party to a transaction by another party to the transaction or by another person.
(2)   The provision of shipping services in respect of a FOB LNG sale by the seller which gives it the appearance of a DES LNG sale. Also called **synthetic DES**.

Slickline   A single strand steel wire which is used to lower tools into a wellbore. *See also* **braided line**.

Slickwater   Water to which certain chemicals have been added in order to increase its ability to flow, so that the water can be pumped downhole at a greater rate during a fracking operation.

Slight show of gas (SSG)   An indication, recorded in a DDR, that drilling has shown the possibility of natural gas in a formation.

Slight show of oil (SSO)   An indication, recorded in a DDR, that drilling has shown the possibility of crude oil in a formation.

Slimhole   A wellbore which is drilled with less than normal diameter tools, using correspondingly smaller-diameter casing and so leading to reduced drilling costs.

Sling   A fabric strop which is used to lift a tubular.

Slip   An alternative term for a **pipe jack**.

Slip forming   A construction method by which concrete is poured into a continuously moving form in order to manufacture a concrete component.

Slipway   An angled ramp on a shoreline from which ships are launched or onto which ships are brought ashore.

Slop   An alternative term for **cutterstock**.

Slop oil   Quantities of oil which are lost during a refining process and are collected and reused as feedstock.

Slops   Water/oil residues which remain in-tank after crude oil has been discharged from the tank. *See also* **crude oil washing**.

Sloshing   The resonant motion of the free surface of a liquid cargo in a moving container (such as a part-loaded cargo of LNG or crude oil in-ship). *See also* **free surface effect**.

Slot
(1)   A defined period of time for the berthing of an LNG carrier at a loading terminal or an unloading terminal.
(2)   A specific area on an offshore drilling unit from which a wellbore is drilled.
(3)   A specific area on an offshore petroleum production platform into which a development wellbore is tied in.

Slot swapping   An inter-shipper mechanism for trading berthing/unberthing slots at a loading port or unloading port.

Slot-type jack-up   A jack-up which has an opening in the drilling deck, over which the drilling derrick is positioned. Also called a **keyway jack-up**. *See also* **canti-jack**.

Sloughing   The disintegration of the wall of a wellbore during or after drilling. Also called **caving**.

Slow steaming   A deliberate reduction in a ship's steaming speed in order to conserve bunkers.

Slow-speed diesel (SSD)   Diesel which does not contain the additives which would transform it to HSD.

Sludge   A viscous, sedimentary residue which forms at the bottom of a storage tank from crude oil mixtures in-tank. Also called **heel**.

Slug   A quantity of liquids and other impurities which has built up within a pipeline which transports wet gas.

Slugcatcher   A filter for the removal of accumulated slug from a pipeline.

Slurry
(1)   A mixture of crushed coal and water which is combusted as a boiler fuel.
(2)   A mixture of cement and water which is used in cementing.

Small scale   An LNG export or import facility which is scaled down from conventional LNG project facility dimensions for improved affordability and operational flexibility.

Small-scale LNG carrier (SSLNGC)   An LNG carrier with a cargo-carrying capacity of 30,000 cubic metres or less.

Smart completion   An alternative term for an **intelligent completion**.

Smart well   An alternative term for an **intelligent well**.

Smoke ingress analysis (SIA)   An assessment and analysis of the ability of a petroleum infrastructure item to withstand an ingress of smoke in the event of an emergency.

Snaglist   A list of defects which are outstanding on the construction and/or commissioning of a petroleum facility or project. Also called a **punchlist**.

Snubbing
(1)   The removal of a debrining tool from a newly created salt cavern in the development of an underground gas storage facility.
(2)   The insertion of casing into a wellbore while the wellhead is still under pressure from the wellbore.
(3)   The forcing of tubulars or tools into a shut in wellbore against the pressure which has built up in the wellbore during a kick.

Soak phase   As part of the CSS process, a pause in the production of petroleum from a wellbore in order to allow even heat distribution from the injected steam.

Social costs   The opportunity costs of a petroleum project when viewed from the perspective of affected individuals and societies.

Social licence to operate   An alternative term for **corporate and social responsibility**.

Social lifecycle analysis (SLA)   A methodology for assessing the internalities and the externalities of the production of petroleum based on social and socio-economic indicators.

Society of International Gas Tanker and Terminal Operators (SIGTTO)   A trade organisation which represents the interests of companies engaged in gas shipping and terminalling (http://www.sigtto.org).

Society of Petroleum Engineers (SPE)   An international industry body which is focused on the technical aspects of petroleum exploration, discovery, development, and production (http://www.spe.org).

Society of Petroleum Engineers—Petroleum Resources Management System (SPE PRMS)   A widely used guide to the classification of petroleum reserves and resources (http://www.spe.org). *See* Appendix 7.

Society of Petroleum Evaluation Engineers (SPEE)   An international industry body which is focused on petroleum evaluation engineering (http://secure.spee.org).

Soft carry   A carry which is repayable by the carried party. Contrast with a **hard carry**, which is not repayable by the carried party.

Soft law   A quasi-legal instrument (such as a voluntary code of regulation) which lacks legally binding force but which persons are accustomed to complying with.

Soft local content   A local content provision which applies a subjective, unquantified local content requirement. *See also* **hard local content**.

Sold cold   LNG which is sold in a refrigerated state to a buyer, rather than that the LNG is regasified and the resultant regas is sold to the buyer.

Sole risk   Under a JOA, a petroleum project activity which is carried out by less than all of the parties to the JOA and paid for only by the participating parties, if the operating committee has not approved the operation as a joint operation. Also called an **exclusive operation**. *See also* **non-consent**.

Sole sourcing   The procurement of goods and services from a supplier which is the only supplier capable of supplying them. Contrast with **single sourcing**, as the procurement of goods and services from a selected single supplier if competitive supply options exist.

Soliton   A single offshore wave which largely retains its shape and strength after colliding with another wave or an object.

Solus tie   A downstream fuel supply contract which is entered into between a petroleum company and an independent dealer (such as in a dealer-owned service station arrangement) and provides for exclusivity in favour of the company for the duration of the contract.

Solution   A mixture of phases in which the individual phases are uniformly distributed.

Solution gas   An alternative term for **associated gas**.

Solution gas drive   Petroleum production through a wellbore which is driven by the expansion of associated gas from a crude oil stream. Also called **dissolved gas drive**. *See also* **capillary pressure** and **depletion drive**.

Solution mining   An alternative term for **leaching**.

Solution oil   Crude oil which contains natural gas in solution. *See also* **associated gas**.

Solvent deasphalting (SDA)   A process whereby a solvent is used to break out asphalt and crude oil from ultra-heavy oil.

Solvent extraction   A process for the manufacture of lubricants in a refinery.

Sonde   A downhole tool which contains various measurement sensors.

Sour gas   Natural gas which contains relatively significant amounts of hydrogen sulphide. Contrast with **sweet gas**, which contains negligible amounts of hydrogen sulphide. Also called **acid gas**.

Sour hole   A wellbore which produces sour gas or sour oil.

Sour oil   Crude oil which contains relatively significant amounts of sulphur. Contrast with **sweet oil**, which contains negligible amounts of sulphur.

Source maturity modelling   A process whereby the maturity of a rock stratum is measured and modelled in order to assess its petroleum-bearing potential.

Source rock   A part of a formation in which petroleum has been formed. *See* Appendix 2A.

South–South   Trade and technical cooperation which takes place between developing nations in the southern hemisphere.

Southern Cone gas network   The interconnection of gas production facilities and gas pipelines across the combined region of Argentina, Bolivia, Brazil, Chile, Paraguay, Peru, and Uruguay.

Sovereign immunity waiver   A provision in a contract whereby a state or a parastatal entity agrees not to apply the defence of sovereign immunity to proceedings which are brought against it in relation to an alleged breach of the contract.

Sovereign wealth fund (SWF)   A state-owned entity which is responsible for investing a state's petroleum project revenues in diversified economic activities.

Spacing   The distance between wellbores which are producing petroleum from the same formation. *See also* **well spacing**.

Spar   A floating chamber which is moored near to an offshore petroleum production platform and provides for the storage of petroleum prior to transfer to a ship for offtake. Contrast with a **single buoy mooring**, which does not provide for storage.

Sparing agreement   An agreement between a group of petroleum companies to pool the ordering and use of common spare parts in order to secure improved efficiencies of scale.

Sparing philosophy   An operational philosophy which a petroleum company applies to determine the extent to which the company will maintain an inventory of spare parts for key equipment items.

Spark spread   The difference between the cost of gas and the price of electric power which is produced from the combustion of the gas. A positive spark spread indicates that the price of the electric power is higher than the cost of the gas and so the spark spread is profitable.

Spear   A particular form of fishing tool for the recovery of a tubular from a wellbore.

Special core analysis (SCAL)   A further set of specialist measurements (compared to routine core analysis) which are carried out on a core in order to determine porosity, permeability, and lithology in respect of a formation.

Special purpose vehicle (SPV)   A limited liability company which is incorporated for the single purpose of participation in a petroleum project and has no interests or assets other than that participation.

Special shut-down (SSD)   An alternative term for **emergency shut-down**.

Special terms and conditions (STCs)   Bespoke terms which apply to a contract between parties. Contrast with **general terms and conditions**, as standard terms which apply between the parties.

Specialties   Petrochemical products which are manufactured for a particular application.

Specific gravity   A measure of the density of the components of a petroleum stream relative to air. Also called **relative density**.

Specific lifting schedule (SLS)   A short-term schedule for LNG deliveries. *See also* **ninety-day programme**.

Spectrum   The range of products which are produced from a refinery.

Speculative seismic   An alternative term for **multi-client seismic**.

Spent gas lift   An alternative term for **recoverable gas lift**.

Sphering   An alternative term for **pigging**.

Spiking   The blending of inerts or higher petroleum fractions into a gas stream in order to modify the calorific value of the gas stream. *See also* **enriching/enrichment**.

Spill point   The structurally lowest point in a formation which can contain petroleum. Petroleum retention in the formation will not be possible once the formation is filled to the spill point. *See also* **filled to spill**.

Spillover   An alternative term for an **externality**.

Spillway   An alternative term for a **diverter**.

Spin-up   The rapid turning of a drill string when one length of drill pipe is joined to another.

Spindle oil   A low viscosity lubricating oil.

Spinning reserve   Extra productive capacity which can be made available by increasing the output of an already-functioning petroleum facility. Contrast with **supplemental reserve**, whereby extra capacity comes from a facility which is not functioning.

Spiral wound heat exchanger (SWHE)   A heat exchanger which consists of a series of tubes that are wound around a central core.

Spirit of mutual understanding and trust (SMUT)   Under a petroleum sales contract, a provision whereby the parties will discuss and seek to agree any necessary amendments to the contract in order to account for the effects of changes in commercial circumstances. *See also* **adaptation clause**.

Splash zone   On an offshore facility, the areas of metalwork which sit above and below the MSL and are exposed alternately to sea and air.

Split and wrap   The disaggregation of a single project contract into a number of smaller, separate contracts (for tax, regulatory, or commercial purposes) which are then administered jointly under an overall single agreement.

Split operator   Under a JOA, the designation of different parties to act as the operator for different operational phases within an overall petroleum project.

Split stream   A sale of petroleum which is made from a single stream by a seller to a number of different buyers.

Sponson block   An extension unit which is fitted to a ship's flank to increase the ship's main deck area, ballast tank volume, and overall buoyancy.

Sponsor   The developer of a petroleum project which relies on third-party finance (such as project finance) in order to meet the related project development costs.

Sponsor support   In a project finance exercise, the support which the lenders require the sponsor to provide for the loan repayment obligation.

Spontaneous potential (SP)   A natural electrical characteristic of a petroleum-bearing formation which is measured by a downhole tool. Also called **self-potential**.

Spot contract   A petroleum trade which is intended for settlement (delivery and payment) on the same day or within two days of the trade date. *See also* **forward contract** and **future contract**.

Spot formation contact   Vertical drilling which interfaces a wellbore with a formation at a single contact point. Contrast with a **spread formation contact**, whereby horizontal drilling interfaces a wellbore with a formation across multiple contact points.

Spot price   A price that is payable for petroleum in a contract which specifies immediate delivery and applies the price prevalent at that time. *See also* **forward price**.

Spot waterflood   The surrounding of a central production wellbore by a number of water injection wellbores which sweep produced petroleum towards the production wellbore (and which could, depending on the aggregate number of water injection wellbores and the production wellbore, be a four spot waterflood or a five spot waterflood).

Spotting   The pumping of oil into a wellbore in order to free a stuck pipe.

Spray cooling   The cool down of an LNG tank by spraying LNG onto its interior surface.

Spread   An assembly of marine vessels, including an offshore drilling unit and a safety boat, which is used for the performance of offshore petroleum exploration or development operations. Also called a **marine spread**.

Spread formation contact   Horizontal drilling which interfaces a wellbore with a formation across multiple contact points. Contrast with a **spot formation contact**, in which vertical drilling interfaces a wellbore with a formation at a single contact point.

Spread mooring   A form of mooring whereby a ship is anchored to the sea-bed using multiple mooring lines, normally attached to the bow and the stern of the ship. *See also* **catenary mooring** and **single point mooring**.

Spread splitting   An alternative term for **benefit sharing**.

Springing guarantee   A guarantee which only comes into effect upon the occurrence of a specific event.

Spud/spud in   The start of the drilling of a wellbore, when a drill bit first penetrates. *See also* **drill-on date**.

Spud can   A support which is placed on the sea-bed in order to underpin a jack-up's legs and is intended to prevent a punch through by the jack-up.

Spudkicker   A kicker which is payable when a well is spudded.

Spur   A small-diameter pipeline which is tied into a larger-diameter pipeline.

Squat   A decrease in a ship's under keel clearance which is caused by an increase in the ship's speed or by the ship passing through a confined channel.

Square kilometre   A square kilometre which is used as a unit of measurement for the acquisition of three D seismic data. Contrast with a **line kilometre**, as the unit of measurement for the acquisition of two D seismic data.

Squeeze
(1) The pressurised injection of cement in order to seal off perforations and fissures in a wellbore. Also called a **cement squeeze**.
(2) In petroleum trading, a lack of actual, deliverable commodity in a market.

Stab   The connection of one piece of drill pipe to another.

Stabilisation   A provision in a concession or a contract whereby an investor will be afforded some measure of protection by a state against expropriatory events which could adversely affect the investor's interests.

Stabilised crude oil (SCO)   Live oil from which solution gas has been extracted.

Stabilised flow   In the testing of a wellbore, the production of petroleum at a sustained rate for a period of time without a drop in production pressure.

Stabilised well   An alternative term for **overbalance**.

Stabiliser   A tool which sets the angle at which a wellbore is drilled. *See also* **deflection tool**.

Stack
(1)   A series of horizontal wellbores which have been drilled at different parallel depths.
(2)   A collection of out-of-use drilling units.
(3)   A vertical arrangement of blowout preventers, in which a ram blowout preventer usually sits below an annular blowout preventer.
(4)   A record of seismic data which incorporates traces taken from different sources in order to improve the quality of the overall data set. *See also* **post-stack** and **pre-stack**.

Stack gas   An alternative term for **flue gas**.

Stacked pay   Payzones which are stacked at different stratigraphic depths.

Stacking   The creation of a seismic data stack.

Stacking location   The location at which an out-of-use drilling unit is laid up.

Stage   An alternative term for an **age**.

Stage cementing   Primary cementing which is conducted in a series of defined stages.

Stand   A pre-assembled section of drill pipe, which is typically made up of two, three, or four lengths.

Stand by   The non-operation of a drilling unit in the drilling of a wellbore for reasons that are not attributable to the drilling contractor.

Stand by costs   The aggregate costs (usually expressed as a daily rate) which are associated with maintaining a spread on station during a stand by.

Stand by rate   Under a drilling contract, the revised dayrate which is payable by the employer company to the drilling contractor during periods of stand by.

Standard bundled unit (SBU)   In a gas storage facility, a combined definition of the injectability, storage, and deliverability services which are purchased by a facility user.

Standard cubic foot   An alternative term for a **cubic foot**.

Standard litre per minute (SLPM)   A unit for measuring the volumetric flow rate of gas at standard temperature and pressure, which is typically used in the United States. *See also* **normal litre per minute**.

Standard metering   An agreed set of base operational conditions by reference to which petroleum quantities are metered.

Standard reach lateral (SRL)   A horizontal wellbore of approximately 4,000 feet, which is a distance less than a mid-reach lateral.

Standard temperature and pressure (STP)   An alternative term for **reference conditions**.

Standardised measure of oil and gas (SMOG)   A protocol which is mandated by the SEC for the standardised reporting of petroleum reserves on an annual basis.

Standby letter of credit (SBLC)   A letter of credit which serves as a secondary payment mechanism for a transaction, whereby an issuing bank promises payment to an identified beneficiary if a principal payer defaults in payment when due under the transaction. *See also* **International Standby Practices 98**.

Standing charge   A tariff which is payable as a fixed charge by a petroleum facility user, irrespective of any variable charges which are incurred by the user.

Standoff location   A designated point at which a ship will wait pending the loading or the unloading of a cargo.

Stand-out right   An alternative term for a **non-consent right**.

Standstill provision   Under an asset or share purchase transaction, a provision which obliges the parties not to replicate or undermine the proposed transaction for a period of time.

Stapling   A requirement that two or more related contractual positions which are held by a contracting party can only be transferred in unison by that party. *See also* **uniform interest clause**.

Start-up   The commencement of commercial operations of petroleum infrastructure item (typically after commissioning has taken place).

Start-up assurance review (SUAR)   An assessment of the safety and functionality of a petroleum project process which is carried out prior to start-up, to ensure that the intended functionality is met. Also called a **pre-start-up audit**.

State-owned enterprise (SOE)   A state-owned or state-linked company. Also called a **government-owned company** and a **government-sponsored enterprise**.

State take   An alternative term for **rent**.

State take in cash (STIC)   The decision of a state to take its produced petroleum entitlements under a concession in cash. *See also* **state take in kind**.

State take in kind (STIK)   The decision of a state to take its produced petroleum entitlements under a concession in kind. *See also* **state take in cash**.

Statement of requirements (SOR)   A documented proposal which states a business problem or opportunity and seeks funding to conduct certain project activities in respect of that problem or opportunity.

Static kill   An alternative term for **top kill**.

Static range   The relationship between undeveloped petroleum resources and reserves and rates of global petroleum production, as an indication of how much petroleum remains for production at any time.

Statute and mandate (S&M)   A legal scheme of arrangement by which a state's entitlement to produced petroleum under a concession is transferred to a person for further on-sale.

Statutory pooling   An alternative term for **forced pooling**.

Statutory unitisation   Unitisation which proceeds as a state-mandated exercise in respect of the participating parties. Contrast with **voluntary unitisation**, which proceeds as a votive exercise between the participating parties.

Steady state equation   A system or process in which variable factors which could influence that system or process do not change over time.

Steam-assisted gravity drainage (SAGD)   An alternative term for **steam flood**.

Steam flood   The injection of high-pressure steam into a heavy oil deposit to reduce viscosity of oil. Also called **cyclic steam stimulation, huff and puff**, and **steam-assisted gravity drainage**.

Steel piled jacket (SPJ)   A steel jacket which is anchored to the sea-bed by deep pilings, typically found in relatively shallow-water depths.

Step-in agreement   An agreement which contains a step-in right.

Step-in right   A right of a person to step in to the contractual position of another person under a contract or a project in protection of its own position. *See also* **step-out**.

Step-out   The end of the exercise of a step-in right by the person holding that right.

Step-out well   A wellbore which is drilled into an identified discovery and is intended to delineate the lateral limits of the discovery. Also called an **appraisal well**, a **confirmation well**, a **delineation well**, and an **outpost well**.

Stick built   A form of construction in which assembly takes place entirely on site. *See also* **modular construction**.

Stiction   The tendency of a valve to close with less efficiency than it is intended to.

Still gas   A mixture of gases which is produced in a refinery by the processes of distillation, cracking, or reforming and is then used as fuel for the refinery. Also called **refinery gas**.

Still water level (SWL)   The average water surface elevation at a particular point, excluding the impact of dynamic local variation such as wave movements.

Stimulation   Any intervention (including tertiary recovery) by which a formation is manually stimulated in order to increase the flow of petroleum.

Stinger   A boom which is used to lower a pipeline onto the sea-bed from a pipelay vessel. *See also* **J-lay** and **S-lay**.

Stock change   A measure of the difference between petroleum stocks in storage at the beginning of a reporting period and at the end of the reporting period. *See also* **build-up** and **drawdown**.

Stock tank barrel (STB)   A measure of the quantity of actual crude oil production from a formation. Contrast with **stock tank oil originally in place**, which measures crude oil in a formation prior to production. Also called **moveable oil originally in place**.

Stock tank oil originally in place (STOOIP)   A measure of the quantity of crude oil which is contained in a formation before any production takes place. Contrast with **stock tank barrel**, which measures the quantity of actual crude oil production from a formation.

Stop card safety   An operational tool which empowers a staff member or a contractor to request an immediate cessation of an ongoing work activity because of health and safety concerns.

Stopcocking   The deliberately intermittent production of petroleum from a low-productivity reservoir in order to allow time for capillary pressure to build up while the reservoir is shut in.

Stoppel   A temporary plug which is inserted into a pipeline under repair.

Storage facility owner (SFO)   An owner of a petroleum storage facility.

Storage facility user (SFU)   A user of a petroleum storage facility.

Storage interruptible   An interruptible gas supply arrangement whereby the seller's interruption will not exceed the buyer's storage or alternative supply position. *See also* **strategic interruptible**.

Storage play   A cargo of petroleum which is held in storage pending the emergence of improved sale terms.

Storage service contract (SSC)   A contract for the provision of petroleum storage services which is made between an SFO and a storage facility user.

Storage service model   A model for the operation of a petroleum storage facility whereby the storage facility owner makes all or part of the storage capacity available to third-party users in exchange for the payment of a tariff. Contrast with an **equity storage model**, whereby the storage facility owner reserves all of the capacity to meet its own commercial needs.

Storm plug   A retrievable tool which is used to suspend the drilling of a wellbore during a storm.

Straddle packer   Packers (often just two) which are spaced apart in a wellbore in order to isolate a section of an open hole.

Straddle plant
(1)   A deep cut plant or a shallow cut plant which is located on a gas transportation pipeline.
(2)   A petroleum-processing facility which provides a tolling service to a user.

Straight hole   A vertically drilled wellbore.

Straight run   Refined petroleum products which are generated from a first round refining process of crude oil before secondary refining and recovery techniques are applied. Also called **virgin stock**.

Straight-up   An alternative term for **heads-up**.

Stranded area power generation (SAPG)   An alternative term for **remote power generation**.

Stranded asset   A petroleum production project which has become incapable of generating an economic return, whether because of inherent project deficiencies or supervening economic circumstances affecting the project.

Stranded gas   Discovered but undeveloped gas deposits which are presently precluded from commercialisation for economic, geopolitical, or operational reasons. Also called **remote gas**.

Stranding   A ship which is grounded by the hull and is not fully afloat and free. Also called **grounding**.

Strat well/stratigraphic well   An alternative term for a **parametric well**.

Strata/stratum   Rock layers/a rock layer within a formation.

Strategic interruptible   An interruptible gas supply arrangement whereby the seller's interruption will not be conditioned by the buyer's storage or alternative supply position. *See also* **storage interruptible**.

Strategic petroleum reserve (SPR)   A petroleum products buffer stock which is maintained by a state for emergency use.

Stratification   An alternative term for **layering**.

Stratigraphic trap   A petroleum-bearing formation which results from changes in rock types or sedimentary unconformities and prevents the further migration of mobile petroleum. *See also* a **structural trap**.

Stratigraphy   The geological study of strata, including lithological and biological content. *See also* **biostratigraphy**.

Stray indemnity   An indemnity provision in a contract which sits outside an expected assembly of indemnity provisions in the contract. Contrast with a **grouped indemnity**, as an indemnity which sits within an assembly of indemnities.

Stream contract   A method of financing the costs of a petroleum project development whereby a producer forward sells a defined percentage of the produced petroleum under a contract which is financially settled. Also called a **flowstream**. *See also* **forward purchase**.

Stream day   A day in a year upon which a petroleum facility is in full operation, free of scheduled and unscheduled maintenance or other interruptions.

Streamer   A long cable, carrying hydrophones, which is towed behind a seismic survey vessel in order to conduct a seismic survey. *See also* **array**.

Strength level blast (SLB)   A high-probability but low-consequence explosion event which by design would not permanently damage a petroleum infrastructure item. *See also* **ductility level blast**.

Stretch target   An objective which is intentionally set to be more difficult to achieve than a base case objective.

Stricto sensu   An alternative term for a **freezing clause**.

Stringer   A support which is placed under the centre of a racked section of pipeline or drill strings in order to prevent the rack from sagging.

Strip   A sequence of ship-delivered petroleum quantities.

Strip mining   The accessing of a formation by the surface removal of overburden.

Strip out   The removal of a drill string from a wellbore which has been shut in because of a kick.

Stripped condensate   Condensate which is recovered from a gas stream in a separator. Also called **plant condensate**.

Stripped gas   Natural gas from which liquids have been removed.

Stripper well   A wellbore which is producing petroleum from a formation that is near to the end of its economic life.

Stripping gas   An inert gas which is used to dehydrate a gas stream.

Structural consequences analysis (SCA)   An assessment and analysis of the ability of a petroleum infrastructure item to withstand a particular emergency event.

Structural trap   A petroleum-bearing formation which results from geologic structures such as folds and faults and prevents the further migration of mobile petroleum. *See also* **stratigraphic trap**.

Stub quantity   A quantity of LNG which is less than a full cargo lot. *See also* **part cargo**.

Stuck pipe   A drill string which has become stuck in a wellbore and cannot readily be rotated or withdrawn.

Stuck point   An alternative term for a **freeze point**.

Sub-allocation   Allocation which is effected in respect of a defined pipeline section which forms part of a wider pipeline network.

Sub-attribution   Attribution which is effected in respect of a defined sub-group of petroleum producers.

Sub-bituminous coal   An alternative term for **lignite**.

Sub-cellar deck   The deck which is immediately below the cellar deck of a drilling unit or of an offshore platform.

Sub-division   The partition of a defined concession area into several smaller, separate concession areas.

Sub-salt   An alternative term for **pre-salt**.

Sub-sea isolation valve (SSIV)   A valve system on a pipeline which is activated to isolate the flow of petroleum in an emergency situation.

Sub-sea tie-back (SSTB)   A sub-sea connection of a satellite to a central processing platform.

Sub-sea umbilicals, risers, and flowlines (SURF)   A composite term for the facilities which are needed to connect sub-surface petroleum production to an offshore petroleum production facility. *See also* **jumper**.

Sub-surface installation (SSI)   A petroleum facility which is installed sub-surface. Contrast with an **above-ground installation**, which is found at or above ground level.

Sub-surface safety valve (SSSV)   An alternative term for a **downhole safety valve**.

Sub-surface sampling   A procedure in which a sampling vessel is lowered into a wellbore in order to take a petroleum sample which is intended to be representative of actual formation conditions.

Subcritical flow   The principle that where gas flows through a choke, the rate of flow will depend on the pressures upstream and downstream of the choke.

Subgeologic map   A map of the formations which lie around an unconformity.

Subjacency   Under a concession, the principle that the contract area will be extended in the concession-holder's favour to include any unlicensed subjacent (higher or lower level) areas into which a discovery could extend. *See also* **adjacency.**

Sublimation   The change of a substance from a solid to a gas without becoming a liquid.

Submerged combustion vaporiser (SCV)   A heat exchanger which is used to convert LNG to regas, using the combustion heat from exhaust gases to raise the temperature of the LNG to the point at which vaporisation occurs. *See also* **ambient air vaporiser** and **open rack vaporiser.**

Submerged turret loading (STL)   A system for the offshore loading of a shuttle tanker through a riser, extending from a sub-sea line from a platform, which is connected to a submerged buoy.

Subordination   In respect of a loan agreement, the promise of a lender to defer its right to enforce security against a borrower in favour of the right of another lender to first enforce its security. *See also* **deed of priority.**

Subrogation   An insurance doctrine whereby an insurer which has made an insurance payment on behalf of an insured person is entitled to take over the rights of action which the insured person has against another person in respect of the insurable event.

Substitution   An alternative term for **attribution.**

Success ratio   An alternative term for a **commercial success ratio.**

Successful efforts accounting   A petroleum accounting convention by which only those operating expenses which are associated with successful exploration are capitalised, and operating expenses which are associated with unsuccessful exploration are offset against revenues. *See also* **full cost accounting.**

Sucker rod pumping   A method of artificial lift in which a pump is located in a bottomhole and a beam jack is used to draw petroleum to the surface. Also called **beam pumping** and **rod pumping.**

Suezmax
(1)   The largest size of ship which can pass through the Suez Canal. *See also* **Capesize.**
(2)   A crude carrier with a deadweight tonnage of 130,000 metric tons.

Sulphate-reducing bacteria (SRB)   Naturally occurring bacteria which can (undesirably) transform sulphur traces in a petroleum stream into hydrogen sulphide. *See also* **biocide.**

Sulphur oxides (SOX)　A pollutant which results from the combustion of fossil fuels. *See also* **nitrogen oxides**.

Sulphur reduction unit (SRU)　A process facility which is used for the extraction of sulphur compounds from a petroleum stream.

Summer valley　A seasonal adjustment which reflects a reduced level of demand and/or prices for certain petroleum grades during summer months. *See also* **winter peak**.

Sun, surf, and breeze (SSB)　A colloquial alternative term for **wind, wave, and solar**.

Sundays and holidays excluded (SHEX)　A methodology for determining which days will count towards allowed laytime (where Sundays and holidays are excluded).

Sundays and holidays included (SHINC)　A methodology for determining which days will count towards allowed laytime (where Sundays and holidays are included).

Sunk cost　An alternative term for **embedded cost**.

Sunset clause　A contractual provision which defines when all or a part of a contract will come to an end.

Super shortfall　Under a petroleum sales contract, an enhanced remedy for a shortfall to which the seller will be liable in certain circumstances.

Supercritical fluid　A substance at a point of temperature and pressure within which gas and liquids do not exist as separate phases but which can diffuse through solids like a gas and can dissolve solids like a liquid.

Superficiary　A person to whom a right of occupation of a surface area is granted, typically in consideration of the payment of a rental payment.

Super-giant field　A new petroleum discovery which is believed to contain more than one billion barrels of recoverable oil. *See also* **elephant field** and **giant field**.

Superintendent　A person who is designated to act as the senior manager of an onshore drilling unit. Also called a **field superintendent**.

Supernumeraries　Under a charterparty, representatives of the charterer who are given passage on the ship.

Superposition　A geological structure in which undisturbed rocks have the oldest layers at the base of a formation and the youngest layers at the top.

Supervisory control and data acquisition (SCADA)　Remotely controlled equipment on a petroleum facility which is used for control purposes and operational data collection. *See also* **telemetry**.

Supplemental reserve　Extra productive capacity which can be made available by applying the output of a petroleum facility which is not then functioning. Contrast with **spinning reserve**, whereby extra capacity comes from increasing the output of an already-functioning facility.

Supplier of last resort (SOLR)　A regulatory principle whereby when an energy supplier fails the affected consumers will be guaranteed continuity of supply by the transfer of their supply rights to another energy supplier (on the same commercial terms).

Supply based contract (SBC)   An alternative term for a **supply contract**.

Supply chain action plan (SCAP)   A wholistic plan for contractor involvement which is prepared in respect of the development of a new petroleum production project, including decommissioning activities.

Supply chain management (SCM)   A series of contracts which are used for the procurement of goods and services for the development of a petroleum project.

Supply contract   A contract for the sale of petroleum to a buyer as a fungible commodity, without an identified source of supply. Contrast with a **depletion contract**, whereby petroleum is sold from a nominated source of supply dedicated to the buyer. Also called a **supply based contract**. *See also* **partial purchase contract**.

Supply led local content   A local content requirement which is developed in isolation from the preferences of a petroleum project participant. *See also* **demand-led local content**.

Support vessel (SV)   A ship which is used to support offshore petroleum exploration and/or production operations. Also called a **work boat**.

Surf n' turf   A colloquial term for a petroleum infrastructure item which is based partially offshore and partially onshore.

Surface casing   The next widest size (after conductor casing) of casing in a wellbore. Also called **surface pipe**. *See* Appendix 2C.

Surface footprint   The spatial surface area of a petroleum facility.

Surface location   The location of a wellhead or of an above-ground installation.

Surface pipe   An alternative term for **surface casing**.

Surface reclamation   The restoration of a surface location after petroleum project operations have ceased.

Surface related multiple elimination (SRME)   A seismic processing technique which eliminates multiple signals and improves data collection.

Surface risk   An economic, environmental, logistical, operational, or political risk which could adversely impact an onshore petroleum project.

Surface to in-seam drilling (SIS)   A wellbore which is drilled from vertical to horizontal, typically to exploit an unconventional petroleum deposit.

Surfactant   A term which is derived from the combination of 'surface', 'active', and 'agent', which is used to describe a compound which lowers the interfacial tension between mixed liquids.

Surfactant flood   A process of waterflooding with surfactants in order to produce higher levels of oil recovery from a formation.

Surge   The forwards and backwards horizontal movement of a ship or offshore petroleum infrastructure item.

Surge tank   A chamber through which a petroleum stream flows to even out pressure surges on a pipeline. *See also* **accumulator**.

Surrender   An alternative term for **relinquishment**.

Surveyed after shipment (SAS)   A quality survey in respect of a cargo of petroleum which takes place after the cargo has reached the unloading port. *See also* **surveyed before shipment**.

Surveyed before shipment (SBS)   A quality survey in respect of a cargo of petroleum which takes place before the cargo has left the loading port. *See also* **surveyed after shipment**.

Suspended discovery   A discovery which is not presently being developed.

Suspended S&W   Sediment and water which are suspended in crude oil and are separated out by a centrifuge test.

Suspension/suspended   A wellbore which has been capped off temporarily. Also called **shut in temporarily abandoned**. *See also* **plug and make safe**.

Sustainability linked loan (SLL)   A form of loan which incentivises the borrower to improve its sustainability profile by measuring performance against targets in the loan documentation (where success could result in improved loan terms for the borrower).

Sustainable natural gas (SNG)   An alternative term for **renewable natural gas**.

Swabbing   The creation of suction in a wellbore by the too-rapid withdrawal of a drill string, possibly leading to the unexpected flow of petroleum or water into the wellbore during drilling which could lead to a kick.

Swaging   The distortion of a tubular by the application of tools to it.

Swamp barge   A flat-bottomed barge which is used as a base for offshore drilling in lakes, lagoons, and other shallow-water situations. Also called an **inland barge rig**.

Swamp gas   An alternative term for **marsh gas**.

Swap   A commercial arrangement whereby petroleum destined for one market is delivered elsewhere, in exchange for a corresponding balancing delivery of petroleum from elsewhere into the original market. *See also* **crosshaul**.

Swath   A particular strip of terrain which is covered during swath shooting.

Swath shooting   A process which is used in a three D seismic survey, using fixed receiver lines.

Sway   The side-to-side horizontal movement of a ship or offshore petroleum infrastructure item.

Sweat equity   A colloquial term for the share of an asset, business, or project which is held by a person in consideration of that person's contribution of resources to the development of the asset, business, or project.

Sweating   The use of a petroleum facility to its greatest possible capacity.

Sweep efficiency   A measure of the effectiveness of an enhanced oil recovery programme which is defined by reference to the rate at which injection fluids displace petroleum.

Sweet gas   Natural gas which contains negligible amounts of hydrogen sulphide. Contrast with **sour gas**, which contains relatively significant amounts of hydrogen sulphide.

Sweet oil   Crude oil which contains negligible amounts of sulphur. Contrast with **sour oil**, which contains relatively significant amounts of sulphur.

Sweetening   A process for the removal of the odour associated with sulphur compounds in a petroleum stream (but without reducing the actual sulphur content) by the application of certain compounds. *See also* **acid treatment**.

Swing
(1) The ability of a petroleum producer to increase the rate of petroleum production from a formation.
(2) Under a gas sales contract, the right of the buyer to increase the offtake of gas beyond an agreed base level.

Swing basin   An alternative term for a **turning basin**.

Swing producer   A petroleum producer which has the ability to manipulate market prices through modulating its production levels.

Switching   A process whereby a dual-fuel electric power-generating facility switches between different feedstocks. *See also* **inter-fuel substitution**.

Swivel   In drilling, a mechanical device which holds the weight of a drill string which hangs from it.

Synbit   A mixture of cleaned bitumen and SCO, which makes the resultant compound sufficiently viscous to be capable of being transported by pipeline.

Syncline   A downwardly concave geological structure in which petroleum could be present at the base of the structure. Contrast with an **anticline**, whereby petroleum could be present at the top of a structure.

Syncrude   Ultra-heavy oil which undergoes processing to improve its API gravity to that of light oil.

Syngas   An alternative term for **synthetic natural gas**.

Synmeth   An alternative term for **synthetic methane**.

Synthesis gas   A mixture of carbon monoxide and hydrogen which is used to make ammonia (which is used to make fertiliser) and methanol.

Synthetic-based mud (SBM)   Drilling fluid which uses man-made chemicals (which simulate oil) as the base component. *See also* **oil-based mud** and **water-based mud**.

Synthetic DES   An alternative term for **sleeving**.

Synthetic equity   An indirect controlling interest in a petroleum project which is effected through contractual rather than direct ownership rights.

Synthetic management structure   A commercial arrangement whereby the prospective purchaser of an asset or a business can exercise a measure of control over the operation of the asset or the business despite not having secured a formal ownership interest.

Synthetic methane   Methane which is artificially created by a biomethanation process using hydrogen and carbon dioxide. Also called **synmeth**.

Synthetic natural gas (SNG)   Synthetic natural gas which is manufactured by the destructive distillation of coal. *See also* **carburetted water gas**. Also called **coal gas**, **hygas**, **manufactured gas**, **syngas**, and **town gas**.

System   An alternative term for a **period**.

T

T-path   Transportation path: the distance which exists between the entry point and the exit point in petroleum pipeline transportation.

Tag-along   In a joint venture, the right of a non-selling co-venturer to tag along with a selling co-venturer in a proposed sale. Contrast with a **drag-along**, as the right of a selling co-venturer to drag along its fellow co-venturers.

Tail   A decline in the production of natural gas from a formation which is caused by the natural depletion of the formation. Also called a **production tail**.

Tail end   The portion of feedstock which vaporises towards the end of an atmospheric distillation process.

Tail gas   The gas stream which exits a petroleum-processing facility after the extraction of liquids has taken place.

Tailgate   The point in a petroleum production facility or a petroleum-processing plant from which petroleum exits the facility or the plant.

Tailgate sale   A sale of petroleum in which the delivery point is the tailgate. *See also* **free in pipe**. Also called a **plant gate sale**.

Tailings   Waste materials which result from the production or the processing of petroleum.

Take   Under a petroleum sales contract, the commitment of the buyer to take delivery of a specified quantity of petroleum over a period of time.

Take and pay (TAP)   Under a petroleum sales contract, a provision whereby the buyer is obliged to take delivery of a defined quantity of petroleum and to pay the seller for the petroleum if it fails to so take delivery. Also called **must-take**.

Take or lose   Under an attribution agreement, a provision whereby a party's underlift transfers (possibly irrevocably) to the benefit of the other lifting parties.

Take or pay (TOP)   Under a petroleum sales contract, a provision whereby the buyer has the option to take delivery of and to pay the seller for a defined quantity of petroleum, or not to take delivery but still to make a minimum payment to the seller.

Take or release   Under a petroleum sales contract, a provision whereby the seller has a termination right, or the right to scale down the quantities in the contract, if the buyer fails to take delivery of a specified minimum quantity of petroleum.

Take point The first point of contact in a horizontal wellbore from which petroleum can be produced.

Taking An alternative term for **expropriation**.

Tallow A substance which is made from animal fats through the process of rendering and is used as a feedstock for the manufacture of biodiesel.

Tally A measure of the total quantity of tubulars which are used in a wellbore.

Tangible drilling costs In drilling, costs which have a recoverable salvage value associated with them (such as equipment and materials costs). Contrast with **intangible drilling costs**, which do not relate to an identifiable asset or interest.

Tank barge A barge which has separated storage compartments that are used for the offshore transportation of petroleum.

Tank bottoms Petroleum fractions and impurities which reside at the base of a tank. *See also* **basic sediment and water**. Contrast with **tank tops**, as petroleum fractions which reside at the surface of a storage tank.

Tank dipping An alternative term for **dipping**.

Tank farm A collection of underground or surface-mounted tanks which are used for the storage of crude oil, products, or petrochemicals. Also called a **battery** and a **tank yard**.

Tank gas An alternative term for **liquefied petroleum gas**.

Tank storage receipt (TSR) A receipt for the payment of a capacity charge or a tariff which is issued by a storage facility owner to a storage facility user.

Tank tops
(1) Petroleum fractions which reside at the surface of a storage tank. Contrast with **tank bottoms**, as petroleum fractions and impurities which reside at the base of a tank.
(2) The situation where in which a petroleum storage tank is full and consequently there is no commercial opportunity for a person storing petroleum to acquire or to produce more petroleum for gain.

Tank yard An alternative term for a **tank farm**.

Tanker Owners Voluntary Agreement concerning Liability for Oil Pollution (TOVALOP) A voluntary agreement between crude carrier owners relating to maritime oil pollution which existed between 1969 and 1997 and was administered by the ITOPF. *See also* **CRISTAL**.

Tanktainer A container tank which is used for the intermodal transportation (by truck or by train) of liquid petroleum or industrial gases.

Tapis A crude oil blend from Malaysia.

Tapping The interconnection of a new petroleum facility to an existing facility. *See also* **cold tap** and **hot tap**.

Tar sands An alternative term for **oil sands**.

Target depth The target depth for the drilling of a vertical wellbore. Also called **objective depth**.

Target distance   The target distance for the drilling of a horizontal wellbore.

Tariff
(1) The fee which is payable by a shipper to a transporter under a quantities-based contract for the shipper's petroleum that is transported by the transporter through a pipeline. *See also* **zero-tariff transportation.**
(2) The fee which is payable by a petroleum facility user to a petroleum facility owner for the use of a petroleum facility.

Tariff in-kind   A tariff which is payable by the intended payer in petroleum rather than in cash.

Tautline referencing   A method of monitoring the position of an offshore drilling unit relative to a sub-sea wellhead by measuring the inclination of a vertical steel line which runs between the drilling unit and the sea-bed.

Tax capacity   The ability of a petroleum company to fund the costs of development of a petroleum project in a state by an offset against tax liabilities which that company owes in that state.

Tax neutrality   A situation in which a petroleum project contract or the project itself neither makes a tax gain nor is exposed to a tax liability as a consequence of its existence.

Tax paid   A provision in a concession whereby the concession-holder is held harmless from the application of certain taxes in the jurisdiction of the concession.

Tax reference price   A posted price for petroleum which establishes the basis for calculating a liability to taxation or royalty payments which are due from a producer.

Teapot refinery   A colloquial term for a crude oil refining and processing facility, which is typically found in eastern China and is owned independently of a state-owned enterprise.

Teaser   A relatively short documentary summary of the key commercial elements of an asset, project, or business which is intended to be sold and is prepared by the proposed seller, which highlights the proposed sales process and the outline investment opportunity, and which often precedes an information memorandum.

Technical analysis   Petroleum price performance analysis which is derived from studying historical market movements and applying predictive tools. Contrast with **fundamental analysis**, which is derived from actual supply and demand factors.

Technical basement   In the drilling of a wellbore, a point of depth at which there is no operational sense in continuing to drill because a technical problem has been encountered which makes drilling impossible or unsafe. Contrast with an **economic basement**, as a point at which there is no economic sense in continuing to drill a wellbore.

Technical committee (TCom)   Under a JOA, a sub-committee which is appointed by the operating committee and is intended to manage certain defined technical issues which arise under the JOA.

Technical committee meeting (TCM)   Under a JOA, a meeting of the technical committee.

Technical evaluation agreement (TEA)   An agreement between a state and a petroleum company whereby the company will undertake defined G&G assessments and evaluations on behalf of the state.

Technical expert report   An alternative term for a **competent person's report**.

Telemetry   The telecommunicated transmission of operational data from a remote petroleum facility. *See also* **supervisory control and data acquisition**.

Temperature correction   An alternative term for **weather correction**.

Temperature gradient   A measure of the rate of change of downhole temperature which takes place as the drilled depth of a wellbore increases.

Temporary abandonment (TA)   An alternative term for **plug and make safe**.

Temporary overnight shelter (TOS)   Personnel accommodation on a normally uninhabited installation.

Temporary refuge   An alternative term for a **refuge**.

Temporary refuge area (TRA)   An alternative term for a **refuge**.

Tenor   The duration of a loan agreement.

Tension leg platform (TLP)   A floating petroleum exploration or production platform, which is stabilised by tendons anchored to the sea-bed and is typically used in water depths up to 6,000 feet. *See* Appendix 2B and Appendix 2D.

Terminal depth   The final depth to which a vertical wellbore is drilled.

Terminal distance   The final distance to which a horizontal wellbore is drilled.

Terminal use agreement (TUA)   A contract which regulates the use of an LNG import or export terminal's capacity by a person other than the terminal owner. Also called a **throughput agreement**.

Terminalling   The business of petroleum storage, processing, and send-out.

Terminus   The end point of a horizontal wellbore.

Tertiary migration   The movement of mobile petroleum from one reservoir rock to another reservoir rock or to a seep. *See also* **primary migration** and **secondary migration**. *See* Appendix 2A.

Tertiary recovery   The third stage of petroleum production, in which petroleum recovery from a formation is effected through steam or water injection. Also called **advanced recovery**. *See also* **primary recovery** and **secondary recovery**.

Test separator   A separator which uses gravity to separate out petroleum constituents and impurities.

Test well   An alternative term for an **exploration well**.

Tethered goat   A colloquial term for a deliberately provocative provision in a draft contract which is intended to be traded away as part of a commercial negotiation. Also called a **sacrificial anode** and **trade-bait**.

Texas deck   The largest load-bearing deck on an offshore drilling unit or offshore production platform.

Texas shoot-out   A colloquial term for a joint venture deadlock resolution provision whereby each party submits a sealed bid price for the purchase of the other party's interest in the joint venture, and the party which submitted the highest-priced bid must buy the other party's interest at the price specified in that bid. Also called a **Mexican shoot-out**. *See also* **Dutch auction** and **Russian roulette**.

Texas two-step   A colloquial alternative term for a **hive-down/hive-up**.

Therm   A unit of calorific value which is equivalent to 100,000 Btus.

Thermal enhanced oil recovery (TEOR)   Any heat-related method of enhanced oil recovery.

Thermal recovery   An alternative term for **fireflood extraction**.

Thermal value   An alternative term for **calorific value**.

Thermalisation   The conversion of a defined volume of petroleum to its equivalent calorific value.

Thermally mature rock   An alternative term for a **kitchen**.

Thermie   A unit of calorific value which is equivalent to one megacalorie.

Thermogenic gas   Natural gas which is formed within a formation at deep depths and high temperatures by the thermal cracking of sedimentary organic matter and crude oil deposits, which is typically wet gas and is often indicative of further petroleum shows. Contrast with **biogenic gas**, which is formed by the anaerobic digestion of sedimentary organic matter.

Thickness map   A geological tool which indicates the thickness of a stratum. *See also* **isochore** and **isopach**.

Thief   A device which facilitates the taking of a sample from a quantity of crude oil in a predetermined location.

Thief zone   A formation which absorbs drilling fluid that has escaped from a wellbore. *See also* **lost circulation**.

Thin cap   A poorly capitalised company, where its debt greatly exceeds its equity capital (so that the company is highly leveraged).

Thinner   An alternative term for **cutterstock**.

Thiol   An alternative term for **mercaptan**.

Third-party access (TPA)   The rights of a third party to access ullage in a petroleum facility which it does not own. Also called **access**. *See also* **infrastructure code of practice, negotiated third-party access, open season,** and **regulated third-party access**.

Third-party inspection (TPI)   An inspection, audit, or assessment of a petroleum facility which is carried out by an independent assessor.

Thixotropy   A property which is exhibited by a material (such as drilling fluid) in a liquid state when flowing and in a semi-solid state when at rest.

Three C (3C)   An estimate of contingent resources. *See* Appendix 7.

Three D (3D)   A seismic survey which creates vertical and horizontal sections of data, thereby allowing a cube of data to be created which gives a three-dimensional image of sub-surface conditions. *See also* **square kilometre.**

Three P (3P)   An alternative term for **possible reserves.** *See* Appendix 7.

Three-phase separator   A separator which separates oil, gas, and water. Also called **free water knock-out.**

Threshold field size   A measure of the minimum amount of petroleum in the ground which is necessary to sustain the development of a petroleum project.

Thribble   A stand of three lengths of drill pipe which have been coupled together. *See also* **double, fourble,** and **single.**

Throttling   *See* **JT valve.**

Throughput   The rate of flow of petroleum through a pipeline or a petroleum facility over a given period of time.

Throughput agreement   An alternative term for a **terminal use agreement.**

Throughput cutback   A forced reduction in the rate of petroleum production which is caused by a shortage of available storage capacity, which is implemented to avoid tank topping. *See also* **production gap.**

Thrust belt   An alternative term for a **fold and thrust belt.**

Thumper truck   A colloquial alternative term for a **vibroseis.**

Tickover   An alternative term for **care and maintenance.**

Tidal current (TC)   An ocean current which ebbs and flows onto a beach. *See also* **longshore current.**

Tidal range   A measure of the difference in height between high and low tides.

Tie-back   The connection of a formation to a petroleum facility. Also called a **tie-in.** *See also* **sub-sea tie-back.**

Tie-back explorer   A petroleum company that limits its exploration activities to areas in which petroleum production infrastructure has already been put in place. *See also* **infrastructure-led exploration** and **frontier explorer.**

Tie-back reach   An agreed radial distance from a central processing platform within which any satellites or remote facilities will be made the subject of a tie-back. *See also* **hub potential.**

Tie-in   An alternative term for a **tie-back.**

Tie-in agreement   A construction and tie-in agreement but without the construction component.

Tie-in platform   An alternative term for a **central processing platform.**

Tiered dispute resolution   A model for dispute resolution in a petroleum contract which applies different forms of dispute resolution to disputes of different monetary values. Also called **multi-tiered dispute resolution.**

Tiered earn-in   A form of farm-out agreement whereby the farmee secures its interests by performing a series of separate steps which correspond to the ongoing performance of obligations.

Tight formation   A formation which has relatively low porosity and permeability.

Tight gas   Natural gas deposits which are characterised by low permeability and porosity, resulting in low flow rates and often requiring fracking and horizontal drilling for extraction.

Tight hole
(1) The drilling of a wellbore in which the resultant drilling data is kept highly confidential.
(2) A constricted section of a wellbore which impedes the free movement of tools.

Tight oil   Crude oil deposits which are characterised by low permeability and porosity, resulting in low flow rates and often requiring fracking and horizontal drilling for extraction. Also called **light tight oil**.

Time & materials contract (T&M)   A contract whereby the contractor is reimbursed for actual time spent and materials used, with a percentage profit element. Also called a **reimbursable contract**.

Time charter   A charterparty for a ship for a defined period of time. Contrast with a **voyage charter**, as a charterparty for a defined voyage. Also called a **time charter party**.

Time charter party (TCP)   An alternative term for a **time charter**.

Time-depth (TD)   A measure of the distance in time which exists between two or more events that are recorded at different stratigraphic depths.

Time–depth curve   As part of a daily drilling report, a schematic which indicates how the drilling of a wellbore is progressing against the axes of time and depth (and which is also indicative of the rate of penetration).

Time losing injury (TLI)   An injury which is suffered by a person which has to be logged as causing the loss of working time. Also called a **lost time injury**.

Time losing injury frequency rate (TLIFR)   A measure of how often a time losing injury occurs within a given period of time.

Time of use tariff (TOUT)   A tariff which is payable by a petroleum infrastructure item user for the use of an item which varies according to the period of time during which the item is being used.

Time stamp   A definition of the exact time at which a petroleum-trading transaction is deemed to have taken place.

Time to depth/time to distance (TTD)   An estimate of the time which will be taken to drill a wellbore to a defined depth or distance.

Time writing   The recording of time which is spent by petroleum company staff on a particular activity, the costs of which can be back charged for payment by another person.

Titanium stress joints (TSJ)   A sub-sea joint on a riser which provides a stronger and more flexible joint when compared to a steel joint.

Title opinion   A formal legal opinion which confirms the content of an abstract.

Title Transfer Facility (TTF)   A notional point on the Dutch gas pipeline network at which shippers trade gas quantities.

Toe-to-heel air injection (THAI)   A process for the production of blown oil.

Tolling   The processing of petroleum through a petroleum facility by person as a facility user in exchange for the payment of a tariff to the facility owner.

Tomography   The use of a penetrating wave in order to create an image of a section of a substance, as part of a seismic survey.

Tonnage   A measure of the size of a fleet of ships.

Tonnage gas   Non-methanous gases (such as oxygen, nitrogen, and argon) which are supplied in large quantities to industrial end-users. *See also* **cylinder gases**.

Tonne (t)   A metric tonne, which is equivalent in weight to 1,000 kilogrammes.

Tonne mile (ton mile)   A measurement which is used in the economics of petroleum transportation to identify the costs which are associated with moving 1 tonne (or ton) of petroleum over 1 mile.

Tonne of oil equivalent (TOE)   A comparative unit of energy which is based on the approximate amount of energy that is released by the combustion of one metric tonne of crude oil.

Tools   A composite term for the various drilling and testing devices which are used downhole.

Top kill   A procedure for controlling a kick from a wellbore, whereby drilling fluid is pumped into the wellbore in order to stop the flow of petroleum by overbalance. Also called **static kill**.

Top slice MHH   A mutual hold harmless liability regime which applies a guilty party pays liability regime to a defined first tranche of liability.

Top stop   An alternative term for a **cap**.

Topdrive drilling   A method of drilling whereby a drive motor is suspended from a derrick above the drill string. Contrast with **rotary drilling**, whereby the drill string is powered by a drive motor at the rotary table level.

Tophole   The topmost vertical section of a wellbore.

Tophole location (THA)   The point at which a tophole sits on a mudline.

Topography   The configuration of a land surface. *See also* **field mapping** and **outcrop**.

Topped oil   Crude oil from which light distillates have been removed. Also called **reduced crude oil**.

Topping refinery   The simplest refinery configuration, which is designed typically to produce unfinished oil products by atmospheric distillation. Also called a **simple refinery**. *See also* **complex refinery**.

Topside umbilical termination unit (TUTU)   A module on an offshore petroleum infrastructure item into which a number of sub-sea umbilicals are gathered.

Topsides
(1) The upperworks of an offshore platform, which is mounted on top of a jacket.
(2) The surface parts of a ship which sit above the hull.

Topsoiling   The placing of topsoil away from other spoil which has been excavated from an onshore trenched pipeline construction project so that it can be replaced in-situ when the pipeline trench is refilled.

Torsional yield strength   The degree of twisting force which a tubular can withstand before being twisted off.

Total acid number (TAN)   A measure of the acidity of a given quantity of petroleum.

Total calculated volume (TCV)   In the measurement of oil in-tank, the total volume of petroleum, sediments, and water which exist at prevailing temperature and pressure but with a correction for reference conditions. TCV is GSV plus free water.

Total dissolved solids (TDS)   A measure of the total amount of solids which are dissolved in a liquid, which is typically measured in milligrammes per litre.

Total energy   An alternative term for **co-generation**.

Total gas in store   The total quantity of gas (working gas and cushion gas) which is contained within a gas storage facility at any time.

Total loss   An insurance term for the situation where the lost value, repair cost, or salvage cost of an insured item exceeds the insured value of that item, such that the item is written off. *See also* **actual total loss** and **constructive total loss**.

Total observed volume (TOV)   In the measurement of oil in-tank, the total volume of petroleum, sediments, and water which exist at reference conditions. TOV is gross observed volume plus free water.

Total organic carbon (TOC)   A measure of the concentration of organic material in source rock which is represented by the weight per cent of organic carbon and is typically revealed by pyrolysis.

Total petroleum hydrocarbons (TPH)   A measure of the total quantity of petroleum compounds which are produced from a given quantity of crude oil.

Total S&W   A measure of the total quantity of sediment and water which resides in a petroleum storage tank. *See also* **basic sediment and water**.

Total suspended solids (TSS)   Solids in water which can be trapped by a filter.

Totaliser   A process step by which a number of monitored inputs are aggregated.

Totally enclosed motor propelled survival craft (TEMPSC)   A form of lifeboat which is installed on a ship or a petroleum infrastructure item and is used for personnel evacuation. *See also* **Brucker capsule** and **Whittaker capsule**.

Tower mooring   A structure which is directly and permanently fixed to the seabed, to which an FPSO or an FSU is anchored. Also called **tower yoke**.

Tower yoke   An alternative term for **tower mooring**.

Town gas   An alternative term for **synthetic natural gas**.

Trace   A single channel of seismic data.

Tracked voting   A mechanism whereby a person's position under a series of contracts to which it is party is applied consistently throughout all of the contracts.

Tracking account   An account through which debits and credits are recorded over time in order to identify a person's overall net entitlement.

Tract/tract participation   In the process of unitisation, the numerical representation of a concession area as a percentage proportion of the overall unit. *See also* **unit interest**.

Trade-bait   An alternative term for a **tethered goat**.

Trade data   Reported data which relates to traded petroleum prices. *See also* **contributed data** and **derived data**.

Trade optionality   An alternative term for a **trade tolerance**.

Trade recap   An alternative term for a **recap**.

Trade tolerance   A percentage tolerance in the size of a petroleum cargo which a trader contracts to buy and to sell, thereby creating an arbitrage opportunity for the trader. Also called **trade optionality**.

Trading limits   Limitations which are specified in a charterparty on the loading and unloading ports to which a ship can be taken by the charterer. Also called **safe trading limits**. *See also* **British Institute Warranties**, **Institute Warranty Limits**, and **International Navigating Limits**.

Trading volume   An alternative term for **churn**.

Train   An individual process facility within a gas liquefaction plant which produces LNG.

Train day   A period of twenty-four consecutive hours (running midnight to midnight) for the operation of a train.

Trajectory   The path of a wellbore.

Tramline   A sale of LNG whereby the LNG carrier traverses between two fixed points without diversion. Also called **point-to-point**.

Tranche   A segment or portion of a whole.

Transaction costs   A measure of the totality of costs (other than the purchase price) which are associated with documenting and completing a particular commercial transaction.

Transfer pricing   An agreed internal (and often non-market) price which is applied to the sale of goods or the provision of services between affiliated companies, which allocates value to different parts of an overall supply chain and can produce taxable profit distortions.

Transhipment   An alternative term for **ship-to-ship**.

Transit fee   A fee which is payable by a shipper or a facility user for the transportation of petroleum through a pipeline or a petroleum facility.

Transit time (TT)
(1) The time which is taken for a quantity of petroleum to travel a defined distance in a pipeline.
(2) The number of days (full or part) which are needed to transport a petroleum cargo from the loading port to the unloading port.

Transition fuel A fuel source which is used in an energy-consuming market in order to bridge the gap between existing fuel sources and envisaged fuel sources.

Transition zone
(1) A zone within a formation from which water and petroleum are capable of being produced together.
(2) A sub-surface area in which pressure increases as the drilled depth of a wellbore increases.

Transmission network An onshore gas transmission pipeline network.

Transmission system operator (TSO) A person who is responsible for operating a pipeline transportation system in isolation from any other activity (particularly the activity of selling or trading petroleum transported through the pipeline). *See also* **independent system operator** and an **independent transmission operator**.

Transmix An alternative term for **cutterstock**.

Transportation and distribution (T&D) A composite term for the activities of petroleum transportation and distribution.

Transportation and processing (T&P) A composite term for the activities of petroleum transportation and processing.

Transportation, processing, and operating services agreement (TPOSA) A contract whereby the owner of a petroleum facility offers a petroleum transportation and processing service to a person as a facility user. Related to a **processing and operating services agreement**.

Transporter A person who contracts to provide transportation rights in a pipeline which that person owns or operates. Contrast with a **shipper**, as a person who contracts to receive those rights.

Trap A stratigraphic trap or a structural trap.

Trap geometry A three-dimensional form of a trap.

Trapped oil Crude oil which is not produced because of a failure to flow through pore spaces in a formation. Also called **residual oil**.

Trapping efficiency The proportion of generated petroleum which remains trapped in a formation (where a high proportion indicates a formation with high development potential).

Treater A chamber in which oil is treated for the removal of sediment and water.

Treatment and transportation (T&T) A composite term for the activities of petroleum treatment (processing) and transportation.

Treatment, storage, and disposal facility (TSDF) A petroleum facility in which hazardous materials are stored and processed and waste components are disposed of.

Trench A scoured channel in a surface area into which a pipeline is lowered for burial.

Trencher A remote-operated vehicle which injects high-pressure water or compressed air into a sea-bed surface area to give a trench.

Trespass well   A vertical or horizontal wellbore which penetrates a formation which is owned by a person other than the person drilling the wellbore.

Trestle   A framework of supporting beams on which a jetty sits.

Tri-ethylene glycol (TEG)   A desiccant, similar to mono-ethylene glycol.

Trigger points   Under a PSC, certain defined conditions which oblige the contractor to pay a production bonus to the state or which alter the agreed profit oil/profit gas splits between the contractor and the state.

Tri-fuel diesel electric (TFDE)   A propulsion system which is used in an LNG carrier. *See also* **dual-fuel diesel electric**.

Trillion   One thousand billion (1,000,000,000,000).

Trip
(1)   The removal of a drill string from a wellbore. Also called **pull back, tripping out**, and **come out of the hole**. *See also* **round trip**.
(2)   An automated process whereby the operation of a petroleum facility or process is suspended upon the occurrence of a certain event. Also called an **automatic shut-down**.

Trip gas   Natural gas which becomes background gas during a trip.

Trip time charter   A time charterparty which is modified for use for a single voyage in a defined time period.

Triple point   A physical state whereby the solid, liquid, and gaseous forms of a substance can all exist in equilibrium.

Tripping out   *See* **trip**.

True amplitude recovery (TAR)   A seismic data processing technique which compensates for certain distortions in order to give a more accurate representation of a surveyed formation.

True horizontal distance (THD)   The furthermost point in the drilled horizontal distance of a wellbore, running from a defined surface point to a defined end point. *See also* **measured distance**.

True sale   An accounting standard which identifies the characteristics of an effective sale of an asset (including the seller's relinquishment of any continuing control rights in respect of the asset) such that the asset is disaggregated from the seller's balance sheet. *See also* **derecognition** and **disaggregation**.

True vapour pressure (TVP)   A measure of the level of pressure which is required to prevent a quantity of petroleum from freely evaporating, which is calculated as a function of temperature. *See also* **Reid vapour pressure**.

True vertical depth (TVD)   The furthermost point in the drilled vertical depth of a wellbore, which runs from a defined surface point to a defined end point. Also called **bottomhole location**. *See also* **measured depth**.

True vertical depth below rotary table (TVDBRT)   True vertical depth which is measured from the bottom of the rotary table on a drilling unit.

True vertical depth sub-sea (TVDSS) True vertical depth which is measured from sea level.

True vertical thickness (TVT) An alternative term for **feet of pay**.

True-up An alternative term for a **reconciliation**.

Trunkline A relatively large diameter pipeline which is used to transport petroleum from production facilities to reception or consumption facilities.

Trust A legal construction whereby legal title to trust property is held by one person (the trustee) for the benefit of another person (the beneficiary), who holds an equitable interest in the property.

Trust property Defined property which is the subject of a trust and is placed in trust by a settlor.

Trustee Under a trust, a person who holds the legal title to trust property.

Trustee and paying agent agreement (TPAA) A contract whereby a person is appointed to receive monies and to disburse those monies to other persons according to an agreed order.

Tubing An alternative term for a **liner**.

Tubing anchor A device which holds the lower end of a tubing string in place in a wellbore.

Tubing hanger A device which suspends a tubing string from a tubing head.

Tubing head A device which supports a tubing hanger and facilitates the attachment of a Christmas tree to a wellhead.

Tubing string A liner which sits within a wellbore, through which petroleum is produced. A wellbore could contain two or more parallel tubing strings for simultaneous petroleum production from different payzones.

Tubulars A composite term for casing, drill pipes, and liners.

Tuck in asset deal An acquisition structure whereby a private equity-backed vehicle acquires a larger operating entity and uses that entity as a foundation for further related asset acquisitions.

Turbidite A sedimentary deposit which is formed by the movement of deepwater currents, which is often found in river deltas.

Turbidity A measure of the cloudiness of water which is caused by suspended sedimentary particles, which is often indicative of water quality and its usefulness for petroleum production and processing operations.

Turbine meter A form of flow meter.

Turbodrill A downhole motor which rotates a drill bit by the action of a flow of drilling fluid on turbine blades within the drill bit.

Turboexpander (TX) A device which converts the energy of a gas or vapour stream into mechanical energy by expanding the gas or vapour stream through a turbine.

Turnaround (TAR)   A period of PSD in a petroleum facility which is intended to allow for maintenance or modification of the facility.

Turnback   The expiry without renewal of a contract for the reservation of capacity in a petroleum facility, leading to the surrender of the capacity to the facility owner.

Turnback right   The right of a buyer to not take delivery of a ship-transported in-bound petroleum cargo.

Turned in line   A wellbore from which gas is channelled into a pipeline for sale.

Turning basin   An area of water in a port which is sufficiently wide to allow a ship to rotate through 180° of its length overall when arriving or departing. Also called a **swing basin**.

Turnkey   An alternative term for **engineering procurement construction installation**.

Turnover   An alternative term for **cycling**.

Turquoise hydrogen   Hydrogen which is produced by the pyrolysis of methane, removing carbon as a solid element.

Turret mooring system   A mooring system which is located on an FPSO or an FSU and which anchors the ship to the sea-bed and allows the ship to freely rotate through 360°. *See also* **weathervaning**.

Twin   A wellbore which is drilled in the same location as a previous wellbore but which produces petroleum from a different payzone.

Twinning   An alternative term for **looping**.

Twist off   The unintended severance of a drill string due to fatigue or excessive torque.

Two C (2C)   An estimate of contingent resources. *See* Appendix 7.

Two D (2D)   A simple seismic survey which creates only a single vertical section of data. *See also* **line kilometre**.

Two P (2P)   An alternative term for **probable reserves**. *See* Appendix 7.

Two-phase pipeline   A pipeline which is intended to transport a mixed dry gas and wet gas stream.

Two-phase separator   A separator which separates gas and oil. Also called a **gas/oil separation plant**.

Two-tier pricing   A price for petroleum which is set by reference to the higher or lower of a choice of different pricing propositions.

Type curve   A representative profile for a well which is to be drilled in a particular play and is used to plot anticipated production rates.

Type curve analysis   A method for quantifying well and formation parameters by comparing actual pressure change data to modelled data.

Typical   A characteristic of crude oil or a product which is attributed to it by the producer but without any formal warranty or guarantee of the existence of the characteristic.

Uberrima fidei   From the Latin for 'utmost (good) faith', the expectation that as a condition of being given insurance cover an insured person will make full disclosure of all the relevant facts which might influence the decision of an insurer to give cover.

Ullage   A measure of the unused capacity in a petroleum-processing, transportation, or storage facility. *See also* **white space**.

Ullage hatch   An opening in the top of an oil or liquids storage tank through which sampling and ullage measurement activities are carried out. Also called a **gauge hatch**.

Ultimate customer   An alternative term for an **end-user**.

Ultimate recovery   The total quantity of petroleum which could be recovered from a formation over the productive lifetime of the formation. *See also* **estimated ultimate recovery**. Contrast with **actual recovery**, as the total quantity of petroleum which is actually recovered from a formation.

Ultra-deep water (UDW)   Water depths of greater than 1,500 metres. Contrast with **deep water**, as water depths in the range of 400 to 1,500 metres.

Ultra-heavy oil   Crude oil which occurs naturally in a solid or semi-solid form which has an API gravity of 10° or lower. Also called **extra-heavy oil**. *See* Appendix 6.

Ultra-large crude carrier (ULCC)   A crude carrier which has a deadweight tonnage in the range of 320,000 to 550,000 metric tons. *See also* **very large crude carrier**.

Ultra-large ethane carrier (ULEC)   A ship which carries approximately 150,000m³ of ethane in a refrigerated state. *See also* **very large ethane carrier**.

Ultra-low sulphur diesel (ULSD)   A form of diesel with low sulphur content.

Ultrasonic metering   The measurement of gas quantities in a pipeline flow which is based on readings that are taken from acoustic pulse deflections.

Umbilical   An assembly of steel tubes and/or thermoplastic hoses which are used variously to control, supply, power, or communicate with sub-sea structures from a platform or a ship.

Umbilical termination assembly (UTA)   A gathering point for the end of a number of umbilicals.

Umbrella agreement   A contract which provides for the consolidation of a number of separate but related disputes within a single proceeding. *See also* **consolidation**.

Unabated petroleum   A quantity of petroleum which has not been produced in accordance with defined carbon reduction standards. *See also* **abated petroleum**.

Unaccounted-for gas (UFG)   An alternative term for **lost gas**.

Unallocated provision (UAP) An alternative term for a **contingency**.

Unbranded fuel Fuel which is the product of white labelling.

Unbroached provisions Provisions which have not been opened for consumption. Contrast with **broached provisions**, which have been opened and partly consumed.

Unbundling The separation and financial disaggregation of different services and the charges which are associated with them within an integrated business. *See also* **accounting separation, functional unbundling**, and **ownership unbundling**. Contrast with **bundling**, as the aggregation of different services and associated charges within an integrated business.

UNCITRAL Rules A set of procedural rules for the governance of arbitration cases (http://www.uncitral.org/uncitral/en/uncitral_texts/arbitration/2010Arbitration_rules.xhtml) which are promulgated by UNCITRAL.

Unconfined compressive strength (UCS) A measure of the strength of a particular substance which is determined by assessing the application of the maximum compressive strength along one axis of the substance under unconfined conditions.

Unconformity A defined lack of continuity and homogeneity between different rock strata which are in contact with each other, around which petroleum-bearing formations could be present. *See also* **contact**.

Unconventional/unconventionals Petroleum which, because of the manner of its physical existence, can only be extracted by (unconventional) recovery techniques. Contrast with **conventionals**, as petroleum which is extracted by traditional (conventional) recovery techniques.

Unconventional gas (UCG) Raw gas which is unconventional in nature, including coal bed methane, hydrates, shale gas, and tight gas.

Unconventional resources operating agreement (UROA) A JOA which is designed specifically for use in an unconventional petroleum project.

Under keel clearance (UKC) The depth between the sea-bed and the lowest part of a ship's keel.

Under vapour The situation whereby an LNG carrier arrives for the loading of a cargo with its tanks containing natural gas vapours and the in-tank temperature is such that cool down will not be required. *See also* **presentation**.

Underbalance The drilling of a wellbore in which the pressure in the formation exceeds the pressure in the wellbore such that petroleum or water in the formation will freely enter the wellbore. *See also* **kick**. Contrast with **overbalance**, whereby the pressure in the wellbore at least matches the pressure in the formation.

Underfunded operation Under a JOA, a planned joint operation for which the operator lacks sufficient funding from all JOA parties. Contrast with an **overfunded operation**.

Underfunding The entry by a petroleum company into a drilling campaign where not all of the anticipated costs have been secured beforehand by the company.

Underground coal gasification (UCG)   A process whereby underground coal is converted into gas by injecting water and oxygen through a wellbore and partially combusting the coal underground.

Underground injection control (UIC)   A wellbore which is used to inject water or produced water into a formation.

Underground waste   Unrecovered petroleum which is lost sub-surface as a result of damage caused to a formation.

Underlift   The lifting of a quantity of petroleum by a producer at a defined point which is less than the quantity of petroleum which the producer is entitled to lift. Contrast with **overlift**, as the lifting of a quantity of petroleum which is greater than the producer's lifting entitlement.

Underreaming   An alternative term for **reaming**.

Undertake   Under a gas sales contract, the offtake of a quantity of gas by the buyer at a defined point which is less than the quantity of gas which the buyer is entitled to offtake. Contrast with **overtake**, as an offtake of gas which exceeds the quantity which the buyer is entitled to offtake.

Underwater inspection in lieu of dry-docking (UWILD)   A method of checking the integrity of a ship's hull without the expense of a full dry-docking. Also called **in-water survey**.

Unexploded ordnance (UXO)   Munitions which have not been properly detonated and present an ongoing risk of detonation during the performance of a petroleum project.

Unfinished oil   A product which requires further processing in order to be commercially useable.

Unfractionated streams   Mixtures of unsegregated natural gas liquid components which are extracted from natural gas.

Unicorn project   A colloquial term for a petroleum project which trials new technology with the capacity to effect significant economic or operational change.

Unification   The combination of lighter petroleum fractions in order to create heavier petroleum fractions.

Uniform Customs and Practices for Documentary Credits (UCP 600)   A set of standard rules and best practices which was published by the International Chamber of Commerce (ICC) in 2007 relating to the use of documentary letters of credit.

Uniform interest clause
(1)   Under a JOA, a provision whereby the parties agree to maintain an agreed level of interests in the underlying concession for a defined period of time after the execution of the JOA. Also called **maintenance of uniform interest**.
(2)   A contractual provision whereby the interests of a person in a contract can only be transferred if a corresponding interest in all other related contracts is also transferred. *See also* **stapling**.

Uniform Rules for Demand Guarantees (URDG 758)   A set of standard rules and best practices which was published by the International Chamber of Commerce (ICC) in 2011 relating to the use of demand guarantees.

Uniformly distributed live load (UDLL)   A live load which is distributed evenly across a petroleum infrastructure item. Contrast with **concentrated live load,** which is unevenly distributed.

Unincorporated joint venture   An alternative term for a **contractual joint venture.**

Uninterruptible power supply (UPS)   Electrical apparatus which provides almost instantaneous emergency power when a principal power supply fails.

Unión Transitoria de Empresas (UTE)   A form of temporary joint venture association between two or more companies which is found in some Spanish-speaking states.

Unique well identifier (UWI)   A unique alpha-numerical or numerical code which identifies a wellbore and the significant operational events which have taken place in the history of that wellbore.

Unit/unit area   The defined area developed under the terms of a UUOA. *See also* **tract/tract participation.**

Unit adjustment   An alternative term for a **redetermination.**

Unit interest   In the process of unitisation, the PPI of an individual interest-holder in the unit.

Unit interval   The cross-block petroleum deposit which results from a unitisation exercise.

Unit operations   Petroleum E&P operations which are carried out in a unit interval. *See also* **non-unit operations.**

Unit operator   The person who is designated and appointed to act as the operator of a unit under the terms of a UUOA.

United Kingdom continental shelf (UKCS)   The petroleum-bearing waters surrounding the United Kingdom, divided into quadrants and blocks and further known by the principal regions of the Central North Sea, East Irish Sea, Northern North Sea, Southern North Sea, and West of Shetland.

United Nations Commission on International Trade Law (UNCITRAL)   A United Nations-sponsored body of international trade law which seeks to reform and to harmonise various rules and trade customs on international commercial law (http://www.uncitral.org/uncitral/index.xhtml).

United Nations Convention on the Law of the Sea (UNCLOS)   An international treaty which defines the rights and responsibilities of the member states with regard to the use of the world's oceans. *See also* **exclusive economic zone.**

Unitisation   The development of contiguous cross-block petroleum deposits as a single unit. *See also* **unit interval, statutory unitisation,** and **voluntary unitisation.** Contrast with **pooling,** as the aggregation of multiple petroleum deposits (which are not necessarily contiguous) for joint development.

Unitisation and unit operating agreement (UUOA) A contract for the conduct of the unitisation process in respect of an identified number of cross-block petroleum deposits and for the operation of the resultant unit by a unit operator.

Universal Transverse Mercator (UTM) A two-dimensional coordinate system which is used to give surface locations on the Earth. *See also* **eastings** and **northings**.

Unless clause A contractual provision whereby the contract will be terminable, or will not move to its next phase of performance, unless a particular person has performed a particular obligation.

Unlevered Petroleum project costs which are met through equity reserves rather than through debt. *See also* **levered**.

Unlevered return The amount of cashflow which a petroleum project generates before the costs of financing have been met. *See also* **levered return**.

Unlined pit An unlined excavation which is used for the storage of produced water from onshore petroleum production operations.

Unliquidated damages Monetary damages which are payable for a breach of contract and are unascertained at the time of entry into the contract. Contrast with **liquidated damages**, whereby the damages payable are agreed at the time of entry into the contract. Also called **damages at large**.

Unloading conflict A conflict which arises between two or more ships that arrive at the same time at an unloading port in order to unload a cargo. *See also* **loading conflict**.

Unmanned production facility (UPF) An alternative term for a **normally uninhabited (unattended) installation**.

Unmanned wellhead platform (UWHP) A wellhead platform which is normally an uninhabited installation.

Unrisked An assessment of resources which does not take into account certain defined geological, commercial, financial, or development risks which would otherwise qualify the estimation of recoverable petroleum. Contrast with **risked**, whereby the assessment does account for the risks.

Unweighted average A calculation of the mean number of a sample with no attribution of values to the sample in order to correct apparent sample irregularities. *See also* **weighted average**.

Unwind costs The transaction costs which are associated with withdrawing from or cancelling a contract or a project. *See also* **break costs**.

Up and over An alternative term for a **processing and operating services agreement**.

Updip The relative positioning of different phases within a formation along a dip. For example, a trap could contain (in ascending order), water then oil then gas—the gas is updip of the oil and the oil is updip of the water. *See also* **downdip**.

Upgrader A facility which processes ultra-heavy oil and heavy oil into lighter oils and syncrude.

Upgrading A process for improving the yield from a quantity of a grade of oil.

Uplift   A quantity of petroleum which can be recovered by a producer from a project before the producer is required to pay tax on the related production revenues.

Upper explosive limit (UEL)   The highest concentration (by percentage) of a gas in air which is capable of combusting with an ignition source, where concentrations of gas above the UEL are too rich to combust. *See also* **lower explosive limit**.

Upper sample   A spot sample which is taken from the upper layer of a petroleum storage tank. *See also* **middle sample**.

Upside   An alternative term for a **payout**.

Upside sharing   An alternative term for **benefit sharing**.

Upstream   The functions of petroleum exploration, project development, and production. *See also* the relative functions of **downstream** and **midstream**.

Upstream government petroleum contract (UGPC)   An alternative term for a **concession**.

Uptime   A period of time during which a petroleum facility is, or can be expected to be, fully operational. Contrast with **downtime**, as a period of time during which a petroleum facility is, or can be expected to be, not operational. Also called a **service factor**.

Upward flexibility quantity (UFQ)   Under a petroleum sales contract, the unilateral ability of the seller or the buyer to increase the quantity of petroleum which is to be supplied during the term of the contract. Contrast with **downward flexibility quantity**, whereby the seller or the buyer can decrease the quantity of petroleum which is to be supplied.

Urgent operational matter (UOM)   Under a JOA, a decision that needs to be made with short notice, usually because of the presence of a drilling rig on site.

Use it or lose it (UIOLI)   A contractual provision or a regulatory principle whereby the holder of ullage in a petroleum facility is obliged to make that ullage available to third parties by a process of open access.

Used laytime   The actual amount of time which is taken for a ship to load or unload its cargo (as appropriate). Contrast with **allowed laytime**, as a defined notional period of time within which a ship is required to load or unload its cargo.

Usufruct   A civil law-founded concessionary right.

Utilisation factor   The proportion of the actual usage of a petroleum facility at any time relative to the maximum useable capacity of that facility.

Utility   A company whose principal purpose is to procure or to generate energy for sale to end-users.

Utility platform   An offshore platform which is used only for the generation of power to be used for adjacent petroleum operations.

Utility rate of return   The rate of return which a service-provider to a petroleum project expects to receive, which is less than an equity rate of return which applies in recognition of the lesser degree of risk which is assumed by the service-provider.

V

Vacuum breaker   A device which prevents the build-up of a vacuum in a petroleum storage tank or a pipeline.

Vacuum degasser   A device which creates a vacuum in order to remove entrained gas from gas-cut mud.

Vacuum distillation unit (VDU)   A crude distillation unit which operates under reduced pressure in order to prevent decomposition of the feedstock.

Vacuum gas oil   A viscous fuel oil which is produced from a vacuum distillation unit.

Valley gas   Under a gas sales contract, gas which is supplied by the seller to the buyer during expected (often seasonal) periods of reduced market demand.

Valorisation   The recovery of money from the commercialisation of an in-kind petroleum entitlement. *See also* **reverse valorisation**.

Value at risk (VAR)   A statistical technique which is used to quantify the risk that a loss might be generated by a trading position. *See also* **cash flow at risk**.

Value chain   An operational and economic continuum through which petroleum is commercialised and its value is enhanced. *See also* **full value chain**.

Value engineering   A petroleum project development process which promotes the use of lowest cost materials and methods in order to reduce overall project costs but without sacrificing basic project functionality.

Value of work done (VOWD)   A petroleum project management technique which is used to measure incurred project costs at a point in time, regardless of whether they have been paid, which is used for overall project management, reporting, and cost control.

Vapour displacement   The release of vaporised petroleum (as gas) from liquid petroleum in a storage tank.

Vapour lock   The displacement by air or gas vapours of liquid fuel in a pipeline.

Vapour loss   *See* **vapourisation**.

Vapour pressure   The in-tank pressure which is exerted by gas vapours which are released in vapour displacement.

Vapour recovery system (VRS)   An item of equipment which is used for vapour return/recovery.

Vapour return/recovery   A process for the capture of vapour displacement from petroleum-processing, transportation, or storage facilities, for conversion to a liquid or for transportation to storage.

Vapour return line (VRL)   A flexible hose which is used to effect vapour recovery from a cargo of petroleum.

Vapour space   The space in a liquid petroleum storage tank which exists above the level of the stored petroleum.

Vapourisation   A transformation of the physical state of a liquid to a gas. Also called **regasification**, where this is an intended industrial process, and **vapour loss**, where it is naturally occurring. *See also* **evaporation loss**. Contrast with **condensation**, as a transformation of the physical state of a gas to a liquid.

Variable deck load (VDL)   A measure of the safe carrying capacity of an offshore drilling unit or a crane barge.

Vegetation survey   A method of detecting leaks in an onshore sub-surface gas pipeline by noting the presence of dead surface vegetation along the pipeline route.

Vehicular natural gas (VNG)   Gas which is used as a vehicle fuel (for example, as CNG).

Velocity meter   An alternative term for a **flow meter**.

Velocity model   A map of the various rock layers within a particular formation and the expected speed of travel of seismic waves in each layer.

Velocity string   Small-diameter tubing which is placed inside the production tubing in a wellbore in order to increase the outflow of accumulated liquids from the bottomhole.

Vena contracta   A point in the flow of a petroleum stream at which the diameter of the stream is at its lowest and the velocity of the stream is at its greatest (such as where the stream passes through a choke).

Vendor due diligence (VDD)   A process in the sale of an asset or a business whereby the seller prepares its own due diligence report in respect of the asset or the business for the benefit of prospective purchasers.

Vent   A deliberate opening point in a petroleum storage or processing facility from which gas or air is permitted to escape to the atmosphere.

Venting   The release of gas into the atmosphere.

Venture information   Confidential information and proprietary data which is generated by the parties to a JOA from the conduct of the joint operations.

Venturi   An alternative term for a **choke**.

Venturi effect   The increased velocity and the reduced pressure which results from the flow of a gas or a fluid through a choke. *See also* **pressure drop**.

Verification   In relation to a loan arrangement, a lender's examination of the existence and the condition of the borrower's assets and of the veracity of the borrower's financial records.

Vertical integration   The participation by an integrated company in each of the upstream, midstream, and downstream petroleum market segments.

Vertical permeability   A measure of the permeability of a payzone, which is assessed from the bottom to the top of the payzone. Contrast with **horizontal permeability**, whereby permeability is assessed according to the payzone's lateral coordinates.

Vertical seismic profile (VSP)   A series of seismic measurements which are obtained from within a wellbore that are used for correlation with surface-sourced seismic data.

Very large crude carrier (VLCC)   A crude carrier which has a deadweight tonnage in the range of 180,000 to 320,000 metric tons. *See also* **ultra-large crude carrier**.

Very large ethane carrier (VLEC)   A ship which carries between 85,000m^3 and 97,000m^3 of ethane in a refrigerated state. *See also* **ultra-large ethane carrier**.

Very large gas carrier (VLGC)   A ship which carries between 50,000 and 80,000m^3 of LPG in a refrigerated state.

Vessel experience factor (VEF)   A measure of the difference between a number of historical ship-side and corresponding shore-side cargo measurements which is applied to more accurately determine a ship's true cargo-carrying capability.

Vessel of opportunity   A ship or a drilling unit which unexpectedly becomes available for hire.

Vessel presentation range (VPR)   A defined range of dates during which a ship is scheduled to undertake a particular activity (such as tendering a notice of readiness).

Vessel-to-vessel (VTV)   An alternative term for **ship-to-ship**.

Vibroseis   A truck with a vibrator plate which is used to conduct seismic surveys on land. Also called a **thumper truck**.

Vienna Convention   The Vienna Convention on the International Sale of Goods 1980: an international treaty which seeks to apply a uniform set of terms to international contracts for the sale of goods. Also called the **Convention on the International Sale of Goods**.

Virgin pressure   The original, inherent capillary pressure which exists within a formation prior to the production of petroleum from the formation.

Virgin stock   An alternative term for **straight run**.

Virgin zone   A part of a formation which is unaffected by invasion.

Virtual data room (VDR)   A data room which exists only through online electronic access and not as a physical location.

Virtual flow meter (VFM)   A process by which a mathematical model is applied to process conditions in order to estimate flow rates instead of using a physical flow meter.

Virtual gas   Gas entitlements which are traded without physical delivery. Also called **notional gas**.

Virtual interconnection point (VIP)   A series of physical interconnection points between two or more pipelines at which the pipeline operators offer capacities at those points through one virtual product.

Vis major   An alternative term for **force majeure**.

Visbreaking   An alternative term for **coking**.

Viscosimeter/viscometer   A device which determines the viscosity of a substance.

Viscosity   The measure of a liquid's resistance to flowing freely (where a higher viscosity reading means a reduced ability of the liquid to flow). *See also* centistoke/centipoise. *See* Appendix 6.

Viscosity gel meter (VGM)   A tool which is used to test the ability of drilling fluid to flow freely during drilling operations.

Viscosity index (VI)   An index which determines the tendency of an oil to become less viscous at increased temperatures.

Vision decision   A commercial or operational decision which is made in support of, and is defensible as being part of, a defined petroleum project vision.

Visioneering   A hypothetical engineering study which is based on envisaged outcomes.

Vitrinite   A form of kerogen which is generally prospective for the presence of petroleum. *See also* **inertinite**.

Voidage replacement   A method of secondary recovery whereby gas or water is injected into a formation in order to boost capillary pressure and to maintain or to increase crude oil recovery from the formation. *See also* **artificial lift** and **waterflood**.

Voidage replacement ratio (VRR)   The ratio, expressed in barrels, of the volume of gas or water which is injected into a formation to the volume of production from the formation during the process of voidage replacement.

Volatile light ends   Liquid phase petroleum fractions which will readily vaporise when exposed to atmospheric temperature and pressure.

Volatile oil   Crude oil that contains fewer heavy fractions and more wet gas than black oil, with higher gas–oil ratios, with a colour generally lighter than black oil (brown, orange, or green), which exists at a certain temperature and pressure but from which entrained gas could emerge if temperature and pressure drop below the bubble point.

Volatile organic compound (VOC)   An organic compound which easily vaporises in air and could be a hazardous air pollutant.

Volume-weighted average price (VWAP)   An averaged price for petroleum sales which is weighted according to sales volumes. *See also* **weighted average price**.

Volumetric gas in place (VGIP)   A measure of gas initially in place which is determined by volume.

Volumetric oil in place (VOIP)   A measure of oil initially in place which is determined by volume.

Volumetric production payment interest (VPPI)   A royalty whereby the payee (the royalty-holder) is entitled to a defined share of the petroleum produced under the concession, which is payable as quantities of petroleum. Contrast with a **dollar-denominated petroleum production interest**, whereby the royalty is payable in cash.

Voluntary relinquishment   Under a concession, an optional right of the concession-holder to surrender all or part of the contract area to the grantor.

Contrast with **mandatory relinquishment**, where relinquishment is a compulsory obligation of the concession-holder. *See also* **relinquishment**.

Voluntary unitisation   Unitisation which proceeds as a votive exercise between the participating parties. Contrast with **statutory unitisation**, which proceeds as a state-mandated exercise in respect of the participating parties.

Vortex breaker   A device which is used to prevent the formation of a vortex when a fluid is drained from a tank, where the existence of a vortex could create entrained vapour in the fluid stream.

Vortex induced vibration (VIV)   The motion which is induced by the interaction of a physical structure with a flow of water around that structure.

Votes by notice (VBN)   Under a JOA, the conduct of the business of the operating committee by the exchange of written communications between the parties rather than through holding a physical meeting.

Voting compact   Under a JOA or a shareholders' agreement, a private arrangement between certain of the co-venturers as to how they will vote their respective interests en bloc.

Voyage charter   A charterparty for a ship which exists for a defined voyage or for a series of voyages. Contrast with a **time charter**, which exists for a defined period of time. *Also called* **voyage charter party**.

Voyage charter party (VCP)   An alternative term for a **voyage charter**. *Also called* **voyage charter**.

Vug   A cavity which exists within a sedimentary rock.

W

Wait-and-see   Under a concession, a provision which allows the concession-holder to not relinquish a part of the contract area which contains a discovery which has not been declared commercial if the concession-holder believes that later circumstances might make that discovery commercial.

Waiting on cement (WOC)   In the drilling of a wellbore, non-productive time which accrues when waiting for curing to take place.

Waiting on weather (WOW)   Under a drilling contract, non-productive time which accrues when a drilling unit is waiting for favourable prevailing weather and/or tidal conditions to occur before engaging in drilling operations.

Walk to work (W2W)
(1) An offshore petroleum production facility in which an accommodation module is bridge-linked to a production module.

(2) A bridge which is deployed from an offshore support vessel in order to allow access to a normally uninhabited installation.

Walkaway vertical seismic profile (WVSP)   A vertical seismic profile in which the source of seismic energy is progressively moved to greater offset from the seismic receivers. *See also* **offset vertical seismic profile** and **zero offset vertical seismic profile**.

Walking rig   An onshore drilling unit which can be moved intact between drill sites, without the need for rig-down and rig-up.

Wall sticking   In the drilling of a wellbore, the situation in which a section of a drill string becomes stuck on accumulated mud cake which has built up on the sides of the wellbore. Also called **differential sticking**.

War risks clause   Under a charterparty, a provision which sets out the course of action which is open to the master of a ship if the ship, its cargo, or its crew would be put at risk because of war affecting the voyage which was due to be made under the charterparty.

Warm and inerted   The situation in which an LNG carrier arrives for the loading of a cargo with its tanks containing an inert gas (such as nitrogen) and where gas up is required. *See also* **presentation**.

Warm rig   A drilling unit which has been mobilised but is not presently being employed, prior to demobilisation. *See also* **cold rig** and **hot rig**.

Warm under vapour   The situation in which an LNG carrier arrives for the loading of a cargo with its tanks containing natural gas vapours and the in-tank temperature is such that cool down is required. *See also* **presentation**.

Warranty and indemnity insurance (W&II)   Under an asset or share sale and purchase agreement, a policy of insurance (taken out by the seller or the buyer) which provides cover for liabilities incurred in connection with warranty or indemnity claims which are made under the agreement against the insured party.

Wash   A refinery process for the purification of crude oil by agitation with water or chemicals.

Wash oil   An alternative term for **absorption oil**.

Wash trade   An instantaneous sale and purchase of a single petroleum cargo, with no change of ownership or market risk (which could be done in order to distort market prices or to trigger commission fees).

Washout   A contractual provision whereby a contract ends when an underlying agreement to which the contract relates also ends. *See also* **anti-washout provision**.

Washover   An operation to free a stuck pipe from within a wellbore by using a larger-diameter pipe which fits over the stuck pipe, grips it and retrieves it.

Waste gas   Gaseous products which are emitted from a petroleum production, processing, or transportation activity.

Waste heat recovery unit (WHRU)   An energy recovery heat exchanger which transfers heat from a process output to another process or for another purpose.

Waste water   An alternative term for **produced water**.

Wat-sat   An alternative term for **water saturation**.

Water alternating gas (WAG)   A process for enhanced oil recovery whereby water and gas are injected alternately into a formation.

Water-based mud (WBM)   Drilling fluid which uses water as the base component. *See also* **oil-based mud** and **synthetic-based mud**.

Water bottom   Water which has accumulated below the lowest level of oil at the base of an oil storage tank. *See also* **basic sediment and water**.

Water breakthrough   The situation whereby water which has been injected to maintain formation pressure, or inherent water within a formation, breaks through to a production wellbore. *See also* **gas breakthrough**.

Water column   A conceptual column of water which runs from the sea surface down to the bottom sediment, which a state could claim jurisdiction over.

Water coning   The point at which a formation starts to produce more water than petroleum, which is caused by pressure depletion from the production of petroleum and the ingress of water.

Water cushion   Water which is injected into a drill string in order to equalise external pressure and to prevent the drill string from deforming when the yield point is reached.

Water drive   Gas production through a wellbore which is driven by naturally occurring water ingress from a formation. Also called **aquifer drive**. *See also* **gas expansion drive**.

Water gas   Synthetic natural gas which is made up mostly of carbon monoxide and hydrogen and is produced by blowing steam through burning coke.

Water–gas ratio (WGR)   The ratio of water to gas in a petroleum stream.

Water leg   A part of a petroleum filtration unit which manages extracted water.

Water lot   A lot for an area which has a maritime frontage.

Water saturation   (WS) A measure of the percentage of total pore space in a formation which is occupied by water. Also called **wat-sat**. *See also* **saturation**.

Water shut-off (WSO)   A sub-surface operation which is intended to reduce the quantities of watercut in a formation.

Water table   A geological level below which rock pores are naturally saturated with water.

Water testing   An alternative term for **hydrotesting**.

Water well   A well which is drilled into an aquifer in order to access water for drilling operations.

Watercut
(1)   The ratio of watercut to petroleum which is produced from a formation. A high degree of watercut indicates a poorly producing formation and can lead to a watered out well.
(2)   Unwanted water in a formation.

Watered out/watering out   The existence of a high degree of watercut in a formation which has developed to the extent that a wellbore is no longer economical to operate and is shut in.

Watered out well (WOW)   A wellbore which has watered out.

Waterfall   A mechanism which is used in the project financing of a petroleum project whereby revenues that are received into the project are collected and are disbursed to a defined progressive series of recipients.

Waterflood   A method of secondary recovery whereby water is injected into a formation in order to displace residual petroleum into a wellbore. *See also* **flooding**, **voidage replacement**, and **waterflood kick**.

Waterflood kick   The first indication of increased oil production which has resulted from waterflood.

Water–oil ratio (WOR)   The ratio of water to crude oil which is produced from a formation.

Water-producing interval   A section of a petroleum-producing formulation from which water is produced.

Wax   A solid or semi-solid mixture which is derived from a petroleum stream.

Wax appearance temperature (WAT)   The temperature point at which wax deposits in crude oil will begin to appear.

Wax disappearance/dissolution temperature (WDT)   The temperature point at which wax deposits in crude oil will begin to dissolve into solution.

Weather correction   A process of estimating adjustments to end-user gas demand which arise from normal seasonal weather conditions. Also called **temperature correction**.

Weather window   A period of time when prevailing weather and/or tidal conditions are the most favourable for the performance of planned petroleum project operations.

Weathered crude   Crude oil which has lost some or all of its entrained gas quantities because of vaporisation which has occurred during transportation or storage.

Weathering   A process whereby a cargo of LNG undergoes a compositional transformation through boil-off.

Weathervaning   The free rotation of an FPSO or an FSU around a turret mooring system, reflective of prevailing weather and/or tidal conditions.

Wedge formula   A mathematical method of calculating the size of a ship's cargo by reference to the ship's trim and the ship's cargo compartment dimensions.

Wedge volumes   LNG cargoes which are available for sale in the ramp-up period between the commissioning of a new export facility and the plateau sales offtakes from the facility.

Weight coating   A physical coating which is applied to a sub-sea pipeline prior to installation, which is intended to give negative buoyancy.

Weight cut   A reduction in the density of drilling fluid which is caused by the presence of entrained gases or liquids.

Weight on bit (WOB)   A measure of the aggregate level of downward force which is transmitted to a drill bit during the drilling of a wellbore as a consequence of the weight of the drill string and drill collars.

Weighted average   A calculation of the mean number of a sample but with a defined attribution of values to certain elements of the sample in order to correct apparent sample irregularities. *See also* **unweighted average**.

Weighted average cost of gas (WACOG)   Under a gas sales contract, a weighted measure of the total cost of gas which is purchased by the buyer, divided by the total quantity of gas that is purchased by the buyer, over a given period of time.

Weighted average price (WAP)   Under a petroleum sales contract, the weighted average of all prices which are paid by the buyer for petroleum over a given period of time. *See also* **volume-weighted average price**.

Weighting   In relation to indexation, a measure of the different percentages which are ascribed to each index in an overall basket of indices to reflect the required pricing proposition.

Well and reservoir management (WRM)   A set of operational and regulatory principles which apply to the effective management of wellbores, reservoirs, and formations.

Well containment package (WCP)   An alternative term for a **well control plan**.

Well control   Operational activities which are undertaken to prevent or to control a kick in a wellbore (including flow check, overbalance drilling, and reduced circulating pressure).

Well control plan (WCP)   A combination of equipment and procedures which are applied to manage a blowout. Also called a **well containment package**.

Well count   The total number of wellbores which are due to be or which have been drilled in a particular place over a particular period of time.

Well-in-progress   Under a drilling contract, a provision whereby the term of the contract will be extended automatically in order to allow the completion of a wellbore which is still being drilled by the drilling contractor at the point when the contract was due to end by the expiry of time.

Well intervention   An operation which is performed on a wellbore in order to manipulate the rate of production from a formation. Also called **intervention** and **well stimulation**.

Well kill   An alternative term for a **kill**.

Well log   An ongoing record of the geological structures which are penetrated during the drilling of a wellbore, including technical details of the drilling operation.

Well logging   The testing of the productive capacity of a wellbore with the drill string removed. Contrast with **drill stem testing**, whereby the wellbore's productive capacity is tested with the drill string still in place.

Well monitoring system (WMS)   A process which estimates and records the flow of petroleum and water from all the individual wells in a producing field.

Well penalty   A cash penalty which becomes payable under the terms of a concession if the concession-holder fails to drill a commitment well.

Well plan   An operator-led plan for the spudding, drilling, completion, and/or workover of a well.

Well spacing   A defined pattern of where wellbores are drilled and the spacing which is required to be maintained between the individual drill sites for those wellbores. *See also* **spacing**.

Well stimulation   An alternative term for **well intervention**.

Well surface location (WSL)   The areal surface coordinates at which a wellbore is to be drilled or has been drilled.

Well symbols   Cartographic symbol standards which are used on maps to indicate current and historical petroleum E&P activity. *See* Appendix 5.

Well to tank (WTT)   A measure of the carbon intensity which is associated with the LNG production to regasification elements of the value chain. *See also* **full lifecycle emissions**.

Well-to-wheel   A method of describing a petroleum project's full life cycle, from the initial production of petroleum to the use of refined petroleum as a vehicle fuel.

Well trajectory   The direction in which a wellbore is being drilled at any time (whether vertical or horizontal).

Wellbore
(1) A borehole which is drilled into a formation in order to explore for petroleum.
(2) A borehole through which discovered petroleum is produced.

Wellhead   The topmost section of a wellbore, above which sits a Christmas tree and below which hangs casing.

Wellhead platform (WHP)   A normally unmanned, remotely operated offshore platform into which multiple production wells are tied in.

Wellhead price   The sales price of petroleum which is required by a producer at the point of production. Also called a **first-hand price**.

Wellstream gas   An alternative term for **raw gas**.

Wentworth scale   A grading scale which is used to classify the diameters of microscopic sediments which are found in a petroleum stream.

West of Suez (WOS)   A notional boundary point which applies to everything due west of the Suez canal (viewed from the south with a longitude of 32.3167E). *See also* **east of Suez**.

West Texas Intermediate (WTI)   A crude oil blend from the United States.

Wet and dry spread   A combined offshore and onshore area in which drilling or seismic works are performed.

Wet barrel   A physical barrel of oil which is intended for delivery, often in a very prompt timeframe. *See also* **paper barrel**.

Wet berth A berth whereby the berthed ship is kept in the water. *See also* **dry berth**.

Wet cargo In a daisy chain, a cargo of petroleum which is actually delivered to an end-user. Contrast with a **paper cargo**, as a cargo which is bought and sold at any time prior to final delivery but without actual delivery of the cargo to an end-user.

Wet gas Natural gas which consists principally of petroleum fractions heavier than methane. Contrast with **dry gas**, which is principally methane.

Wet oil Crude oil which contains basic sediment and water. *See also* **dry oil**.

Wet pre-commissioning Pipeline pre-commissioning with the use of hydrotesting. *See also* **dry pre-commissioning**.

Wet shipping Shipping issues which are associated principally with addressing maritime misadventures and incidents. Contrast with **dry shipping**, which is associated principally with contract drafting and negotiation issues.

Wet tow The transportation of an offshore petroleum infrastructure item to site by tug towage as a floating item. *See also* **dry tow**.

Wet tree A sub-sea Christmas tree. Contrast with a **dry tree**, which is installed above-water or is installed sub-sea but in a waterproof cell.

Wetheader A pipeline system which is designed to transport wet gas.

Wheeling The transfer of physical gas quantities between different pipelines in order to ensure optional overall system performance. *See also* **operational balancing agreement**.

Whipstock
(1) A sidetrack which is drilled from an already-existing wellbore.
(2) A steel wedge which is applied in a wellbore to start the drilling of a sidetrack.

White labelling The production of a fuel on an anonymous basis by a producer which is then sold to a reseller and rebranded prior to on-sale to an end-user. *See also* **unbranded fuel**.

White oil
(1) Crude oil which contains lighter fractions when compared to black oil.
(2) A highly refined, colourless, and odourless mineral oil which is used as a blending base for pharmaceutical products. Also called **albino oil**.

White products Light-end refinery products such as gasoline and kerosene. Contrast with **black products**, as heavy-end refinery products such as residual fuel oil.

White space Ullage which exists in a petroleum storage facility.

Whittaker capsule A particular form of lifeboat which is used in offshore facilities for escape and survival. *See also* **totally enclosed motor propelled survival craft**.

Whole mud A measure of the entirety of the components of a drilling fluid, including solid and liquid phases.

Wide azimuth An alternative term for **multi-azimuth**.

Wild well   An alternative term for a **runaway well**.

Wildcat well   An alternative term for an **exploration well**.

Wilful arbitrage   Under a petroleum sales contract, a deliberate decision of the seller to sell and deliver the buyer's petroleum entitlement to a third party (and to accept the agreed contractual liability to the buyer for that delivery failure) in order to secure a greater profit from the third party.

Will o' the wisp   A colloquial alternative term for **marsh gas**.

Wind, wave, and solar (WWS)   The principal sources of renewable energy. Also called **sun, surf, and breeze**.

Windfall tax   A tax which is levied by a state on an investor's profits from a petroleum project, typically as a reaction to the state's perception that the investor's profits are excessive.

Window
(1) A hole which is made in casing as the first stage of creating a whipstock.
(2) A time stage which results from the windowing process.

Windowing   A process by which a start date is arrived at in a petroleum project contract through a progressively reducing series of time stages. Also called **funnelling**.

Winner's curse   In a licensing round, the risk to a bidder that a lack of available data or insight about the target concession is compensated for by overpayment to win the auction. *See also* **bidder's remorse**.

Winter peak   A seasonal adjustment which reflects an increased level of demand and/ or prices for certain petroleum grades during winter months. *See also* **summer valley**.

Wireline logs   Measurements which are taken from within a wellbore by wireline tools as part of well logging.

Wireline tools   Tools which are lowered into a wellbore by wire.

Withdrawal   Under a JOA, the right of a party to withdraw from continuing participation in the JOA and the underlying concession (subject to certain conditions for doing so).

Withering interest provision   Under a JOA, the forfeiture by a defaulting party of part of its petroleum project interests proportionate to the degree of that party's payment default. Also called a **dilution provision**. Contrast with **absolute forfeiture**, by which the defaulting party forfeits the entirety of its interests. *See also* **forfeiture**.

Within-pipe blending (WPB)   The blending of petroleum quantities with different compositions in a pipeline in order to give a single commingled stream.

Wobbe Index   An index which is used to compare the calorific value of gas of different compositions when combusted in order to determine the compatibility of gas compositions with gas-burning equipment. *See also* **enriching/enrichment**.

Wobbe quality adaptation (WQA)   The treatment of a gas or LNG stream in order to alter the Wobbe Index through spiking.

Won and saved (W&S)   A measure of the gross quantity of petroleum that is produced from a defined interest or area, prior to the processing, consumption, storage, or transportation of that petroleum, which is often used in the context of UKCS licensing.

Work boat   An alternative term for a **support vessel**.

Work breakdown structure (WBS)   A methodology for the definition of individual responsibilities in a petroleum project.

Work obligation payout   Under a concession, a monetary payment which is due for payment from the concession-holder to the grantor as compensation for the concession-holder's non-performance of a minimum work commitment under the concession. *See also* **drill or pay**.

Work order (WO)   An instruction which is given by an employer or by an operator in order to permit the carrying out of a particular activity.

Work programme and budget (WP&B)   Under a JOA, an agreed activity and costings plan which applies to intended joint operations for a year or for multiple years.

Workaround   A colloquial term for an improvised legal, contractual, operational, or technical solution to an issue which is otherwise impeding progress in the development of a petroleum project.

Working gas   Gas which is injected into and withdrawn from a gas storage facility during the ordinary course of operations. Contrast with **cushion gas**, which stays in the gas storage facility.

Working interest   Under a concession, the aggregate percentage participating interests of all the concession-holders (adding up to 100 per cent). Contrast with a **net working interest**, which represents the percentage participating interest of an individual concession-holder.

Working interest barrel (WIB)   A barrel of oil (or oil equivalent) which is attributable to a petroleum company's percentage entitlements (net of state take) under a concession.

Working load limit (WLL)   An alternative term for **safe working load**.

Workover   The modification, repair, or further stimulation of a wellbore which is carried out in order to restore or to prolong petroleum recovery from an underlying formation. Also called a **re-entry** and a **re-opening**.

Workover fluid   A special drilling fluid which keeps a wellbore under control during a workover.

Workover rig   A drilling unit which is used to carry out a workover.

World Association for Waterborne Transport Infrastructure   *See* **Permanent International Association of Navigation Congresses**.

World geodetic system (WGS)   A widely accepted standard which is used to determine coordinates in cartography and navigation, managed by the US National Geospatial-Intelligence Agency.

World Petroleum Council (WPC) An international industry body which is focused on the sustainable management and use of global petroleum resources (http://www.world-petroleum.org).

Worldscale index A published index (New Worldwide Tanker Nominal Freight Scale) of the costs of chartering a ship, where the actual charter costs of a ship will be expressed as a premium to or a discount from the index (http://www.worldscale.co.uk).

Worst credible metocean conditions (WCMC) A design and/or operational parameter which applies the worst possible estimate of metocean conditions to the operation of a ship or port facility.

Wrap An alternative term for a **construction management agreement**.

Wrong pockets A colloquial term for a contractual provision whereby a person who receives a monetary sum to which they were not entitled will account for it to the person who was entitled to receive it.

XL *See* **exploration licence.**

XRL *See* **extended reach lateral.**

X-O *See* **crossing.**

X-OV *See* **crossover valve.**

X-ray diffraction (XRD) A process which provides detailed information about the physical properties, chemical composition, and crystallographic structure of a substance.

X-tree *See* **Christmas tree.**

Yarn creep The irrecoverable increase in length over time of a rope or cable which results from its subjection to a sustained load.

Yaw On a ship or an offshore drilling unit, a side-to-side bow and stern motion which occurs because of prevailing weather and/or tidal conditions.

Year-on-year (YOY) A comparison of data which applies to a year with comparable data which applies to a previous year.

Yellow tipping   The incomplete combustion of gas, which could result in the production of unacceptable levels of carbon monoxide.

Yield   The total quantity of the products (of all forms) which result from the refining of a given quantity of crude oil. Also called **refinery yield**. *See also* **gross product worth** and **net product worth**.

Yield analysis   A measure of the value of a given quantity of crude oil which is determined by the yield which results from that crude oil.

Yield point/yield pressure (YP)   The point at which the level of pressure in or upon a section of a pipeline or a tubular is sufficient to cause it to deform. Also called **collapse pressure**.

Yield shift   An alteration which is made to the processing configuration of a refinery in order to alter the balance of the resultant products and to improve the yield.

Yoo-hoo buoy   A colloquial alternative term for a **pop-up buoy**.

Z

Zeebrugge trading point (ZTP)   A notional point on the Belgian gas pipeline network at which shippers trade gas quantities.

Zero gas   A calibration gas which contains no flammable gas components.

Zero offset vertical seismic profile (ZOVSP)   A vertical seismic profile in which the seismic receivers are located directly below the source of seismic energy. *See also* **offset vertical seismic profile** and **walkaway vertical seismic profile**.

Zero-tariff transportation   An arrangement for the transportation of petroleum through a pipeline whereby the shipper does not pay a tariff but does pay a proportionate part of the transporter's operating expenses. Also called **opex-covered transportation**.

Zombie partner   A colloquial term for a party to a JOA which cannot undertake the full performance of its contractual rights and/or obligations because it is subject to some form of suspension of its interests under the JOA.

Zombie party   A colloquial term for a contracting party which cannot undertake the full performance of its contractual rights and/or obligations because it is subject to some form of suspension of its interests under the contract or by applicable law.

Zonal tariff   Under a pipeline transportation agreement, a tariff which is payable by the shipper to the transporter which reflects distinct transportation zones to which distinct tariffs apply. *See also* **distance tariff** and **postage stamp tariff**.

Zone   An interval of rock within a formation which is capable of being differentiated from its surrounding rock intervals on the basis of its petroleum-bearing composition.

Zone isolation   The separation of different zones within a formation by the use of casing and packers in order to maintain separated zonal pressures and to prevent the commingling of different phases within the formation.

Zone of influence   The operational and environmental relationship which exists between the spatial footprint of a petroleum facility and its surrounding habitat.

Zone of lost circulation   A formation which contains natural fissures which are sufficiently large to allow drilling fluids or cement to flow into the formation rather than to stay within a wellbore. Also called a **lost circulation zone**.

Zone time (ZT)   The local time which is selected by the master of a ship to apply on board and in respect of a ship that is traversing a number of different time zones on a voyage.

PART C

APPENDICES

PETROLEU M UNITS AND CONVERSIONS

Energy

The principal units of energy used in the measurement of petroleum are the British thermal unit (Btu), the calorie (Cal), the Joule, and the Watt/hour (W/hr):

1 Btu	251.995 Cal	1 Joule	0.0009478 Btu
	1,055.06 Joules		0.2388467 Cal
	0.293071 W/hr		0.000277778 W/hr
1 Cal	0.0039683 Btu	1 W/hr	3.41214 Btu
	4.1867865 Joules		859.8453502 Cal
	0.001163 W/hr		3,599.988 Joules

Pressure

The principal units of pressure used in the measurement of petroleum are the atmosphere (atm), the bar, the Pascal (Pa), and the pound per square inch (psi):

1 atm	1.01325 bar	1 Pa	0.0000099 atm
	101,325 Pa		0.00001 bar
	14.6959 psi		0.000145 psi
1 bar	0.986923 atm	1 psi	0.068046 atm
	100,000 Pa		0.0689476 bar
	14.5038 psi		6,894.75 Pa

Volume

The principal units of volume used in the measurement of petroleum are the cubic foot (ft^3) and the cubic metre (m^3):

$1\ ft^3$	$0.0283168\ m^3$
$1\ m^3$	$-35.3146667\ ft^3$

TECHNICAL ELEMENTS

2A—petroleum formation

The existence of recoverable quantities of petroleum below the Earth's surface depends on the presence of several combined conditions:

Source material: dead and decaying organic plant and animal matter ('kerogen') was continuously deposited on sea-beds over a period of time of between 30 and 570 million years ago (see Appendix 4).

Compaction and conversion: the kerogen was compacted with sediments over time and subjected to increasing temperature and pressure. Through anaerobic bacterial decomposition in this source rock the kerogen underwent chemical conversion to form petroleum.

Migration, seal, and reservoir rock: once formed, gaseous or liquid phase petroleum would migrate upwards through microscopic fractures and fault lines in the source rock until it seeped out at the Earth's surface or became trapped in reservoir rock (typically sandstone, limestone, shale, or mudstones) beneath an impervious rock seal.

It is the combination of kerogen, converted into petroleum within source rock and the migration of that petroleum to a reservoir rock, which determines the existence of a viable petroleum system.

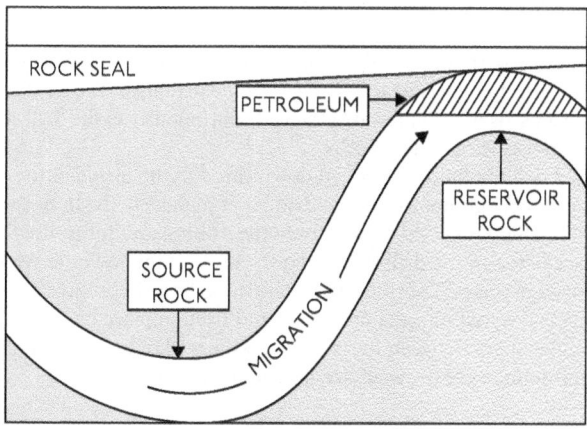

Petroleum formation also depends on depth and temperature. Crude oil tends to form in the oil window of between 1 and 5 kilometres below surface level, in a range of 120°F (50°C) to 350°F (175°C). Natural gas tends to form in the gas window of between

3 and 7 kilometres below surface level, beyond 350°F (175°C). Sub-surface temperatures exceeding 500°F (260°C) at depths beyond 8 kilometres below surface level tend to make the kerogen overmature—it becomes carbonised and not worthy of recovery.

2B—exploration drilling

In offshore exploration drilling, the type of drilling unit used to explore for petroleum is determined by the applicable water depth.

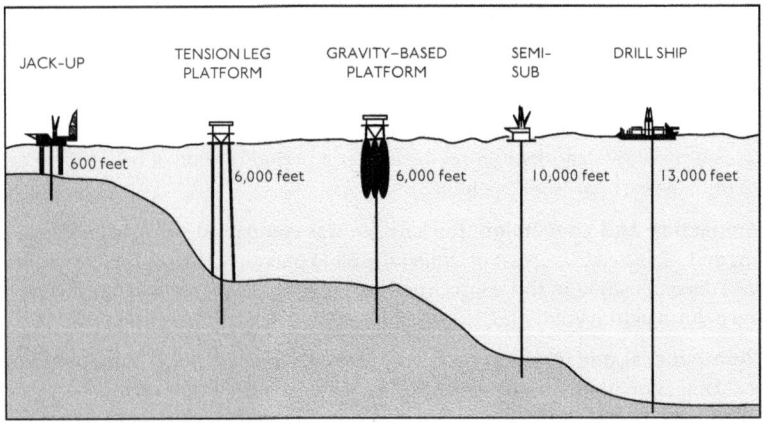

2C—drilling a wellbore

In drilling a wellbore a drill bit is attached to the hanging end of a drill string, made up of lengths of hollow steel drill pipe which are screwed together in two, three, or four unit lengths. The drill string is assembled until it has the length necessary to enable the drill bit to meet the surface point (which, offshore, is the sea-bed). At this point the wellbore is 'spudded'—the drill string is rotated from the drilling unit and the drill bit begins to penetrate the surface. As the drilling of the wellbore gets underway extra lengths of drill pipe are attached continuously to the drill string as the drill bit penetrates ever-greater depths.

Specialised drilling fluid is pumped down through the inside of the drill string and comes out through nozzles in the drill bit at high velocity. In normal circulation the drilling fluid is cycled back up onto the drilling unit for re-use through the annular space between the drill string and the walls of the wellbore. Wellbores are typically drilled in stages. To protect the integrity of the wellbore, as the drilling proceeds, casing is lowered into the open hole and fixed in place by pumping cement into the annular space between the casing and the wall of the wellbore. As drilling continues the diameter of the wellbore, and of the casing, reduces.

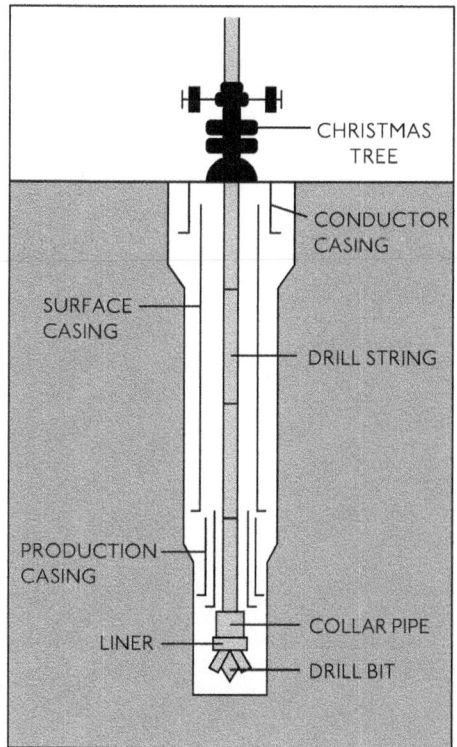

2D—petroleum production

Produced petroleum could be taken through a sub-sea tie-back to an existing facility or could be taken onto a newly installed platform. Offshore, the type of platform used will depend on the applicable water depth.

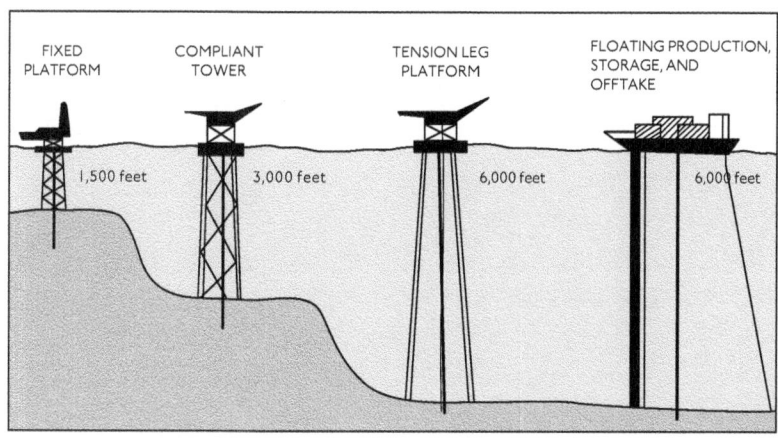

2E—refining

In atmospheric distillation crude oil is heated to boiling point in a furnace and passes into a crude distillation unit, where different petroleum fractions are formed by evaporation, separation, and condensation and exit the unit at different output points.

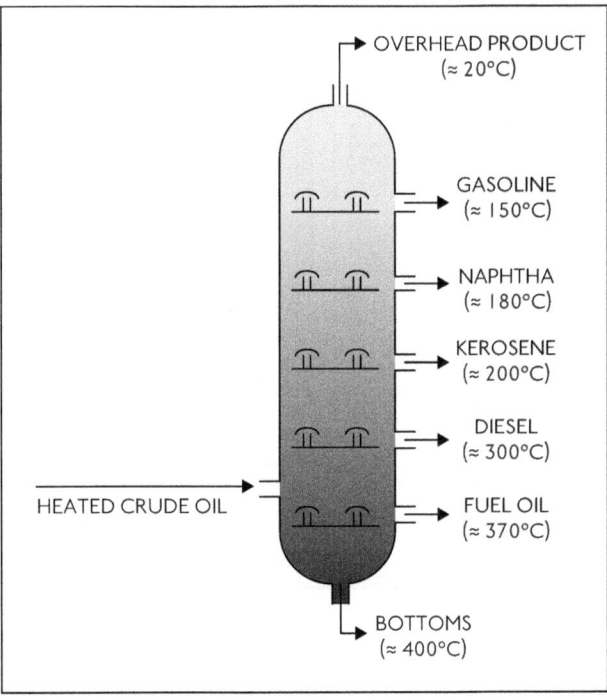

Further refinery processes which can take place, depending on how complex a refinery is, include alkylation, catalytic cracking, coking, hydrocracking, hydrotreating, reforming, and solvent extraction.

PROMOTE LEVELS

Under an earn-in agreement or a farm-in agreement, the relationship between the percentage of the farmor's total required work costs which are assumed by the farmee and the net working interest sold down by the farmor to the farmee, expressed as a percentage of the working interest, is known as the 'promote level'. The most commonly used promote levels are set out below.

The promote level definition	The amount payable by the farmee (expressed as a percentage of the farmor's total required work costs)	The net working interest earned by the farmee (expressed as a percentage of the working interest)
Four for one (4:1)	100%	25%
	75%	18.75%
	50%	12.5%
	25%	6.25%
Three for one (3:1)	100%	33.3%
	75%	25%
	50%	16.6%
	25%	8.3%
Two for one (2:1)	100%	50%
	75%	37.5%
	50%	25%
	25%	12.5%
Ground floor (1:1)	100%	100%
	75%	75%
	50%	50%
	25%	25%
One for two (1:2)	12.5%	25%
	25%	50%
	37.5%	75%
	50%	100%
One for three (1:3)	8.3%	25%
	16.6%	50%
	25%	75%
	33.3%	100%

The promote level definition	The amount payable by the farmee (expressed as a percentage of the farmor's total required work costs)	The net working interest earned by the farmee (expressed as a percentage of the working interest)
One for four (1:4)	6.25%	25%
	12.5%	50%
	18.75%	75%
	25%	100%
Quarter for a quarter	25%	25%
Quarter for a third	25%	33%
Quarter for a half	25%	50%
Third for a quarter	33%	25%
Third for a third	33%	33%
Third for a half	33%	50%
Half for a quarter	50%	25%
Half for a third	50%	33%
Half for a half	50%	50%

GEOLOGICAL TIME PERIODS

The geological time scale (GTS) is a system which relates the Earth's geological strata to the passage of time since the Earth is believed to have been first formed approximately 4.6 billion years ago. The geological history of the Earth is defined according to a number of hierarchies, with several elements used to describe that history:

The period: this is the basic unit of geological time which reflects when a particular rock system was formed.

The epoch: some periods are further divided into epochs.

The age: some epochs are further divided into ages.

The era: this describes the combination of two or more periods.

The eon: this is the largest division of geologic time and describes the combination of two or more eras.

Thus the overall sequence is **eon–era–period–epoch–age**. This hierarchy is best understood by relying on the assumptions that rock layers have been laid down horizontally over time without deformation; that the oldest layers of rock reside at the base of a formation; and that progressively younger rock layers lie above ('superpositioning').

From a petroleum perspective the Phanerozoic eon (which encompasses the Cenozoic, Mesozoic, and Paleozoic eras) is key as it represents the time during which the majority of macroscopic organisms lived. This eon is generally assumed to cover the period from today to around 541 million (mm) years before present (ybp). As a rough estimate, approximately 20 per cent of the Earth's petroleum deposits existing today were formed in the Cenozoic era, 70 per cent in the Mesozoic era, and 10 per cent in the Paleozoic era. Despite its importance in the creation of petroleum, however, the Phanerozoic eon represents only 12 per cent of the assumed total history of the Earth.

The time before the Phanerozoic eon is called the Precambrian eon, which encompasses the Proterozoic, Archean, and Hadean eras. The Precambrian eon is of little interest from a petroleum perspective because it pre-dates the existence of plant or animal life on Earth.

All of the above elements can be expressed diagrammatically as follows, reflecting the geological time scale conventions established by the International Commission on Stratigraphy:

Eon	Era	Period	Epoch	Age	Years before present
Phanerozoic	Cenozoic	Quaternary	Holocene	Meghalayan	0–4,200
				Northgrippian	4,200–8,270
				Greenlandian	8,270–11,700
			Pleistocene	Upper	11,700–129,000
				Middle	129,000–774,000
				Calabrian	774,000–1.8m
				Gelasian	1.8–2.58m
		Neogene	Pliocene	Piacenzian	2.58–3.6m
				Zanclian	3.6–5.33m
			Miocene	Messinian	5.33–7.24m
				Tortonian	7.24–11.63m
				Serravallian	11.63–13.82m
				Langhian	13.82–15.97m
				Burdigalian	15.97–20.44m
				Aquitanian	20.44–23.03m
		Paleogene	Oligocene	Chattian	23.03–27.82m
				Rupelian	27.82–33.9m
			Eocene	Priabonian	33.9–37.8m
				Bartonian	37.8–41.2m
				Lutetian	41.2–47.8m
				Ypresian	47.8–56m
			Paleocene	Thanetian	56–59.2m
				Selandian	59.2–61.6m
				Danian	61.6–66m
	Mesozoic	Cretaceous	Upper	Maastrichtian	66–72.1m
				Campanian	72.1–83.6m
				Santonian	83.6–86.3m
				Coniacian	86.3–89.8m
				Turonian	89.8m–93.9m
				Cenomanian	93.9–100.5m
			Lower	Albian	100.5–113m
				Aptian	113m–125m
				Barremian	125–129.4m
				Hauterivian	129.4–132.9m
				Valanginian	132.9–139.8m
				Berriasian	139.8–145m

				Tithonian	145–152.1m
Phanerozoic	Mesozoic	Jurassic	Upper	Kimeridgian	152.1–157.3m
				Oxfordian	157.3–163.5m
			Middle	Callovian	163.5–166.1m
				Bathonian	166.1–168.3m
				Bajocian	168.3–170.3m
				Aalenian	170.3–174.1m
			Lower	Toarcian	174.1–182.7m
				Pliensbachian	182.7–190.8m
				Sinemurian	190.8–199.3m
				Hettangian	199.3–201.3m
		Triassic	Upper	Rhaetian	201.3–208.5m
				Norian	208.5–227m
				Carnian	227–237m
			Middle	Ladinian	237–242m
				Anisian	242–247.2m
			Lower	Olenekian	247.2–251.2m
				Induan	251.2–251.9m
	Paleozoic	Permian	Lopingian	Changhsingian	251.9–254.1m
				Wuchiapingian	254.1–259.1m
			Guadalupian	Capitanian	259.1–265.1m
				Wordian	265.1–268.8m
				Roadian	268.8–272.9m
			Cisuralian	Kungurian	272.9–283.5m
				Artinskian	283.5–290.1m
				Sakmarian	290.1–295m
				Asselian	295–298.9m
		Carboniferous	Pennsylvanian (Upper)	Gzhelian	298.9–303.7m
				Kasimovian	303.7–307m
			Pennsylvanian (Middle)	Moscovian	307–315.2m
			Pennsylvanian (Lower)	Bashkirian	315.2–323.2m
			Mississipian (Upper)	Serpukhovian	323.2–330.9m
			Mississipian (Middle)	Visean	330.9–346.7m
			Mississipian (Lower)	Tournaisian	346.7–358.9m

Phanerozoic	Paleozoic	Devonian	Upper	Famennian	358.9–372.2m
				Frasnian	372.2–382.7m
			Middle	Givetian	382.7–387.7m
				Eifelian	387.7–393.3m
			Lower	Emsian	393.3–407.6m
				Pragian	407.6–410.8m
				Lochkovian	410.8–419.2m
		Silurian	Pridoli		419.2–423m
			Ludlow	Ludfordian	423–425.6m
				Gorstian	425.6–427.4m
			Wenlock	Homerian	427.4–430.5m
				Sheinwoodian	430.5–433.4m
			Llandovery	Telychian	433.4–438.5m
				Aeronian	438.5–440.8m
				Rhuddanian	440.8–443.8m
		Ordovician	Upper	Hirnantian	443.8–445.2m
				Katian	445.2–453m
				Sandbian	453–458.4m
			Middle	Darriwilian	458.4–467.3m
				Dapingian	467.3–470m
			Lower	Floian	470–477.7m
				Tremadocian	477.7–485.4m
		Cambrian	Furongian	Stage 10	485.4–489.5m
				Jiangshanian	489.5–494m
				Paibian	494–497m
			Miaolingian	Guzhangian	497–500.5m
				Drumian	500.5–504.5m
				Wuliuan	504.5–509m
			Series 2	Stage 4	509–514m
				Stage 3	514–521m
			Terreneuvian	Stage 2	521–529m
				Fortunian	529–541m
Precambrian	Proterozoic				541–2500m
	Archean				2500–4000m
	Hadean				4000–4600m

WELL SYMBOLS

Well symbols are cartographic symbol standards which are used on maps to indicate current and historical petroleum exploration and production activity. There is no universally accepted single set of well symbols, however, and so the following guide represents the well symbols which are most commonly encountered in practice.

Producing Wells

- ●   oil
- ✿   gas
- ✹   oil and gas
- ⊗   coal bed methane
- ▦   tar/bitumen

Suspended/Abandoned Wells

- ✛   oil (abandoned)
- ✺   gas (abandoned)
- ⬤   oil (suspended)
- ✿   gas (suspended)

Other

- ○   permitted location
- ∅   expired permitted location
- ⊙   water injection wellbore
- ⊕   gas injection wellbore
- ⊠   tight hole
- -○-   dry hole
- ⏀   well being drilled
- ◆   gas storage
- ★   well with unknown status

APPENDIX 6

API GRAVITY

The API gravity scale is a measure of the density of petroleum fractions relative to water, measured in degrees (°), used to define petroleum fractions and which also indicates the relative viscosity of petroleum fractions. A higher API gravity value typically indicates a lighter and more commercially valuable petroleum fraction.

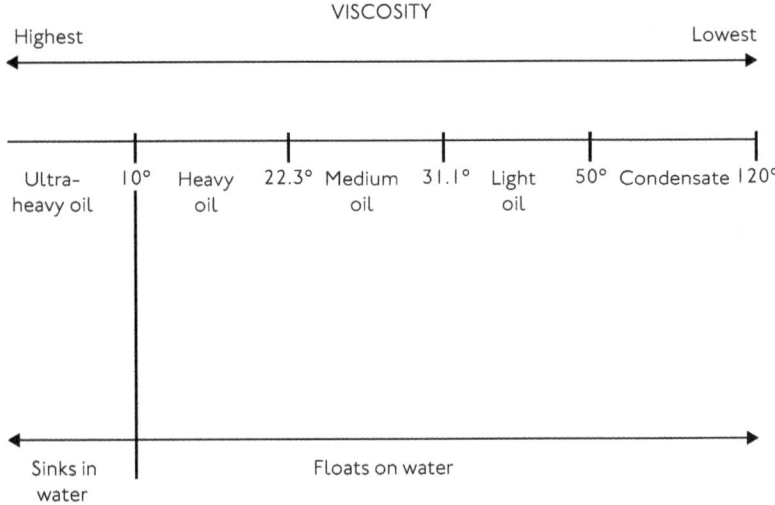

APPENDIX 7

SPE PRMS

The SPE PRMS (Society of Petroleum Engineers—Petroleum Resources Management System) is a standardised series of petroleum reserves and resources definitions and classification guidelines which have been in continuous evolution since 1997 (and which were most recently revised in 2018).

The SPE PRMS represents a collaboration between the Society of Petroleum Engineers (SPE), the American Association of Petroleum Geologists (AAPG), the Society of Exploration Geophysicists (SEG), the Society of Petroleum Evaluation Engineers (SPEE), and the World Petroleum Council (WPC).

At the heart of the SPE PRMS is the following two-axis categorisation of reserves and resources (reproduced here under licence from the SPE) which classifies a petroleum project according to its individual status and maturity by reference to three principal classes of reserves, contingent resources, and prospective resources (and within-class sub-divisions) across the axes of project uncertainty and the prospects for project commerciality.

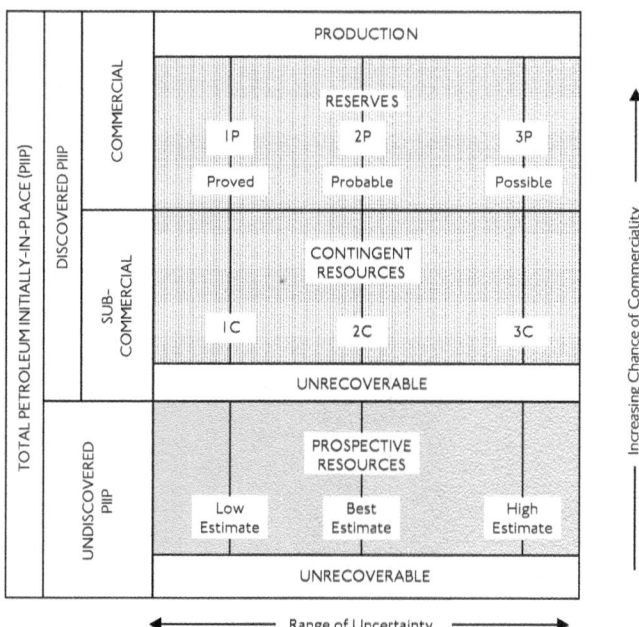